ADVANCES IN WATER MODELLING AND MEASUREMENT

THE FLUID ENGINEERING CENTRE

Acknowledgements

The valuable assistance of the Technical Advisory Committee, Corresponding Members and Panel of Referees is gratefully acknowledged.

Technical Advisory Committee

Dr A C Flemming (Chairman)	Halcrow Maritime
Dr P C Barber	Ceemaid Division, British Maritime Technology
Mr D H Cooper	ABP Research & Consultancy Ltd
Prof K R Dyer	Plymouth Polytechnic
Prof R A Falconer	University of Bradford
Mr D Fiddes	Water Research Centre
Mr M H Palmer (Editor)	BHRA, The Fluid Engineering Centre
Mr R Thomas	Posford Duvivier
Mr G Thompson	Binnie & Partners

Overseas Corresponding Members

Prof M B Abbott	International Institute for Hydraulic and Environmental Engineering, The Netherlands
Mr J A Cunge	SOGREAH, France
Prof R E Nece	University of Washington, USA
Prof E Todini	University of Bologna, Italy
Dr A C E Wessels	Delft Hydraulics Laboratory, The Netherlands
Prof R L Wiegel	University of California, USA

ADVANCES
IN WATER MODELLING
AND MEASUREMENT

M H PALMER (EDITOR)

Published by
BHRA (INFORMATION SERVICES)
The Fluid Engineering Centre
Cranfield, Bedford MK43 0AJ, UK

First published by

BHRA (Information Services)
The Fluid Engineering Centre
Cranfield, Bedford MK43 0AJ, UK

Tel. (0234) 750422 Telex: 825059 Fax: (0234) 750074

© BHRA 1989

British Library Cataloguing in Publication Data

Advances in water modelling and measurement.
1. Coastal waters. Oceanography
I. Palmer, M. H.
551. 46
ISBN 0-947711-75-9 ✓

6038

ADVANCES IN WATER MODELLING AND MEASUREMENT

CONTENTS

PART 1 ESTUARINE WATERS

PART II COASTAL WATERS

PART 1
Estuarine Waters

Chapter 1
MONITORING WATER QUALITY IN THE MARINE ENVIRONMENT – PRESENT METHODS AND SOME LIKELY FUTURE IMPROVEMENTS

R F Rayner and R M Pagett
Marine Environmental Sciences Division
Wimpol Limited, UK

Summary

Recent UK and European legislation coupled with greater public awareness of the problems of coastal pollution has led to a considerable increase in the requirements for monitoring of water quality in Britain's estuaries and coastal waters. This paper provides a review of the methods that are currently applied to the routine monitoring of marine and estuarine water quality. Consideration is given to some of the problems and limitations of present techniques both in terms of the practicality of their application and the quality of the results that may be achieved. Problems of calibration and standardisation of methodologies are also discussed. Having reviewed the present day approach the paper considers a variety of new techniques that may, once fully developed, permit more representative, accurate and cost effective measurements to be made.

1. Introduction

Historically, UK pollution legislation has tended to focus on issues related to land, freshwater and atmospheric pollution. However, UK public awareness of coastal pollution problems coupled with growing legislative pressures from the European Community has resulted in the coastal environment receiving increased attention in recent years. A summary of the current principle items of UK and European legislation relating to control of coastal pollution in the UK is given in Table 1. For further details of legislative controls the reader is referred to Bates (1) and Haigh (2).

The indications are that the pressures for tighter controls on coastal discharges, dumping of dangerous substances at sea and general management of coastal areas are likely to increase. This is evident from a number of recent national and international meetings notable amongst which was the Intergovernmental Conference on the North Sea held in London during November of 1987. These pressures are coupled with yet greater public awareness as a result of the growing levels of activity of specialist environmental pressure groups. A good recent example is the publication by the Marine Conservation Society of 'The Good Beach Guide' (3) which describes the environmental status of Britain's beaches. Publications of this type further stimulate public awareness and the pressure on legislators to act to alleviate and control coastal pollution problems.

All of these legislative and public pressures increase the need for routine monitoring of water quality parameters in the marine and estuarine environment. Addressing this requirement for increased monitoring can pose considerable problems. At first sight it would seem that there should be considerable commonality with techniques applied to freshwater monitoring. Their are, however, differences in the physics, chemistry and biology of marine systems that greatly influence the types of techniques that may be applied. Consideration of access and the practical problems of working at sea also have profound effects on the types of strategy that may currently be applied to water quality monitoring in seas and estuaries.

2. The physical environment

Methods of monitoring the physical environment in coastal seas are relatively well developed and a variety of instrumentation is available for the measurement of physical oceanographic parameters such as waves, tides and currents. There are a wide range of publications which consider this area. For a good summary review see (4) and (5). In the context of this paper the importance of the physical environment lies in its effects on the measurement of water quality parameters and hence on the quality of monitoring itself.

In contrast to measurements in the riverine environment where flows are predominantly unidirectional and generally subject to only gradual change, water movement in the marine and estuarine environment is usually subject to rapid change over short timescales. This is particularly true of waters dominated by tidal forcing where currents can vary by orders of magnitude in speed and be subject to reversals in direction over the course of a single tidal cycle. The majority of the UK's coastal waters are dominated by such tidal currents. A detailed description of the processes of coastal circulation is outside the scope of the present paper but it is important that the influence of large scale rapidly changing horizontal water movement on water quality sampling strategies is appreciated. Such changes mean that determination of the concentration of a particular contaminant from discrete samples at a site are of little value since the advection and dilution of a pollutant will be critically dependent on tidal state. It is therefore necessary to sychronise sampling with the tidal cycle and to determine the temporal variability of a particular water quality indicator throughout the tidal cycle. Sampling strategies which meet these needs give rise to the need for large numbers of measurements at each site of interest. This is compounded by the fact that the marine and estuarine environment often exhibit considerable vertical structure and hence there is usually a need for sampling at the surface and through depth as well as in any underlying sediments.

The need to monitor water quality parameters spatially and temporally is less of a problem when in situ techniques are available. However, for many water quality determinands in situ techniques do not currently exist and so there is a need to collect, preserve and subsequently analyse large numbers of samples.

The physical environment of coastal waters also poses practical and logistic problems which increase the difficulty associated with meaningful sampling strategies. It is rare that sampling of a contaminant on land or in fresh water is disrupted by changes in physical conditions whereas such disruptions are common in programmes of sampling at sea!

3. The chemical and biological environment

The majority of routine monitoring of coastal and estuarine water quality is based on measurements of the concentration of particular chemical and microbiological determinands, the determination of parameters indicative of the effects of pollutant inputs (eg dissolved oxygen concentration or chemical oxygen demand), and the sampling of biota which may be sensitive to the effects of particular pollutants.

A typical list of chemical entities that would need to be considered in routine monitoring of water quality at a particular site might include nutrients, toxic metals, pesticides and herbicides. Microbiological monitoring would typically include an assessment of the distribution of bacteria (notably coliforms and certain pathogens such as salmonella) and viruses.

Bearing in mind the need for spatial and temporal sampling outlined above it would be ideal if all of the chemical entities could be measured by in situ techniques and preferably using instrumentation that may be moored for periods of sampling thus reducing the risk of sampling disruption as a result of unworkable conditions for vessels. In practice few of these measurements are possible by in situ methods and fewer still may be addressed by self contained instrumentation capable of collecting time histories of particular determinands. Table 2 summarises the present state of the art for a range of determinands that require routine monitoring.

From Table 2 it is clear that sampling strategies are still largely dependent on collection of discrete water samples. However, this is not the end of the problem. Because of the labile nature of many of the chemical and microbiological parameters of interest sample preservation is not a simple matter. Great care is needed in order to ensure that the concentration of a particular chemical species or the numbers of a particular bacterium or virus remain unaffected by the method of sample collection or storage. Long term storage of samples is not possible for some of the determinands (eg faecal coliforms which must be processed within six hours of sample collection) or poses particular problems (eg the need for immediate analysis of nutrients or freezing of samples to -20 degrees Centigrade for transport to the laboratory).

Having successfully collected and transported the samples from a particular monitoring exercise there remains the not inconsiderable problem of standardising the method of laboratory analysis. This is essential to ensure adequate knowledge of the accuracy and sensitivity of particular analyses and to allow comparability of measurements from different locations and from programmes conducted by different organisations. Here again there are difficulties.

In the water industry the HMSO publications, familiarly known as the 'Blue Books' (6) have sought to lay down firm guidelines on the analytical procedures to be followed when assaying waters, effluents and tissues for a range of determinands. Although comprehensive in their coverage of certain areas these publications do not adequately cover the analysis of determinands in sediments which are now more likely to be a vital component of coastal and estuarine water quality investigations. The development of appropriate standards for this area has now become an urgent requirement. Such standards take a considerable time to produce in view of the consultation that must accompany their preparation. For the present, the handbooks produced by the International Biological Programme and the Estuarine and Brackish Water Sciences Association (7 and 8) provide a

3

useful digest of present approaches and methodologies for the examination of sediments.

Sampling of benthic communities has posed a set of problems that have endured through the years. Soft and hard substrata require totally different approaches. the former traditionally uses remote techniques for which there is a wide choice of equipment and an equal number of opinions concerning their relative merits. Hard substrata are best examined in situ. Although the use of video is becoming increasingly common there is still no real substitute for the diving marine biologist. The subtleties of floral and faunal distribution cannot be fully resolved by underwater photography or television. The two handbooks referred to above distil in a very accessible form the different ways in which a benthic survey may be conducted and the relative merits of the various approaches.

3. The future

A key area with potential for future improvements in techniques is the routine measurement of chemical determinands in marine and estuarine waters. In order to overcome dependency on the collection of discrete water samples the principal development must be the perfection of in situ methods. Information generated by such techniques needs to be of a quality equal to or better than that which is typically expected of more traditional collection/storage techniques. Any new analytical or measurement technique will need to be validated and be acceptable as a robust technique by those who administer the UK and European legislation.

A recent approach which is under evaluation in the UK is the use of columns with chelating resins which can provide a time-integrated sample (9). Whilst this in situ approach to collection may go some way to solving some of the problems outlined above, the list of determinands that may be sampled in this way is not likely to be comprehensive. For specific determinands and for specific locations, though, it may prove to be of great utility.

The use of biological indictors of water quality has had a chequered history (10). There are several long running mussel watch programmes which have produced interpretable results. The freshwater mussel has been used to detect organochlorine compounds and this development continues. Research also indicates that there is potential for assessing sublethal stress induced by pesticides by means of enzyme assay techniques (11). Qualitative indicators of water quality have been gaining in popularity and are used to provide rapid indications of a decrease in water quality and for triggering trouble shooting of water supply quality by the appropriate authorities. Rainbow trout and the bizarre-looking elephant fish are on active service in the UK (12).

The use of in situ probes to measure the concentration of a particular determinand has many attractions. Ion selective electrodes are available which are specific to particular ions. When immersed in solutions containing the specific ion of interest a potential is developed across the electrode - solution interface that is related to the activity of the ion. The use of such ion-selective electrodes is likely to increase, although a number of problems still need to be resolved with regard to use in saline waters and ensuring adequate long term stability (13).

A further technique which is gradually advancing in robustness and utility is voltammetry (14). Voltammetry is an electrochemical technique which is based on measuring voltage changes at an electrode according to presence (and amount) of a particular metal. Its particular application is in trace metal analysis. A benefit of voltammetry is the simultaneous measurement of metals in the presence of organic compounds with little pre-treatment of samples. A further development is stripping or inverse voltammetry. Inverse voltammetry is a two stage process in which the metal of interest

4

is concentrated onto an electrode and subsequently stripped off by changing the potential at the electrode. During this second stage the measured current is directly proportional to the metal concentration. This technique has been shown to be a powerful tool for elucidating metal speciation which is of importance in determining the toxicity of a particular metal.

In the relatively near future these types of in situ techniques coupled with existing data telemetry technology may allow the provision of information on the variability of chemical entities in real time in the office of the environmental manager.

Problems to be resolved for monitoring the biological environment seem to revolve around the conceptual notion of health. What is a healthy community? Many cyclic patterns are found in the marine and estuarine environments on a variety of timescales ranging from a few hours to several years. Devising monitoring programmes that can sensibly cope with such variability is the challenge that faces those who seek to establish environmental management plans that include the monitoring and assessment of water quality. A conventional measure of the well being of a given benthic community is based on the collection of samples, counting the species and individual and relating them with respect to physical or chemical parameters. The determination of environmental status (or, well being) is then largely based on a comparison with other areas and assemblages. This determination relies on the skill and experience of the assessor.

In order to overcome this rather subjective approach, attempts have been made to quantitatively measure the health of certain individual species by a scope for growth test which provides an index of sublethal stress (15). This measurement is a quantification of the energy available for growth and reproduction after the demands of basal metabolism have been satisfied. Scope for growth measurements provide physiological information at the individual level. Biochemical responses can also be obtained which provide information at the cellular level. This technique is now moving from the research tool domain to limited routine use. However, there is still much work left to be done to go from the level of the individual to the level of the population and then to the community level. From there, true ecosystem modelling is required. These are the kind of developments that need to be done and which we might see in the future.

A further aspect which should not be neglected is the importance of how the results of environmental monitoring are made available to those who are obliged to make decisions and, necessarily, are unlikely to be involved in the monitoring process itself. Indices of environmental quality have been developed for limited routine use in the freshwater environment (16) and a similar approach could be adopted for coastal and estuarine waters.

A number of indices should be available which should be sensitive to certain factors and not to others. An index, unlike the direct measurement on which it is based, reflects some technical comprehension and this underlying understanding can thus be provided to the non-specialist in a form where it may be used for decision making. Provided appropriate thresholds are incorporated, indices could be the means by which deterioration in water quality could be monitored by decision-makers. Such deterioration could be on a scale which features various categories such as "normal", "cause for concern", "action" and "alarm". Refinements such as early warning trigger could also be incorporated.

A range of indices could be used to indicate various aspects of environmental decline. These could include: pollutant stress in sediments, pollutant stress in the water column, coliform risks, viral risks, benthic species composition and abundance, fish disease, shellfish disease, oxygen depletion effects and so on.

Indices cannot be predictive but should assist in evaluating large data sets and therefore should provide a potential framework for evaluating and comparing different environmental measures that must be weighed not only against each other, but also against other considerations such as economic and aesthetic constraints. This differential weighting enables the decision making process to reflect the integration of the complexity of ecosystems with socio-economic needs.

The difficulty of distinguishing natural change from change brought about by human activity is of fundamental importance in deriving solutions to environmental problems. This difficulty underpins the inevitable compromise between scientific uncertainty, economic constraints, conflicts of multiple use and the sometimes differing priorities of scientists and environmental managers. Temporal and spatial scales and subtleties of heterogeneity are immense challenges for the scientist to understand and quantify and for the environmental manager to incorporate into management plans.

Present environmental monitoring strategies are based almost entirely on the collection of discrete samples, be they water samples, sediment samples, photographs, profiles of physical data or written descriptions of aesthetic condition. These samples may be singly collected or collected in conjunction with other samples and with the recording of various environmental parameters (such as temperature, salinity and dissolved oxygen concentration). The basis for the collection is that they can, if collected over a given time basis (which may vary from determinand to determinand) provide information about the changes that occur during a period of time. This period could be from high to low water, from Spring to Neap tides or could be seasonal or in some cases annual. Whatever the time base, it is generally cyclic or regular. That is to say, discrete samples are collected in order to define given parameters during a particular cyclical period.

The present legislation calls for long term planning in dealing with water quality. Incorporated into this planning are data collection programmes that are short in their duration and conservative in their prediction. They provide information on regular sequences of events. They provide little information to indicate how episodic events may change a given sequence, and therefore, provide less criteria for a longer term strategy. Long term planning and design are very often based on very short term data sets.

The questions that we must ask are, firstly, do episodic events influence general patterns and, secondly, if they do what is the magnitude of their effect?

Water quality modelling may be able to incorporate this to a certain extent by, say, adding a particular margin of conservatism. These margins cost money because they can have substantial impacts on the costs of schemes to alleviate coastal pollution problems. Simulation models of marine and estuarine systems can be valuable tools for exploring the inherent variability of dissolved constituents and for predicting annual mean concentrations (17). These models are typically driven by measured river run-off and predicted tidal range with dispersion coefficients that may be used as a basis for simulating the fate of particular pollutants (18). Their predictive ability may often be weakened by the impact of infrequent and poorly understood events such as sediment resuspension and consequent contaminant remobilisation during storm events.

One way of assessing the effectiveness of present approaches to the prediction of performance of coastal environmental management schemes is to conduct comprehensive monitoring after particular schemes have been implemented. Such monitoring can allow the comparison of actual performance with predicted performance. Monitoring of this type is rarely conducted in any comprehensive way usually for economic reasons. The availability of more reliable and cost effective techniques for water quality monitoring would make monitoring of this type more attractive.

Adequate monitoring of water quality must take account of the physics, chemistry and biology of the marine environment, the temporal and spatial heterogeneity of the water and sediment fabric and cyclic and episodic events. This is necessary in order to formulate management plans that respect the environment, protect human health and that can be effected within the inevitable constraints of time and money. This is the challenge for the future.

4.   References

1.   Bates, J.H. (1985). United Kingdom Marine Pollution Law. Lloyd's of London Press 461 pp.

2.   Haigh, N (1987). EEC Environmental Policy and Britain, 2nd Edition. Longman, Harlow 380 pp.

3.   Marine Conservation Society (1988). The Good Beach Guide. Ed. A Scott. Ebury Press, London 192 pp.

4.   Society for Underwater Technology (1985). Evaluation, comparison and calibration of oceanographic instruments. In: Advances in underwater technology and offshore engineering 4. Graham and Trotman, London 267 pp.

5.   BHRA (1980). International conference on Measuring Techniques of Hydraulics Phenomena in Offshore, Coastal and Inland Waters, BHRA, The Fluid Engineering Centre, Cranfield 451 pp.

6.   HMSO (1980). Methods for the Examination of Waters and Associated Materials, London.

7.   IBP Handbook (1984). Methods for the Study of Marine Benthos 2nd Edition. Ed. Holme, N.A. and McIntyre, A.D. Blackwell Scientific Publications, London 387 pp.

8.   EBSA Handbook (1987). Biological Surveys of Estuaries and Coasts Ed. Baker, J.M. and Wolff, W.J. Cambridge University Press, Cambridge 449 pp.

9.   Green, D.R. (1986). Determination of chlorinated hydrocarbons in coastal waters using a moored in situ sampler and transplanted live mussels (unpublished).

10.   White, H.H. (1983). Mussel Madness. Use and misuse of biological monitors of marine pollution. In: Concepts in marine pollution measurements. Maryland Sea Grant, Maryland p.325-337.

11.   Anon. Musseling in on monitoring pollutants. In: Water and Waste Treatment, November 1987 p.6.

12.   Anon. Trouble trout pick up pollution. In: New Civil Engineer, July 1987 p.4.

13.   Whitfield, M. (1985). ion-selective electrodes in estuarine analysis In: Practical Estuarine Chemistry. Ed. Head, P.C. Cambridge University Press p.201-272.

14.   Anon. One Step ahead for water analysis. In: Water and Waste Treatment, September, 1987 p.16.

15. Lack, T. and Widdows, J (1986). Physiological and cellular responses of animals to environmental stress - case studies. In: The role of the oceans as a waste disposal option. Ed. G Kullenberg. p.647-665.

16. Department of Environment (1970). Report of a River Pollution Survey in England and Wales 1970, Vol.1, London, HMSO 39 pp.

17. Radford, P.J. and West, J (1986). Models to minimise monitoring. Water Research 20, (8), 1059-1066.

18. Radford, P.J., R.J. Uncles and Morris, A.W. (1981). Simulating the impact of technological change on dissolved cadmium distribution in the Severn Estuary. Water Research 15, 1045-1052.

TABLE 1 : Principal UK and European Community Environmental Legislation Relating to Coastal and Estuarine Waters

Legislation	Summary of application to marine and estuarine Waters
UK Acts of Parliament:	
Water Act, 1973	General provisions for the maintenance of the quality of coastal waters.
Control of Pollution Act Part II 1974 (part II implemented 1986).	Controls discharges to coastal waters.
Food and Environment Protection Act, 1985.	Controls dumping at sea.
E C Directives:	
Directive on the quality required for shellfish waters, 1979.	Controls maintenance of an environment suitable for shellfish growth.
Directive concerning the quality of bathing water, 1975.	Sets quality objectives for bathing waters and prescribes sampling requirements for monitoring of bathing water quality.
Directive on pollution caused by certain substances discharged into the aquatic environment of the Community, 1976.	Sets the framework for the elimination of pollution in coastal and territorial waters by particularly dangerous substances. Subsequent daughter Directives set standards for particular substances.

TABLE 2 : Techniques for Obtaining Water Quality Data

Determinand	Conventional Technique
Salinity	In situ probe or laboratory analysis of discrete water sample
Dissolved Oxygen	In situ probe
Biochemical Oxygen Demand	Laboratory analysis of discrete samples
Chemical Oxygen Demand	Laboratory analysis of discrete samples
Nutrients	Laboratory analysis of discrete samples
Suspended solids	In situ optical measurement of turbidity to infer solids content or laboratory analysis of discrete samples
Turbidity	In situ probe
Metals	Laboratory analysis of discrete samples
Organics	Laboratory analysis of discrete samples
Chlorophyll	In situ fluorimetric technique or laboratory analysis of discrete samples
Bacteria	Laboratory culture from discrete samples
Viruses	Laboratory culture from discrete samples

Chapter 2
TIME VARIANT APPROACH TO THE MODELLING OF WATER QUALITY IN STREAMS

A James and D J Elliott
University of Newcastle upon Tyne, UK

Summary

Water quality standards have recently been framed in terms of percentage compliance. This has focussed attention on the need for appropriate models of stream quality.

Steady-state models have considerable attraction in simplicity, modest data requirements and moderate computational times and can by Monte-Carlo simulation be used to generate frequency distributions of pollutant concentrations in streams. However such an approach has serious limitations since it gives no indication of the duration of failures and cannot handle complex interaction of time-dependent processes.

A dynamic model of river water quality is described which has been used to overcome these limitations. The calibration and validation of model are discussed and the method of synthetic data generation is explained. The paper concludes with a comparison of the two approaches.

1. Introduction

Water quality in streams needs to be specified in relation to use or potential use and may be characterised by a large number of parameters, but as indicated in Table 1 quality requirements generally centre about dissolved oxygen (DO) and biochemical oxygen demand (BOD).

The concentration of DO is usually the resultant of a large number of processes operating simultaneously, e.g.
i.   Photosynthesis of algae or macrophytes
ii.   Respiration of algae or macrophytes
iii. Bacterial oxidation of organic matter in solution or suspension
iv.   Bacterial oxidation of organic matter on the bed
v.   Oxygen exchange with the atmosphere

The rates of all of these vary in time and the result of this complex interaction is a constantly varying DO. From an ecological viewpoint it is important to specify the following
i.   the level of DO required
ii.   the percentage of the time that this level should be achieved
iii. the maximum frequency, the maximum duration and minimum levels during any departures from (i.)

River models may be used to explore these departures as described below.

2. Use of Steady-state models

These have already been described by Warn (1,2) and Brown (3) and therefore do not require detailed coverage.

River Class	Quality Criteria	Current & Potential Use
1A	(i) DO > 80% (ii) BOD ≯ 3 mg.l^{-1} (iii) Ammonia ≯ 0.4 mg.l^{-1} (iv) Where abstracted must conform to EEC requirements (v) Non-toxic to fish in EIFAC terms	(i) Water of high quality suitable for potable abstraction (ii) Game or high class fishery (iii) High amenity value
1B	(i) DO > 50% saturation (ii) BOC ≯ 5 mg.l^{-1} (iii) Ammonia ≯ 0.9 mg.l^{-1} (iv) Where abstracted must conform to EEC requirements (v) Non-toxic to fish in EIFAC terms	Water of less high quality but usable for substantially the same purposes
2	(i) DO > 40% saturation (ii) BOD ≯ 9 mg.l^{-1} (iii) Where abstracted must conform to EEC requirements (iv) Non-toxic to fish in EIFAC terms	(i) Water suitable for supply after advanced treatment (ii) Supporting good coarse fishery (iii) Moderate amenity value
3	(i) DO > 10% saturation (ii) Not likely to be anaerobic (iii) BOD ≯ 17 mg.l^{-1}	(i) Suitable for only low grade industrial abstraction (ii) Not suitable for fishery

95% percentage compliance

Table 1   Classification of river water quality proposed by the Fish Committee

The essential idea, as shown in Fig. 1, is to select a flow and a load from appropriate frequency distributions and then to route these down the river using a steady-state model to calculate the DO and BOD at various stations. This process is repeated in the Monte Carlo simulation a large number of times so that frequency distributions of DO and BOD are obtained for each of the downstream stations.

The advantages of this approach are mainly reflected in the minimal data requirement and the modest computational effort. Warn (4) argued that measurement errors are so large that the use of all but the simplest model is unjustifiable. This has the great attraction that the small computational effort involved allows several hundreds of runs to be followed thus generating   statistically sound distributions of downstream concentrations.

The main drawbacks of the steady-state simulation are as follows:

a)   During the routing of flows down the river there are many changes e.g. in light intensity, which affect the oxygen balance.

Such changes are responsible for diurnal fluctuations, which, since they do not appear explicitly in steady-state models, are a large part of what is called measurement error.

Such variations especially where they are cyclical will affect the frequency distribution at the downstream stations.

b) Steady-state models are incapable of providing information on the duration of pollution episodes. As illustrated in Fig. 2 the biological consequences of any failure to meet standards depends upon the combination of intensity and duration. The frequency distribution at downstream stations is therefore not an adequate guide to whether or not an effluent may damage a river.

3. Time-Varying Model

Most dynamic models of water quality in rivers are based upon a finite-difference solution to the 1-dimensional convective-diffusion equation (Rinaldi (5))

$$A \frac{\partial AC}{\partial t} = U \frac{\partial AC}{\partial x} - \frac{\partial}{\partial x} (AE \frac{\partial C}{\partial x}) - KAC + La$$

where C = Concentration of water quality parameter
       A = Area
       U = Velocity
       E = Dispersion coefficient
       K = Decay coefficient
       La = Pollution source
       x = Distance downstream
       t = time

Various explicit and implicit schemes have been used and have been shown to give adequate numerical stability (McBride, 6). The NUT* model used in this work incorporated the QUICK** routine.

The structure of the NUT model as summarised in Fig. 3, shows a modular construction. The main elements of the model may be briefly characterised as follows:

a) Alternative Hydraulic Inputs - Data in the form of

Flow		Flow		Velocity
&		&		&
Velocity	or	Cross-Sectional Area	or	Cross-Sectional Area
&		&		&
Depth		Depth		Depth

are read in at known sections for the times of observation. Intermediate values are obtained by interpolation in space and time.

Coefficients of dispersion, reaeration and re-suspension are calculated from the hydraulic data.

b) Quality Inputs - Data on water quality at upstream boundary and initial conditions are read in together with loads. Loads and boundary conditions are specified at particular times and interpolated.

c) Coefficients for processes like BOD, SOD, Photosynthesis etc. are read in at selected points and times and interpolated. Temperature corrections are applied.

d) Subroutines for handling all the main operations like interpolation, finite difference scheme, source and sink routines, output routines.

The NUT model can be operated in the interpolation mode as described above or it can be used in the synthetic data model. In the latter some short sequence or series of short sequences of data are subjected to time-series analysis by first removing the trend, secondly removing the principal harmonics and finally computing the mean and variance of the resuidual.

---

* Water Quality Model developed at the University of Newcastle upon Tyne
** Numerical Algorithm developed by Leonard (1979).

These components can then be recombined to generate longer sequences of records which can be analysed to determine the following characteristics:

i. Frequency distributions of quality parameters like DO and BOD with associated percentages of compliance. These can be determined over an annual period or shorter period e.g. summer when critical conditions are more likely to occur.

ii. Duration/concentration relationships for each failure with compliance together with the time intervals between.

The operation of the various parts of the model is illustrated in Fig.4. The dynamic model approach clearly provides much more useful data for judging the consequences of alternative pollution control strategies.

The main drawback to this approach is the additional data requirement for the model calibration and validation. The extent of the additional effort can be gauged from the calibration/validation example described in the next section.

The other disadvantage of the dynamic model is the great computational effort, but this is not a serious problem even for micro-computers.

4. Calibration and Validation of NUT Model

Figs. 5a and 5b show the preliminary results of a calibration exercise carried out on the interpolation version of the NUT model using a data set from a Scottish river. The data set contained results of DO, BOD and Ammonia concentrations on samples taken at 60-minute intervals over a 120-hour period at 5 stations in the river together with flows, temperatures and ancillary data on SOD, plant biomass, etc. Parallel data on the BOD load from the only discharge was also collected. Stream sections were surveyed for cross-sectional area and depth at a variety of measured flows.

The calibration phase was carried out using the first 60 hours of data. In the initial runs as shown in Fig. 5, the model failed to reproduce the marked diurnal fluctuations in DO and the predicted ammonia concentrations at the downstream stations were much higher than those observed. Both of these discrepancies appear to have been connected with the nature of the plant biomass. This contained very little in the way of macrophytes but consisted almost entirely of attached algae (mainly Cladophora), which has a higher rate of biochemical activity per unit biomass. Increasing the photosynthetic and respiratory coefficients aboe the range generally quoted in the literature for macrophytes brought the amplitude of the diurnal variations in line with observation. The higher values used are more appropriate for algal activity.

One further difficulty in the DO calibration was caused by the phase difference between observed and predicted, as shown in Fig. 5a. There appeared to be 3 possible explanations, namely:

a) A lag in the oxygen production being reflected in the oxygen concentration;

b) Incorrect velocity data causing DO changes to be advected downstream too quickly.

c) Incorrect biomass specification causing distorted DO patterns to be advected to the downstream station.

Examination of ammonia and BOD patterns indicated that advection appeared to be correct and no lag effect had previously been reported in DO response to photosynthesis or respiration.

The biomass in the upper 2 km was adjusted and provided a response which was then in phase with the observation, as shown in Fig. 6a.

It was also apparent that such a biomass of algae would cause a significant removel of ammonia from solution which accounted for the discrepancy between predicted and observed concentrations at the downstream station. The introduction of a new subroutine for ammonia uptake brought the model predictions into line.

(See Fig. 6b). Attempts at adjusting the ammonia concentration by increasing the rate of nitrification were unsuccessful as this distorted the DO pattern.

The second phase in the exercise was to validate the model using the second part of the data set (hours 61 - 120). The agreement achieved for DO is shown in Fig. 7.

5. Time-Series Analysis

Time series analysis confirmed the dominance of a diurnal cycle. Fig. 8 shows the normalised autocorrelation before and after the removal of the principal harmonics.

The statistics obtained from the time series analysis can be combined as shown in Fig. 4 with the parameters estimated by the calibration/validation exercise to generate synthetic records. The synthetic data generation of the NUT model can be used for the same purpose as the Monte Carlo simulations using steady-state models. But as shown in Fig. 9 the dynamic record yields much more ecologically useful information than the percentage compliance which is the sole result of the steady-state approach.

References

1. WARN, A.E. & BREW, J.S. (1980) Mass Balance Water Research, 14, 1427-34.
2. WARN, A.E. (1982) Calculating Consent Conditions to Achieve River Quality Objectives Effluent & Water Treatment Journal, 22, 152-5.
3. BROWN, S.R. (1987) TOMCAT : A Computer Model Designed Specifically for Catchment Quality Planning Within the Water Industry, pp 37-50 in International Conference on Water Quality Modelling in the Inland Natural Environment. Bournemouth, June 1986.
4. WARN, A.E. (1986) Planning Investment for River Quality Using a Catchment Simulation Model, pp 1-8 in International Conference on Water Quality Modelling in the Inland Natural Environment. Bournemouth, June 1986.
5. RINALDI, T. (1979) Modelling for the Control of River Water Quality, McGraw Hill.
6. McBRIDE, G.B. & RUTHERFORD, C.J. (1986) Accurate Modelling of River Pollutant Transport ASCE J. Env.Eng., 110, 808-27.
7. JORDAN, D.H.M. & LLOYD, R. (1964) The Resistance of Rainbow Trout & Roach to Alkaline Solutions, Int. J. Air Water Pollution, 8, 405-29.
8. LEONARD, B.P. (1979) A Stable and Accurate Modelling Procedure based on Quadratic Upstream Interpolation, Computer Methods in Applied Mechanics and Engineering, 19, (1) pp. 59-98.

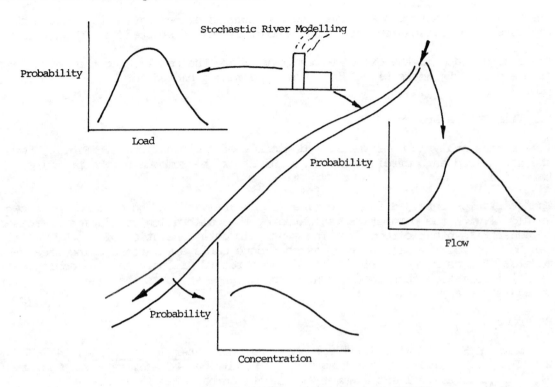

Fig. 1 Monte-Carlo simulation to determine compliance using a steady-state model

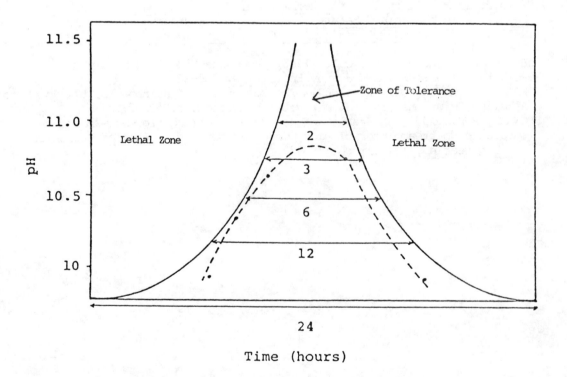

Fig. 2 Method of evaluating fluctuating toxicity (after Jordan & Lloyd (1964)

Solid line indicates boundary of zone of tolerance.
Dotted line indicates fluctuation of pH in the environment.
If the environmental fluctuation crosses the boundary some
degree of mortality will occur.

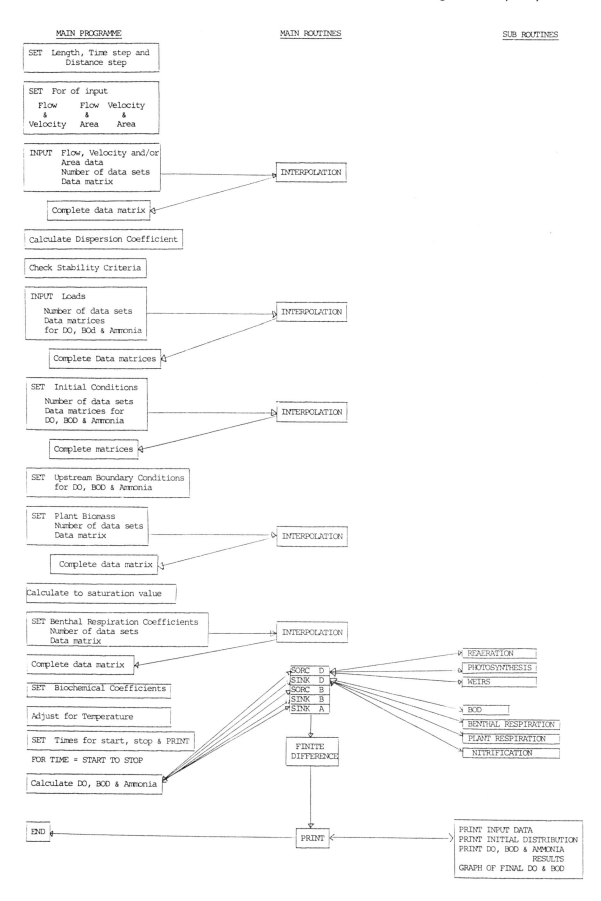

Fig.3 Modular Structure of NUT Model

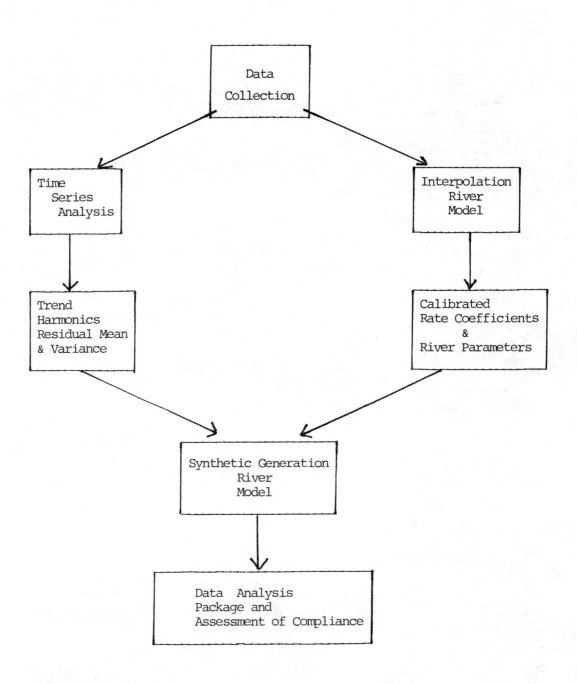

Fig 4 Flow diagram of data processing

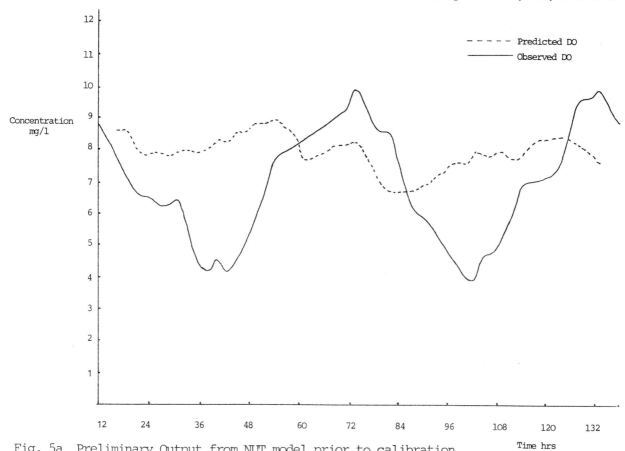

Fig. 5a   Preliminary Output from NUT model prior to calibration
          DO Results

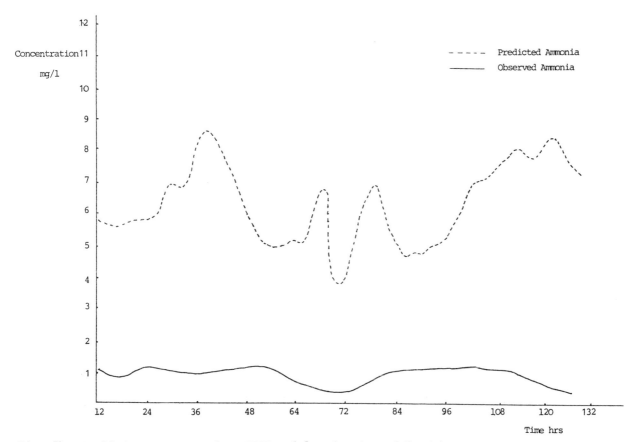

Fig. 5b   Preliminary Output from NUT model prior to calibration
          Ammonia results

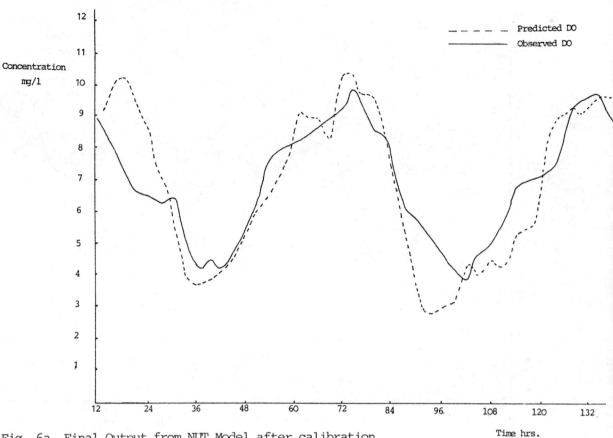

Fig. 6a   Final Output from NUT Model after calibration
          DO results

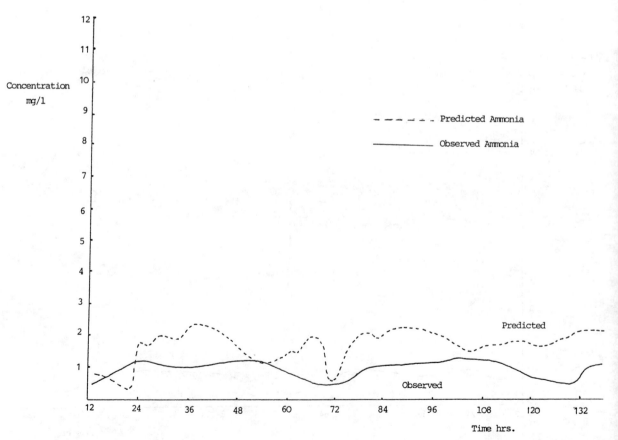

Fig 6b   Final Output from NUT model after calibration
         Ammonia Results

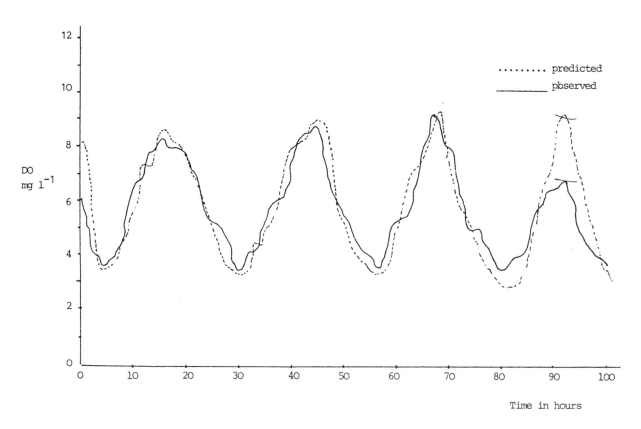

Fig. 7 Results of the Validation exercise on DO

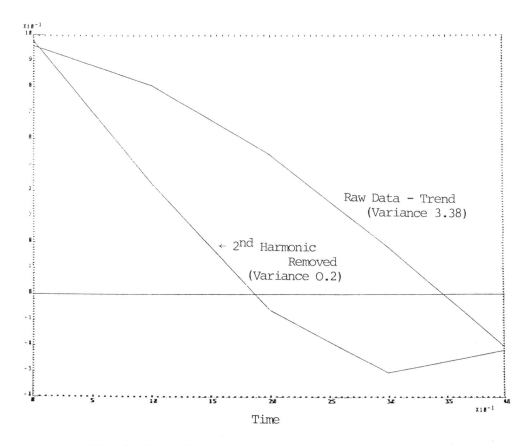

Fig. 8 Normalised Autocorrelation Function

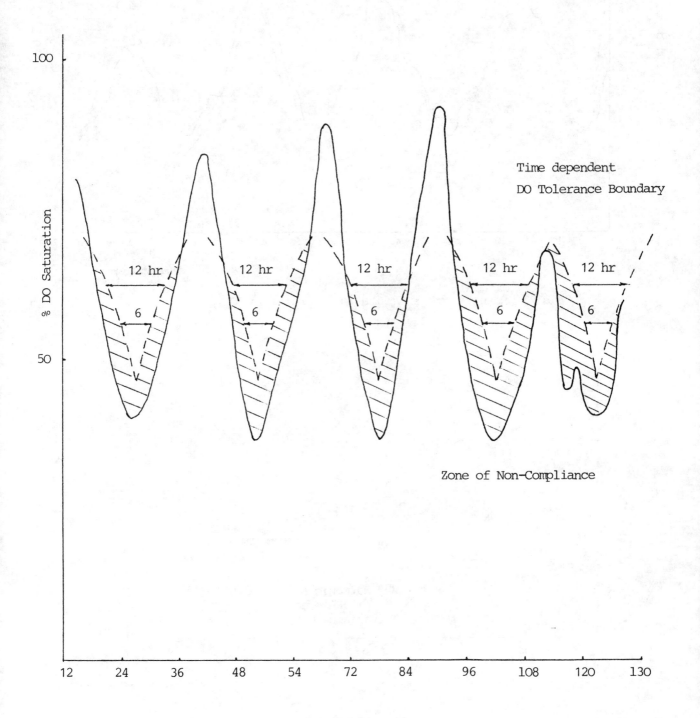

Fig. 9 Data analysis of synthetic DO record showing the following features:

1. No. of times when non-compliance occurs
2. Antecedent conditions prior to non-compliance
3. Length of time for each non-compliance episode

Chapter 3
CONSIDERATION OF A ONE DIMENSIONAL TRANSPORT MODEL OF THE UPPER CLYDE ESTUARY

S G Wallis
Heriot-Watt University, UK
J M Crowther
University of Strathclyde, UK
J C Curran, D P Milne and J S Findlay
Clyde River Purification Board, UK

Summary

This paper describes a one-dimensional time-dependent mathematical model for solute transport in estuaries. The model is applied to an 11 km reach of the Clyde estuary in Scotland and has been developed as part of a larger study into the dispersal and survival of indicator bacteria in coastal and estuarine waters. The larger study has been stimulated by the recent failure of many UK designated bathing waters to meet the recommended European Community quality standards. The model consists of a finite element solution to the advection-dispersion equation with decay which interacts with a finite difference solution to the hydrodynamic equations. The hydrodynamic model has been verified using water level data and measurements of flow velocity, and the conservative version of the solute model has been calibrated approximately using salinity measurements. The hydrodynamic model is shown to perform well, but the solute model remains unverified at the time of writing due to a shortage of appropriate salinity data. The values of the dispersion coefficient used in the model, however, are shown to be consistent with values estimated from measurements of the flow structure. Other work has yielded values for the mortality or decay rate appropriate to Escherichia coli bacteria in estuarine waters which, when used with the non-conservative version of the solute transport model, will enable a bacterial concentration model to be calibrated.

© BHRA, The Fluid Engineering Centre,
   Cranfield, Bedford, MK43 0AJ, 1989

Nomenclature

A = cross-sectional area of flow.
$\tilde{A}D, A1, A2, A3, A4$ = coefficients in solute model.
C = cross-sectional area average solute concentration.
C_O = solute concentration in lateral inflow.
$c_1, c_2, c_3, c_4, c_5, c_6, c_7$, = coefficients in hydrodynamic model.
D = longitudinal dispersion coefficient.
$\tilde{DT}1, \tilde{DT}2, \tilde{DT}3, \tilde{DT}4$ = coefficients in solute model.
F = any function in finite difference operators.
f = Darcy-Weisbach resistance coefficient.
g = gravitational acceleration.
h = water level relative to a datum.
i = spatial index (subscript).
K = channel conveyance.
k = solute mass decay rate.
L = lateral inflow per metre run.
n = number of space steps (or elements) in the models.
Q = discharge.
$\tilde{Q}1, \tilde{Q}2, \tilde{Q}3, \tilde{Q}4$ = coefficients in solute model.
R = hydraulic radius.
S = salinity.
S_f = friction slope.
t = time.
W = water surface width
x = longitudinal co-ordinate direction.
Δt = time step.
Δx = space step (or element length).
λ = constant in equation of state for salt.
θ = weighting coefficient.
$*$ = denotes function value at time $t + \Delta t$ (superscript).
$\wedge$ = denotes latest estimate of converged value (superscript).
$\alpha_i, \beta_i, \gamma_i, \delta_i, \epsilon_i, q_i, t_i, s_i, p_{i+1}, r_{i+1}$ = coefficients in hydrodynamic model.
$a_i, b_i, d_i, e_i, f_i, g_i, \ell_i, m_i, x_i, y_i, z_i$ = coefficients in solute model.

NB. f_i, g_i, t_i, x_i, do not refer to ith value of f, g, t, x,: they are different parameters.

1. Introduction

Recent European Community directives have defined certain microbiological standards for bathing and shellfish-growing waters. This now allows the appropriate regulatory authorities to develop a standardised and scientifically based rationale for the licensing of potentially polluting discharges. Although some techniques are currently available for modelling the mixing of effluents in receiving waters (Fischer (1)), the resulting models often suffer from a lack of knowledge of the mixing processes. When concerned with bacteria in coastal waters the problem is augmented because their non-conservative behaviour is poorly understood and hence difficult to quantify.

The authors are currently engaged in an investigation funded by the Department of the Environment which is aimed at remedying some of these deficiencies. Microbiological experiments in the laboratory using the indicator bacteria Escherichia coli (Milne et al.(2)), field work and mathematical modelling of the mixing and bacterial survival dynamics are being combined in a study based on the Clyde estuary in Scotland. In this paper, we describe the initial modelling work which, to date, has involved the development of a one-dimensional transport model for an 11 km reach of the upper estuary.

The model is based on numerical solutions to the one-dimensional hydrodynamic and solute transport equations and employs a combination of finite difference and finite element techniques. This enables the use of the same non-uniform spatial discretisation of the estuary in both the hydrodynamic model and the solute model. In the larger study the solute model will satisfy a number of needs. Its non-conservative version will describe the advection, dispersion and mortality of the bacteria. Since the mortality of the bacteria is a function of salinity, the conservative version of the solute model is required to give the salinity field, and since the study team's laboratory work (Milne et al. (2)) is demonstrating the importance of suspended solids on the survival of the bacteria, a slightly modified version will probably be required to yield the suspended solids field. In this paper, however, we only consider results from the hydrodynamic and conservative solute models.

Model results are compared with field data collected by the study team and co-operating institutions, and some aspects of model calibration are discussed. Measurements of the flow structure are also presented which indicate some deficiencies in the model and the type of modifications which will be considered in future work.

2. The Clyde Estuary

The Clyde estuary lies on a north-west to south-east axis at the western end of Scotland's central valley. It extends for about 36 km between the city of Glasgow and the town of Greenock, seawards of which it becomes the Firth of Clyde (See Fig. 1). Major dredging operations during the last two hundred years enabled the development of Glasgow as a port, and have left the estuary as a navigable waterway for most of its length (Riddell (3)). The tidal limit for neap tides is a tidal weir situated about 1 km upstream of the centre of Glasgow. On spring tides, however, tidal action can occur as far as Dalmarnock about 3 km further upstream.

The model described in this paper covers the upper 11 km of the estuary between Rothesay dock and the tidal weir (See Fig. 2). This part of the estuary conforms well to the concept of a one-dimensional estuary being largely canalised, narrow and of simple cross sectional shape. For about 10 km upstream of Rothesay dock there is virtually no bed slope, and tidal mean water depths are of the order of 10 m in the centre of the channel. For about 1 km below the tidal weir, however, siltation has occurred with a consequent rapid decrease in water depths and the exposure of mud banks at low water. The width changes little with tidal state but varies with location. The average width is about 130 m with maxima and minima of about 200 m and 110 m. Both banks of the estuary contain various open docks and basins which serve as dynamic storage areas.

Fresh water inputs to the upper estuary (CRPB (4)) come from the river Clyde itself (50% exceedance flow = $26m^3/s$) and from the river Kelvin (50% exceedance flow = $5m^3/s$), which enters the estuary about 4 km below the tidal weir. The tidal response approximates well to a standing wave with a small (1%) amplification of the tidal range between Rothesay dock and the tidal weir, and virtually no phase lags. Mean tidal ranges at Broomielaw (Glasgow) are 3.9 m (springs) and 2.5 m (neaps) (Admiralty Tide Tables (5)). Tides in the estuary are not simple, however, because high frequency components distort spring flood tides in particular (Thomson (6)) and create unusual variations in flow velocity. The flow structure appears to be very complex (Thomson (6), West et al (7)) with a tendency for stratified conditions to prevail.

3. Model formulation

3.1 Hydrodynamic model

The movement of water in an estuary may be described mathematically in a number of ways. In a one-dimensional scenario it is commonly described by the following continuity and dynamic equations:

(continuity) $$W \frac{\partial h}{\partial t} + \frac{\partial Q}{\partial x} = L \qquad (1)$$

(dynamic)
$$\frac{\partial Q}{\partial t} + \frac{\partial}{\partial x}\left[\frac{Q^2}{A}\right] + gA\frac{\partial h}{\partial x} + gAS_f + \frac{gA\lambda R}{2(1+\lambda S)}\frac{\partial S}{\partial x} = 0 \quad (2)$$

where W is the water surface width, h is the water level relative to a datum, Q is the discharge, L is the lateral inflow per metre run, A is the cross-sectional area of flow, g is the gravitational acceleration, S_f is the friction slope (= Q^2/K^2, where K is the conveyance), R is the hydraulic radius, λ is the constant in the equation of state for salt, S is salinity, x is the longitudinal co-ordinate direction and t is time.   As well as the usual simplifications and assumptions associated with a one-dimensional approach, equations (1) and (2) ignore the effects of wind stresses, Coriolis forces and cross-sectional density gradients, and also imply that the lateral inflow imposes no momentum drag on the flow in the estuary and that the Boussinesq (or momentum) coefficient is unity (Cunge et al. (8)).   The last term in equation (2), which represents the effects of the longitudinal density term, assumes vertically well mixed conditions and that the depth of flow may be approximated by the hydraulic radius.   Since analytical solutions are not available to these equations recourse is usually made to numerical techniques such as characteristics (Abbott (9)), finite differences (Abbott (9), Cunge et al. (8)) or finite elements (Chung (10), Baker (11)).   In this work, the equations have been solved using the Preissman finite difference scheme (Cunge et al. (8)).   The main advantages of the scheme are that it is implicit, which allows the use of large time steps;   it caters for a non-uniform spatial discretisation of the estuary, which allows the computational grid to reflect the longitudinal non-uniformity of estuary geometry; and its practical and theoretical performance is well documented (Abbott (9), Cunge et al. (8), Samuels (12)).

In applying the scheme to equations (1) and (2) the following difference operators are used to replace temporal gradients, spatial gradients and coefficients, respectively:

$$\frac{\partial F}{\partial t} = \frac{1}{2\Delta t}(F^*_{i+1} - F_{i+1} + F^*_i - F_i) \quad (3)$$

$$\frac{\partial F}{\partial x} = \frac{1}{\Delta x_i}\left\{\theta(F^*_{i+1} - F^*_i) + (1-\theta)(F_{i+1} - F_i)\right\} \quad (4)$$

$$F = \frac{1}{2}\left\{\theta(F^*_{i+1} + F^*_i) + (1-\theta)(F_{i+1} + F_i)\right\} \quad (5)$$

where F is a function of x and t, i is the spatial index, Δx_i is the ith space step, Δt is the time step and θ is a weighting coefficient.   The superscript * denotes values of F at time t + Δt while the absence of the superscript denotes values at time t.   The result of using these operators is a pair of difference equations which are centred in space but weighted over time by θ.   A linear Fourier series analysis shows that the scheme is stable for $0.5 \leq \theta \leq 1$ and that θ determines the numerical behaviour of the scheme for any combination of Δt, Δx_i and flow conditions.   In practice, local instabilities may occur during the solution of the complete (non-linear) difference equations.   This is most likely to happen when the cross sectional area of flow varies rapidly with x (Samuels (12)) and/or when θ approaches 0.5 and/or when the friction term is small.   If, however, the friction term dominates the dynamic equation, then the resulting physical damping will usually dissipate any instabilities which may be generated. Alternatively, θ may be adjusted so that any instabilities are controlled by the numerical damping.

After substitution of equations (3)-(5) into equations (2) and (1), the following equations may be obtained for the ith space step:

$$h^*_{i+1} - h^*_i + \alpha_i Q^*_{i+1} + \beta_i Q^*_i = \gamma_i \tag{6}$$

$$\delta_i h^*_{i+1} + \delta_i h^*_i + Q^*_{i+1} - Q^*_i = \epsilon_i \tag{7}$$

where the coefficients $\alpha_i - \epsilon_i$ are defined in Appendix 1. For n space steps we have 2n equations and 2n + 2 unknown elevations and discharges. By specifying two of these as measured boundary conditions we reduce the system of equations to a tri-diagonal matrix of unknowns which is easily solved using a double sweep algorithm (Abbott (9), Cunge et al. (8)). This involves the successive solution of the following pair of recurrence equations for i=n to 1:

$$Q^*_i + q_i h^*_{i+1} + t_i Q^*_{i+1} = s_i \tag{8}$$

$$h^*_{i+1} + p_{i+1} Q^*_{i+1} = r_{i+1} \tag{9}$$

where the coefficients $p_{i+1} - t_i$ are defined in Appendix 2. The normal choice of boundary conditions is h_1 and Q_{n+1} for which $p_1 = 0$ and $r_1 = h_1^*$.

The solution starts from a prescribed initial condition and advances through each time step iteratively. This is necessary because the coefficients α_i, β_i and γ_i are functions of unknown variables and coefficients which derive from the non-linear nature of equations (1) and (2). In Appendix 1 these unknowns are denoted by the superscript $\wedge$ which indicates that they are the latest estimates of the values obtained when the iteration has converged. For each time step, the first solution is based on values of these unknowns pertaining at the convergence of the solution over the previous time step, and with each iteration they are updated according to the latest solution.

3.2 Solute transport model

One dimensional solute transport in estuaries is usually assumed to be described by the following advection - dispersion equation:

$$\frac{\partial}{\partial t}(AC) + \frac{\partial}{\partial x}(QC) = \frac{\partial}{\partial x}\left[AD\frac{\partial C}{\partial x}\right] - kAC + LC_o \tag{10}$$

where A, Q, L, x and t are as previously defined, C is the cross-sectional area average solute concentration, D is the longitudinal dispersion coefficient, k is the solute mass decay rate and C_o is the solute concentration in the lateral inflow. In this work, to date, the last term in equation (10) has been omitted (See section 4). The most important assumption implied in this equation is that the effects of all mechanisms responsible for longitudinal dispersive transport are lumped together and represented by a Fickian type gradient diffusion term employing an effective diffusion coefficient.

Advances in water modelling and measurement

Various numerical methods have been used to solve equation (10) (Stone & Brian (13), Pinder & Gray (14), Cunge et al. (8), Sauvaget (15)). Here, the Galerkin finite element method (Becker et al. (16), Pinder & Gray (14), Chung (10)) has been used for the spatial discretisation, with the solute concentration and all coefficients being assumed to vary linearly within each element (or space step). The Preissman finite difference operators (equations (3) and (5)) are used to replace the temporal gradient and to weight the terms in the discretised equations over the time step. The scheme is similar to that used by O'Connor & Thompson (17) and Nassehi & Williams (18) but is more rigorous in its treatment of the coefficients. The main advantage of the method is that the computational grid is exactly the same as for the hydrodynamic model from which values of discharge and cross-sectional area are supplied. It is also a robust scheme which performs encouragingly well even when advection dominates dispersion.

After the application of the Galerkin method to equation (10) the following pair of simultaneous equations may be obtained for the ith space step:

$$a_i C_i^* + b_i C_{i+1}^* = d_i C_i + e_i C_{i+1}$$

$$\left. \begin{array}{c} \\ \\ \end{array} \right\} \quad (11)$$

$$f_i C_i^* + g_i C_{i+1}^* = \ell_i C_i + m_i C_{i+1}$$

where the coefficients a_i - m_i are defined in Appendix 3. On assembling the contributions of all the elements a tri-diagonal global element matrix is obtained which is solved via a double sweep algorithm using the following recurrence equation for i = n to 2:

$$x_i C_i^* + y_i C_{i+1}^* = z_i \quad (12)$$

where the coefficients x_i, y_i and z_i are defined in Appendix 4. Assuming C_1 and C_{n+1} are given boundary conditions, we have $x_2 = g_1 + a_2$, $y_2 = b_2$ and $z_2 = \ell_1 C_1 + (m_1 + d_2)C_2 + e_2 C_3 - f_2 C_1^*$. As before, the weighting parameter θ determines the numerical behaviour for any combination of Δt, Δx_i, Q, A and D, but here we have found it preferable that $\theta = 0.5$ because simulations showed that solutions tend to be over dispersed when $0.5 < \theta \leq 1$ (see also Pinder & Gray (14)).

4. Application of the models to the Clyde estuary

The upper estuary was represented by 30 grid points in the computational plane. The space step varied between about 300 m and 600 m in accordance with the longitudinal variations of channel geometry. During computations, the geometrical coefficients (water surface width, cross-sectional area and hydraulic radius) at each grid point were evaluated by interpolation of a stored table of their values as a function of water level, which was compiled from hydrographic charts of the estuary.

Data for the boundary conditions of the hydrodynamic model were supplied by an automatic tide gauge at Rothesay dock (operated by the Clyde Port Authority) and the Daldowie gauging station on the river Clyde (operated by CRPB). Lateral flows, into and out of the dynamic storage areas, were estimated from the product of plan area and water level change during a time step and were distributed over the lengths of the space steps in which they occurred. Other inputs were treated similarly: flows from the river Kelvin being supplied by the Killermont gauging station (operated by CRPB) and the discharge from the Shieldhall sewage treatment works being specified as 2.5 m^3/s throughout (CRPB (4)). In all runs the longitudinal salinity gradient was held constant and was based on salinities predicted by regression equations developed from routine surveys (CRPB (19)).

The solute model will fulfil a number of requirements in the larger microbiological pollution study. So far, the conservative (k = 0) version has been used to model the salinity variations in the upper estuary. As well as providing the salinity field, for use in other parts of the fully integrated pollution model, its calibration yields values of the longitudinal dispersion coefficient for use elsewhere. Data for the Rothesay dock boundary came from measurements undertaken by the study team, and the salinity at the tidal weir was assumed to be zero. The salinity of all the other inputs was also assumed to be zero and the effects of the dynamic storage areas were neglected.

5. Calibration and verification

5.1 Hydrodynamic model

The hydrodynamic model was calibrated using a 12 hour water level data set measured by the Clyde Port Authority in October 1987. As well as records from two automatic tide gauges (Rothesay dock and Broomielaw) readings from tide boards at King George V dock and Prince's dock were also used. After a number of simulation exercises, a time step of 15 minutes was selected and θ was chosen to be 0.6. The model was run for a number of constant values of the Darcy-Weisbach friction coefficient, f (for which $K^2 = 8gRA^2/f$) and the root mean square error (based on the three calibration locations) was evaluated. The minimum error (0.076 m) occurred for a friction coefficient of 0.03.

Figure 3 shows the results of a verification exercise using this calibration, and compares the predicted water level at Broomielaw with measurements from the tide gauge. Figure 4 compares predicted velocities with measurements of the centre line depth average velocity, undertaken by the study team and co-operating institutions (West et al. (7)), at Shieldhall for the same day. Both figures show that the model is in good agreement with the measurements. Figure 5 shows another water level comparison at Broomielaw, on this occasion for an extremely high tide which was caused by unusual meteorological conditions of very low pressure accompanied by high river flow, and Figure 6 shows a second velocity comparison at Shieldhall. Again the agreement is good.

5.2 Salt transport model

The salt transport model has been calibrated using data collected by the study team. Surveys (of varying degrees of success) were conducted during which an approximately instantaneous longitudinal distribution of centre line depth average salinity was measured, followed by similar salinity measurements every 30 minutes at Rothesay dock and at certain times at other locations.

Figure 7 shows model output and measurements of the longitudinal distribution of salinity at about 4 hours after high water for a neap tide in August 1987. The hydrodynamic model was run from the beginning of the previous tidal cycle and the salt model was started at about high water. Within the salt model θ was specified as 0.5 and the time step was 15 minutes. The model results were obtained with dispersion coefficients held constant in time but varying spatially in a linear fashion from 100 m^2/s at the tidal weir to 900 m^2/s at Rothesay dock. The agreement is promising, but since at present we have no other reliable data sets with which to verify the calibration, it must be considered as preliminary and approximate.

6. Discussion

One approximation in the application of the models is the treatment of the landward boundary, i.e. assuming that the Daldowie gauged flows are applicable at the tidal weir and that the salinity there is always zero. For neap tides this is probably good enough, but for spring tides when flow reversal occurs at the weir it is not a very good representation of reality. Nevertheless, the performance of both the hydrodynamic and the solute models is encouraging, and there is no evidence from the comparison of predicted and measured velocities that the neglect of storage upstream of the weir is too serious an omission. Of course, there is an element of uncertainty in these velocity comparisons because we are comparing cross-sectionally averaged model values with depth average centre line measurements.

The water levels and velocities shown in Figures 3 - 6 illustrate well the effects of the high frequency components in the tidal oscillation, with the variation of velocity during the flood tide being particularly characteristic of the estuary. They also show that using a constant resistance coefficient in the model is perfectly adequate in this part of the estuary. Indeed, there is little justification for allowing it to vary since the channel is relatively uniform and the water is deep.

The calibration of the solute model, on the other hand, is less well founded. It was anticipated that calibrating it would be more difficult than calibrating the hydrodynamic model because the longitudinal dispersion coefficient is known to be strongly dependent on flow conditions, see e.g. theoretical (Fischer (1), Smith (20)) and practical (James and Park (21), West and Mangat (22)) studies. There is also the additional complication of the complex flow structure of the Clyde estuary (West et al. (7)) which is associated with the stratified nature of the flow. Unfortunately, the collection of the amount of field data necessary to elucidate the variation of the dispersion coefficient throughout the estuary is beyond the study team at present, and therefore the calibration used is simple and approximate. Indeed, our attempts to collect sufficient good quality salinity data for simple calibrations have not yet been particularly successful.

A number of other simple calibrations were investigated but the one given in the previous section produced the best all round agreement. The form of the spatial variation of the dispersion coefficient is based on the assumption that vertical shear is the primary source of dispersive transport, combined with the knowledge that velocities increase in magnitude in a seawards direction. Paradoxically, but for reasons of simplicity, the effects of temporal changes in velocity have not been taken into account.

The results of flow structure surveys lend support to the magnitude of the dispersion coefficients used and illustrate well the stratified flow conditions which tend to prevail in the estuary. Figure 8 shows measurements of the centre line vertical profiles of salinity and velocity on a number of occasions near the Shieldhall sewage treatment works. Following West and Mangat (22) depth average longitudinal dispersive salt fluxes were calculated from these profile data and were typically of the order of 0.5 kg/m^2s (landward). For a typical longitudinal depth average salinity gradient of 10^{-3} kg/m^4, this yields a dispersion coefficient of 500 m^2/s, which is similar to the value used in the model at this location.

Since dispersion coefficients are large and the time scale is short, the results presented here do not do justice to the wider ability of the model to describe constituent transport in estuaries. Testing of the model, however, has shown that it has favourable characteristics with regard to mass conservation and numerical dispersion, and that it simulates the expected features when run over longer time scales under a variety of tidal, fluvial and constituent concentration regimes.

Figure 7 illustrates an important characteristic of ebb tides in the upper Clyde estuary which is also apparent in other data sets, namely the approximately steady state salinity field. This is consistent with the large dispersion coefficients since a steady state salinity field implies a balance between advective and dispersive transport. On flood tides, when advective and dispersive fluxes are expected to be both landward (assuming a Fickian type of process is applicable for dispersive transport), field data confirm that depth average salinities do tend to increase, albeit by only ~ 2 kg/m^3. Flow structure measurements have usually yielded landward dispersive fluxes during flood tides, but on one occasion seaward fluxes were found (see James and Park (21) for details of similar results for the Tyne estuary). Interestingly, on this occasion the depth average salinity remained constant during the three hours of flood tide measurements.

Although salinity field is approximately steady state on the time scale of a tidal cycle, over longer time scales it is greatly influenced by the river flow. During high river flow events the salt is pushed seaward, and in the most landward reaches, vertically homogeneous conditions can result and depth averaged velocities may be seawards throughout the tidal cycle, as shown in Figure 5.

To some extent, the prevalent stratified flow conditions call into question the propriety of using a one-dimensional model. Clearly, there are some important vertical variations in parameters which are averaged out in the one-dimensional approach. This will probably be of greater significance in the microbiological model because the two parameters which most effect the mortality of the bacteria, namely the intensity of ultra-violet radiation and salinity, are both functions of depth. Also most of the bacterial inputs to the estuary occur in the top few metres, and with vertical mixing being weak, the bacteria tend to stay in the upper layers. Evidence from a number of field work activities suggests that transverse mixing is much more efficient than vertical mixing, so that either a simple two-layer or a more complex, multi-layer, width averaged model would seem to offer a sensible avenue for future work. In the meantime, however, the fully integrated one-dimensional microbiological model will be completed and assessed.

7. Conclusions

A one-dimensional transport model, comprising a hydrodynamic and a solute model, has been applied to the upper 11 km of the Clyde estuary as the first part of a larger study of microbiological pollution of coastal waters. The hydrodynamic model has performed well, but the solute model remains unverified, at the time of writing, because of a shortage of field data. The calibration proposed for an ebb tide, however, although approximate, is consistent with estimates of the longitudinal dispersion coefficient made from flow structure measurements. Bearing in mind previous work and the complex flow structure of the estuary, it is likely that a different calibration may be required for flood tides. The models represent well both the temporal changes in water level and velocity and the approximately steady state salinity field, which appears to be characteristic of ebb tides.

In the near future a non-conservative version of the solute model will be applied to E. coli concentrations in the estuary. This will require the dispersion coefficients used in the calibration of the salinity model (assuming them to be the same for both constituent species) and bacterial mortality rates. The latter are currently being investigated by the study team through carefully controlled laboratory experiments. The final model will be fully integrated, allowing the longitudinal density term in the hydrodynamic model and the mortality rate in the bacteria model (which is a function of salinity, among other things) to be updated from the salinity model. Some account of suspended solids concentrations will also need to be included.

In the longer term, it is likely that a two-dimensional, width averaged model will also be developed. This should offer considerable improvements since the vertical flow structure and variation of model parameters will be retained explicitly, rather than being averaged as in the present one-dimensional model.

8. Acknowledgements

The work was carried out under a contract with the Department of the Environment and publication of the paper is with their agreement.

The authors would also like to thank the following for their assistance: Mr. D.

Thomson of the Clyde Port Authority for the supply of tide level and hydrographic data, and many useful discussions; the Sewerage Department, Strathclyde Regional Council for access to Shieldhall sewage treatment works and use of the quay; and Dr. J.R. West, Dr. I. Guymer and Mr. N.M. Lynn for their assistance and advice with field work.

9. References

1. Fischer, H.B. et al: "Mixing in inland and coastal waters". London, UK, Academic Press, 1979.

2. Milne, D.P. et al: "The effect of estuary type suspended solids on survival of E. coli in saline waters". Presented at IAWPRC International Conference (Brighton, UK: July 17-23, 1988).

3. Riddell, J.F.: "Clyde navigation - a history of the development and deepening of the river Clyde". Edinburgh, UK, John Donald Publishers, 1979.

4. Clyde River Purification Board: "Water quality - a ten year review 1976-1985". Glasgow, UK, 1986.

5. Admiralty tide tables, volume 1, 1988. Published by the hydrographer of the navy.

6. Thomson, A.: "A hydraulic investigation of the Clyde estuary". PhD Thesis, University of Strathclyde, Glasgow, UK, 1969.

7. West, J.R. et al: "A preliminary investigation into the flow structure of the upper Clyde estuary". Dept. of Civil Engineering, University of Birmingham, Birmingham, UK, 1988.

8. Cunge, J.A. et al: "Practical aspects of computational river hydraulics". London, UK, Pitman, 1980.

9. Abbott, M.B.: "Computational hydraulics - elements of the theory of free surface flows". London, UK, Pitman, 1979.

10. Chung, T.J.: "Finite element analysis in fluid dynamics". New York, USA, McGraw-Hill, 1978.

11. Baker, A.J.: "Finite element computational fluid mechanics" (International student edition). Singapore, McGraw-Hill, 1985.

12. Samuels, P.G.: "Computational modelling of open channel flow - an analysis of some practical difficulties". Report IT 273, Hydraulics Research, Wallingford, UK, 1984.

13. Stone, H.L. and Brian, P.T.: "Numerical solution of convective transport problems". Am.Inst.Chem.Engng.J., No. 9, 1963, pp 681-688.

14. Pinder, G.F. and Gray, W.G.: "Finite element simulation in surface and subsurface hydrology". London, UK, Academic Press, 1977.

15. Sauvaget, P.: "Dispersion in rivers and coastal waters - 2. Numerical computation of dispersion". Chapter 2 of Developments in hydraulic engineering - 3. P. Novak (ed). London, UK, Elsevier, 1985.

16. Becker, E.B. et al: "Finite elements: an introduction". New Jersey, USA, Prentice-Hall, 1981.

17. O'Connor, B.A. and Thompson, G.: "A mathematical model of chloride levels in the Wear estuary". In: Proc. International Symposium on Unsteady Flow in Open Channels (Newcastle-upon-Tyne, UK: April 12-15, 1976), Cranfield, UK, BHRA, The Fluid Engineering Centre, 1976, Paper H1-1.

18. Nassehi, V. and Williams, D.J.A.: "Mathematical model of upper Milford Haven - a branching estuary". Estuarine, Coastal and Shelf Science, 23, 1986, pp.403-418.

19. Clyde River Purification Board: "Topographic, hydrographic and loading data for the Clyde estuary". Technical report No. 48, Glasgow, UK.

20. Smith, R.: "Mixing and dispersion in estuaries: which mathematical model and when?". In: Proc. International Conference on Water Quality Modelling in the Inland Natural Environment (Bournemouth, UK: June 10-13, 1986), Cranfield, UK, BHRA, The Fluid Engineering Centre, 1986, Paper K1, pp. 375-385.

21. James, A. and Park, J.K.: "Modelling of pollutant dispersion in estuaries". In: Proc. International Conference on Water Quality Modelling in the Inland Natural Environment (Bournemouth, UK: June 10-13, 1986), Cranfield, UK, BHRA, The Fluid Engineering Centre, 1986, Paper J2, pp. 345-357.

22. West, J.R. and Mangat, J.S.: "The determination and prediction of longitudinal dispersion coefficients in a narrow, shallow estuary". Estuarine, Coastal and Shelf Science, 22, 1986, pp. 161-181.

Appendix 1

$$\alpha_i = \left[c_2 + 2\hat{Q}_{i+1}/\hat{A}_{i+1}\right]/c_1 + |\hat{Q}_{i+1}|/c_4$$

$$\beta_i = \left[c_2 - 2\hat{Q}_i/\hat{A}_i\right]/c_1 + |\hat{Q}_i|/c_4$$

$$\gamma_i = c_3\left[(h_i - h_{i+1}) - (|Q_i|Q_i + |Q_{i+1}|Q_{i+1})/c_4\right]$$

$$+ \left[c_2(Q_{i+1} + Q_i) - 2c_3(Q_{i+1}^2/A_{i+1} - Q_i^2/A_i)\right]/c_1$$

$$- c_6\left[S_{i+1}^* - S_i^* + c_3(S_{i+1} - S_i)\right]/\left[2(2+c_7)\right]$$

$$\delta_i = c_2 c_5/4$$

$$\epsilon_i = \delta_i(h_{i+1} + h_i) + c_3(Q_i - Q_{i+1}) + \hat{L}_i \Delta x_i/\theta$$

where:

$$c_1 = g[\theta(\hat{A}_i + \hat{A}_{i+1}) + (1-\theta)(A_i + A_{i+1})]$$

$$c_2 = \Delta x_i/(\theta \Delta t) ; \quad c_3 = (1-\theta)/\theta$$

$$c_4 = \left[\theta(\hat{K}_i^2 + \hat{K}_{i+1}^2) + (1-\theta)(K_i^2 + K_{i+1}^2)\right]\Delta x_i$$

$$c_5 = \left[\theta(\hat{W}_i + \hat{W}_{i+1}) + (1-\theta)(W_i + W_{i+1})\right]$$

$$c_6 = \lambda\left[\theta(\hat{R}_i + \hat{R}_{i+1}) + (1-\theta)(R_i + R_{i+1})\right]$$

$$c_7 = \lambda\left[\theta(S_i^* + S_{i+1}^*) + (1-\theta)(S_i + S_{i+1})\right]$$

Appendix 2

$$P_{i+1} = \left[1 + t_i(1 + \delta_i P_i)\right]/\left[\delta_i + q_i(1 + \delta_i P_i)\right]$$

$$q_i = 1/(\beta_i + P_i)$$

Advances in water modelling and measurement

$$r_{i+1} = \left[\epsilon_i - \delta_i r_i + s_i(1 + \delta_i p_i)\right] / \left[\delta_i + q_i(1 + \delta_i p_i)\right]$$

$$s_i = (\gamma_i + r_i)/(\beta_i + p_i)$$

$$t_i = \alpha_i/(\beta_i + p_i)$$

Appendix 3

$$a_i = \Delta x_i \hat{A}1/(12\Delta t) + \tilde{Q}1\theta/6 + \widetilde{AD\theta}/(6\Delta x_i) + \Delta x_i \widetilde{DT1\theta}/60$$

$$b_i = \Delta x_i \hat{A}2/(12\Delta t) + \tilde{Q}2\theta/6 - \widetilde{AD\theta}/(6\Delta x_i) + \Delta x_i \widetilde{DT2\theta}/60$$

$$d_i = \Delta x_i A1/(12\Delta t) - \tilde{Q}1(1-\theta)/6 - \widetilde{AD(1-\theta)}/(6\Delta x_i) - \Delta x_i \widetilde{DT1(1-\theta)}/60$$

$$e_i = \Delta x_i A2/(12\Delta t) - \tilde{Q}2(1-\theta)/6 + \widetilde{AD(1-\theta)}/(6\Delta x_i) - \Delta x_i \widetilde{DT2(1-\theta)}/60$$

$$f_i = \Delta x_i \hat{A}3/(12\Delta t) + \tilde{Q}3\theta/6 - \widetilde{AD\theta}/(6\Delta x_i) + \Delta x_i \widetilde{DT3\theta}/60$$

$$g_i = \Delta x_i \hat{A}4/(12\Delta t) + \tilde{Q}4\theta/6 + \widetilde{AD\theta}/(6\Delta x_i) + \Delta x_i \widetilde{DT4\theta}/60$$

$$\ell_i = \Delta x_i A3/(12\Delta t) - \tilde{Q}3(1-\theta)/6 + \widetilde{AD(1-\theta)}/(6\Delta x_i) - \Delta x_i \widetilde{DT3(1-\theta)}/60$$

$$m_i = \Delta x_i A4/(12\Delta t) - \tilde{Q}4(1-\theta)/6 - \widetilde{AD(1-\theta)}/(6\Delta x_i) - \Delta x_i \widetilde{DT4(1-\theta)}/60$$

where,

$$A1 = 3A_i + A_{i+1}; \quad A2 = A3 = A_i + A_{i+1}; \quad A4 = A_i + 3A_{i+1}$$

$$\hat{A}1 = 3\hat{A}_i + \hat{A}_{i+1}; \quad \hat{A}2 = \hat{A}3 = \hat{A}_i + \hat{A}_{i+1}; \quad \hat{A}4 = \hat{A}_i + 3\hat{A}_{i+1}$$

$$\tilde{Q}1 = \theta(-4\hat{Q}_i + \hat{Q}_{i+1}) + (1-\theta)(-4Q_i + Q_{i+1})$$

$$\tilde{Q}2 = \theta(\hat{Q}_i + 2\hat{Q}_{i+1}) + (1-\theta)(Q_i + 2Q_{i+1})$$

$$\tilde{Q}3 = \theta(-2\hat{Q}_i - \tilde{Q}_{i+1}) + (1-\theta)(-2Q_i - Q_{i+1})$$

34

$$\tilde{Q4} = \theta(-\hat{Q}_i + 4\hat{Q}_{i+1}) + (1-\theta)(-Q_i + 4Q_{i+1})$$

$$\tilde{AD} = \theta\hat{AD} + (1-\theta)AD$$

$$\tilde{DT1} = \theta\hat{DT1} + (1-\theta)DT1 \quad ; \quad \tilde{DT2} = \theta\hat{DT2} + (1-\theta)DT2$$

$$\tilde{DT3} = \theta\hat{DT3} + (1-\theta)DT3 \quad ; \quad \tilde{DT4} = \theta\hat{DT4} + (1-\theta)DT4$$

$$AD = 2A_iD_i + A_{i+1}D_i + A_iD_{i+1} + 2A_{i+1}D_{i+1}$$

$$\hat{AD} = 2\hat{A}_i\hat{D}_i + \hat{A}_{i+1}\hat{D}_i + \hat{A}_i\hat{D}_{i+1} + 2\hat{A}_{i+1}\hat{D}_{i+1}$$

$$DT1 = 12A_ik_i + 3(A_ik_{i+1} + A_{i+1}k_i) + 2A_{i+1}k_{i+1}$$

$$\hat{DT1} = 12\hat{A}_i\hat{k}_i + 3(\hat{A}_i\hat{k}_{i+1} + \hat{A}_{i+1}\hat{k}_i) + 2\hat{A}_{i+1}\hat{k}_{i+1}$$

$$DT2 = DT3 = 3A_ik_i + 2(A_ik_{i+1} + A_{i+1}k_i) + 3A_{i+1}k_{i+1}$$

$$\hat{DT2} = \hat{DT3} = 3\hat{A}_i\hat{k}_i + 2(\hat{A}_i\hat{k}_{i+1} + \hat{A}_{i+1}\hat{k}_i) + 3\hat{A}_{i+1}\hat{k}_{i+1}$$

$$DT4 = 2A_ik_i + 3(A_ik_{i+1} + A_{i+1}k_i) + 12A_{i+1}k_{i+1}$$

$$\hat{DT4} = 2\hat{A}_i\hat{k}_i + 3(\hat{A}_i\hat{k}_{i+1} + \hat{A}_{i+1}\hat{k}_i) + 12\hat{A}_{i+1}\hat{k}_{i+1}$$

Appendix 4

$$x_i = g_i + a_{i+1} - (f_iy_{i-1}/x_{i-1})$$

$$y_i = b_i$$

$$z_i = \ell_iC_{i-1} + (m_i + d_{i+1})C_i + e_{i+1}C_{i+1} - (f_iz_{i-1}/x_{i-1})$$

35

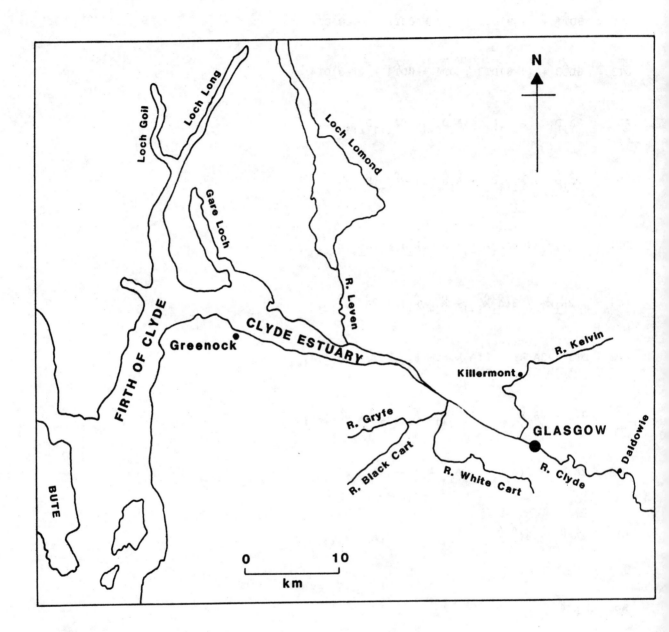

Figure 1   Location map.

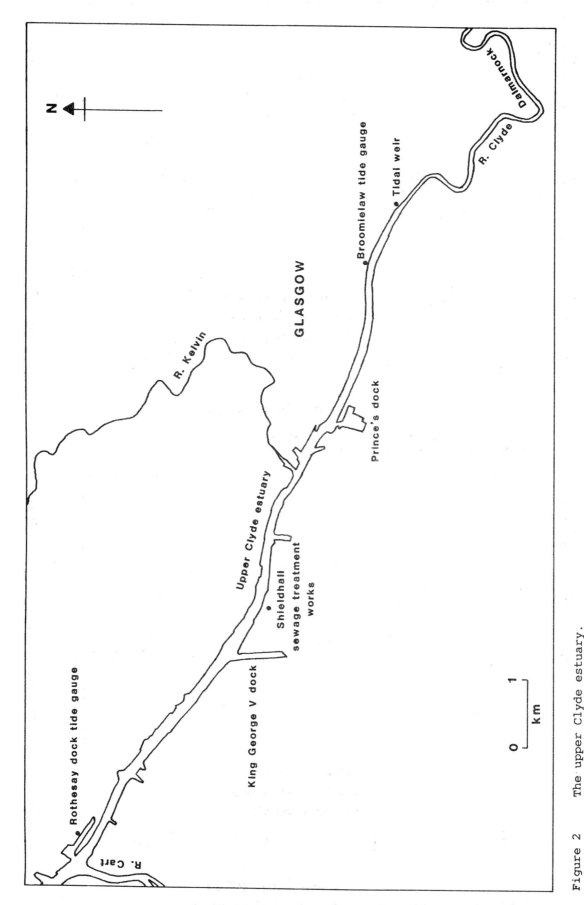

Figure 2   The upper Clyde estuary.

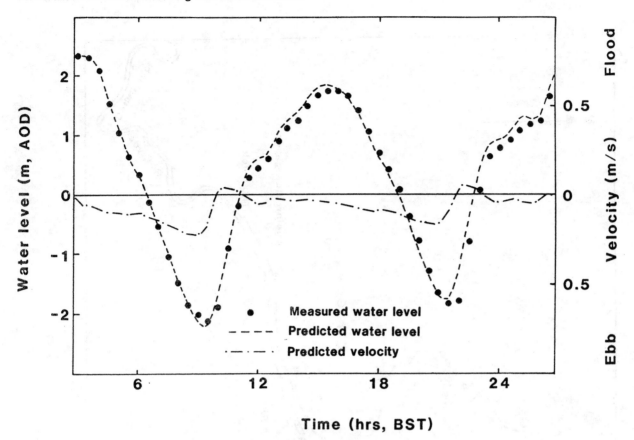

Figure 3   Comparison of predicted and measured water levels at
            Broomielaw on 13th July, 1987.

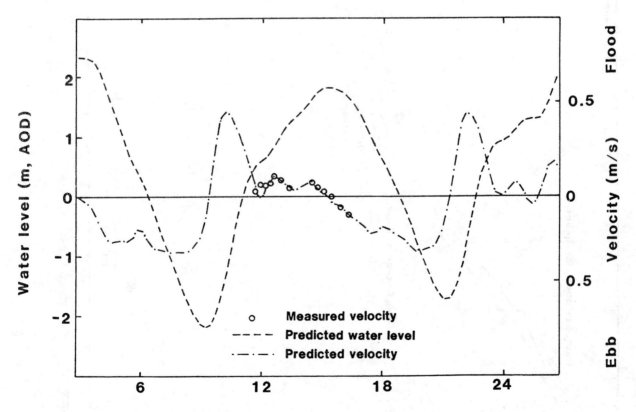

Figure 4   Comparison of predicted and measured velocities at
            Shieldhall on 13th July, 1987.

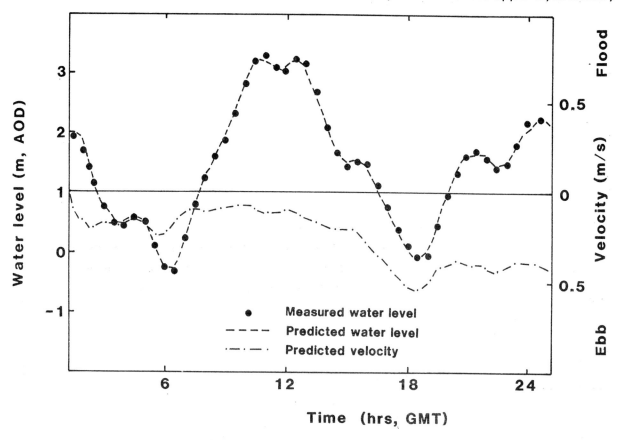

Figure 5   Comparison of predicted and measured water levels at
             Broomielaw on 1st February, 1988.

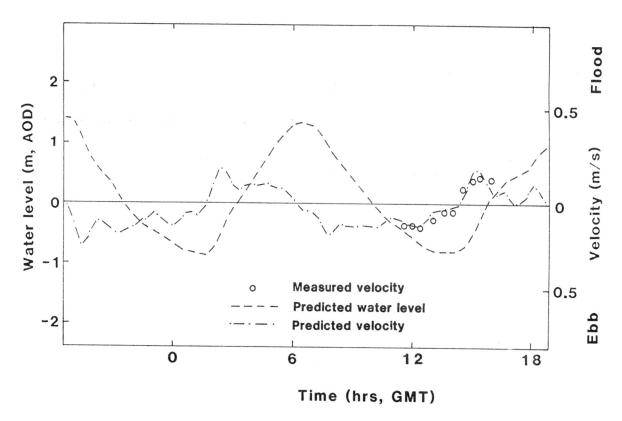

Figure 6   Comparison of predicted and measured velocities at
             Shieldhall on 26th February, 1988.

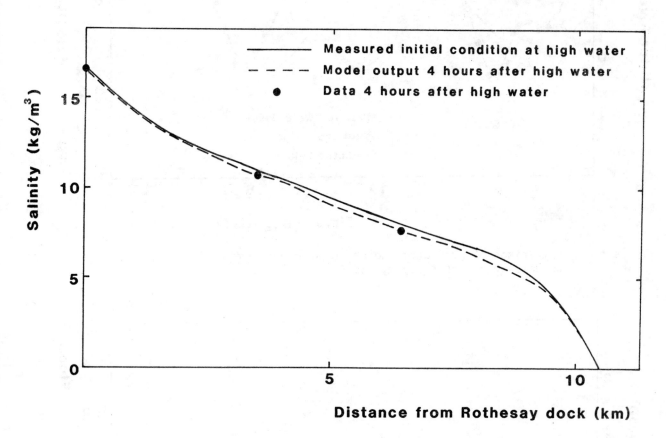

Figure 7   Comparison of predicted and measured salinities in the
            estuary on 7th August, 1987 (calibration run).

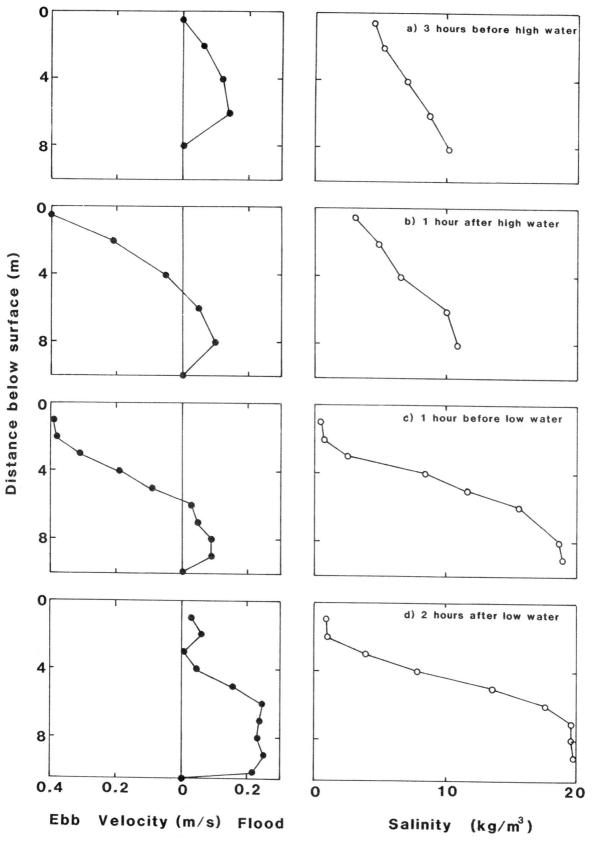

Figure 8   Measurements of centre line vertical profiles of velocity
            and salinity on 13th July, 1987 (a,b) and 26th February,
            1988 (c,d).

Chapter 4
MODELLING MUD TRANSPORT IN A MACROTIDAL ESTUARY

P Le Hir, P Bassoulet and J L'Yavanc
Institut Français pour l'Exploitation de la Mer (IFREMER)
France

Summary

In order to determine the fate of fine particles and adsorbed contaminants in a
short macrotidal estuary, a mathematical modelling system for cohesive suspended
mater transport and bottom evolution, including a new consolidation algorithm, is
developed. An analysis of successive erosion, transport and deposition processes
due to tidal forcing is given and comparisons with field data are provided. A
description of mean-term bottom variations has been attempted and is partially
validated by differential soundings.

1. Introduction

The urban sewage of Morlaix, a 30 000 inhabitants town on the northern coast of
Brittany (France) is discharged in the upper part of an estuary. In order to know
the fate of the effluent and its impact downstream, especially over shellfishing
areas, a multidisciplinary environmental study has been carried out. As many
pollutants may be adsorbed on the particles, turbidity variations and mud
transport have to be considered.

For a better understanding of different integrated phenomena and to enable further
calculations of concentrations for several adsorbed and/or dissolved constituents
interacting with each other, mathematical modelling seems the best way to study
the sedimentary processes. However, to face the lack of basic knowledge on the
phenomena involved in erosion and sedimentation, many field data have to be
collected and taken into account, either to calibrate the model, or to validate
it.

2. The study area

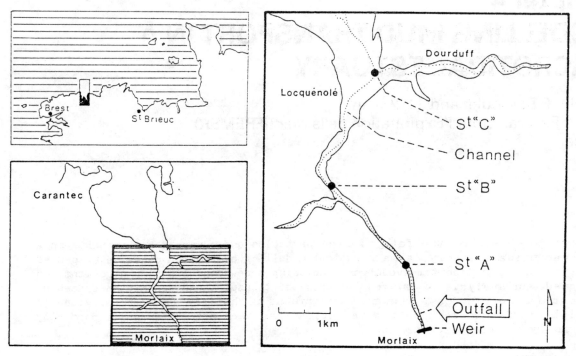

Figure 1 : Location of the Estuary of Morlaix.

2.1 Geographical presentation

The estuary of Morlaix stricto sensu is very short (about 6 km long) with a mouth situated in a bay opened to the English Channel (Figure 1). Upstream a weir closes a dock with a little harbour. An affluent so called Dourduff river joins the Morlaix river just before the bay. Figure 2 shows off typical cross-sections with extensive intertidal mudflats. Actually, at low tide, channel widths vary from 10 to 100 m in the two rivers, whereas at high tide the estuary is 50 m wide upstream and 1 000 m at the beginning of the bay. More often than not, cross-sections present a series of flat mud benches seperated with steep slopes (until 20 %).

The two rivers (Morlaix and Dourduff) have a small drainage basin (270 km^2) so that their mean discharges are very low, respectively 3 $m^3.s^{-1}$ and 0,8 $m^3.s^{-1}$.

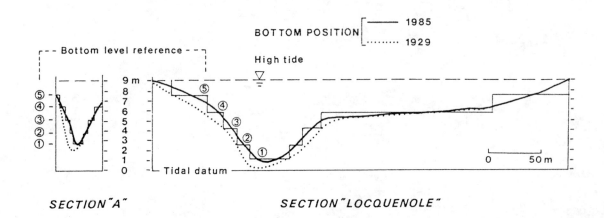

Figure 2 : Typical cross-sections of the Estuary of Morlaix. The intertidal area is represented by 5 bottom level references that are spread over the tidal variation.

2.2 Hydrodynamics

The main hydrodynamic forcing in the area is the tide : the level variation is 7,6 m for a mean spring tide and as the estuary is very shallow, which leads to empty and fill it up at each tide. Flood and ebb currents reach 1,2 m.s^{-1} and 0,8 m.s^{-1} respectively, with a classical but strong asymmetry induced by the tide propagation over depth varying area (see figure 3).

Water mass trajectories have been computed by the means of a mathematical model (Salomon and Breton, 6) : on spring tide, water leaving the weir upstream at high tide can reach the bay at low tide and come back within 500 m from the starting point during the next flood, depending on the river discharge.

Except on neap tide and for high freshwater input, the estuary is nearly well mixed and density currents are weak. However, current speed induced by the river flow at low water is never negligible, as the channel is very narrow.

Generally speaking, wind induced currents and short waves effects are weak because the bay is protected by a prominent peninsula and a series of little islands off the outlet. But some exceptional events may contribute to a redistribution of particles inside the bay.

2.3 Sedimentology

From the weir to the confluence with the Dourduff river, the channel mainly consists of pebbles. On both sides mudbanks are essentially constituted of particles less than 30 μm. However, near the outlet of the Dourduff river, the sediment becomes more silty. Everything goes on as though inputs from Morlaix river were "selected" by the dock located upstream the weir, most coarse particles settling in that dock (mean input to the estuary : 12 g.m^{-3}) [Bassoullet et al., 1]. Concentration ranges of that surficial sediments expressed in dry weight reach 400 to 650 kg.m^{-3} at flood period and 300 to 480 kg.m^{-3} at ebb period according to the position in the transect.

Concerning the microgranulometry of suspended particulate matters, along the 6 first kilometres, particles are less than 10 μm for 90 % except at low water slack

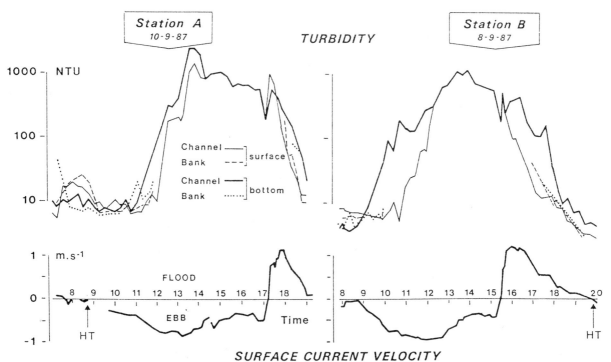

Figure 3 : Estuary of Morlaix. In situ measurements at fixed stations. Spring tide - low river discharge (1 NTU $\simeq$ 3 g.m^{-3}). [See Fig. 1 for location].

of spring tide situation where particles greater than 10 µm reached 70 to 80 per cent. Maxima contents of these suspended particulate matters are on average in the region of 40 g.m⁻³ (dry weight) but for low water slack on spring tide situation they can reach some grams per litre within the turbidity maximum (figure 3).

3. Models description

3.1 Hydrodynamics

Considering the tidal forcing as the main factor of the estuarine circulation, vertically averaged models have been run in the area : a two dimensional one in the bay and a one dimensional model to represent the two estuaries and their connection, with realistic variable cross-sections computed from a refined soundings field work (Salomon and Breton, 6). With a mesh size of 250 m, the one dimensional grid is 41 meshes long.

The downstream boundary condition is the tide level deduced from a previous running of the two coupled models (bay + estuary). The model is calibrated with a constant friction factor in order to simulate good flow velocities at station A and B (see figures 3 and 4).

3.2 Sedimentary processes

The principle of mud transport models is mass conservation of sediment. Two compartiments can be distinguished, according to the environment of the particles :

3.2.1 Processes in the water column

Suspended matter can be calculated by solving an advection/diffusion equation that can be written in a one dimensional form :

$$\frac{\partial \sigma C}{\partial t} + \frac{\partial U \sigma C}{\partial x} = \frac{\partial}{\partial x}\left(K \sigma \frac{\partial C}{\partial x}\right) + (E + D).\ell \qquad (1)$$

where : C = concentration of suspended matter
σ = cross section area
ℓ = bed width
U = mean current velocity
K = diffusion coefficient
E and D = erosion and deposition rate

Solving the equation (1) can fit the computation of U (x,t) and σ (x,t) with the one dimensional hydrodynamic model.

The lateral averaging of the equation is justified by the weakness of lateral gradients of concentration that have been measured (see figure 3).

As for the vertical profile, figure 3 shows off high gradients, especially at mid-tide. To take them into account, a vertical distribution of suspended matters has been introduced to fit the deposition term D to the bottom concentration. This distribution results from the balance between the vertical exchanges due to the eddy diffusion and the settling velocity Ws. If a logarithmic velocity profile and a same diffusivity for mass transfert and momentum are assumed, the concentration profile can be written :

$$C(z) = C_{bottom} . e^{-\frac{W_s z}{K.u^*}} \qquad (2)$$

Boundary conditions

Upstream, as for any other outfall along the estuary, the concentration of suspended sediment has to be given, as well as the water discharge. Downstream, we only need a flood condition for C. As the model results are sensitive to this condition (see 5 Discussion), the following one has been selected : at the beginning of the flood period, the boundary concentration is the same as at the end of the ebb, and then it decreases as the local cross-section increases, until the influx of suspended matter balances the previous outflux. By this way, the estuary system cannot be fed downstream, which is better for understanding the processes inside.

Deposition law

Most of authors use the Krone formula :

$$D = \frac{\partial C}{\partial t} = W_s . C . \left(1 - \tau/\tau_d\right) \qquad (3)$$

where $\tau = \rho u*$ is the bottom shear stress (u* : friction velocity) and τ_d is the critical stress of deposition, which is often a constant.

The bottom stress can be deduced from the computed vertically averaged velocity U and a hypothetic rugosity length according to a logarithmic profile.

The settling velocity formulation has to take into account the floculation phenomena. When the concentration is higher than 0,3 kg.m^{-3} , this settling velocity is found to increase simultaneously until the concentration value of 3,5 kg.m^{-3} , then it is hindered when aggregates form a continuous network (Krone, Thorn in Mehta (4) ; Migniot (5)). Besides floculation occurs in salted waters when salinity S exceeds 1 to 8 p.p.m. (Migniot, 5). Thus a general formula has been computed, with the following form :

$$\left[\begin{array}{ll} W_s = W_o \, f(S) \;\; C^{n_1} & \text{for } C < 3,5 \text{ kg.m}^{-1} \\[2mm] W_s = W_o \, f(S) \; (1 - mC)^{n_2} & \text{for } C > 3,5 \text{ kg.m}^{-3} \end{array} \right. \qquad (4)$$

$$\text{with } f(S) = S/(S + \alpha_1 + \alpha_2 C^{-\alpha_3})$$

α_1, α_2, α_3, n_1 and n_2 are constant to be calibrated.

However the lack of experimental data on in situ settling velocities and the will to simplify the interpretation of the model results lead to first considering a constant settling velocity.

Erosion law

Following many authors as Owen, Onishi, Ariathurai and Krone, Baeyens et al (see De Nadaillac, 2) we use the classical formula :

$$E = \frac{\partial C}{\partial t} = k_1 \left(\tau/\tau_e - 1 \right) \qquad (5)$$

where k_1 is a constant.

The critical bottom shear stress of erosion, τ_e, is bound to the concentration of surficial sediment Cs according to the formula : $\tau_e = k . C_s^n$ (6) which is used by several authors (Owen, Thorn and Parsons, Hayter in Mehta, 4 ; Migniot, 5). The knowledge of Cs needs modelling the soil behaviour and its consolidation.

Note that when $\tau_d > \tau_e$ [= f (Cs)], erosion and deposition can occur at the same time : this point will not be discussed in this paper.

3.2.2. Processes in surficial sediment

Consolidation algorithm

From the modelling point of view, consolidation means time variation of the sediment concentration depending on the effective stress (weight of overlying sediment) and on the permeability which rates the expulsion of pore water.

Instead of considering typical sediment concentration profiles in several layers as many authors do, De Nadaillac (personal communication) suggested to manage a stock of elementary quantities of sediment ("quanta") which number varies with successive erosion and deposition events. Every quantum is identified by its concentration and may be followed along the time. Thus consolidation may be represented by a differential equation relative to the concentration, which is closer to the processes than a given profile of soil concentrations.

Up to now, the knowledge of consolidation processes does not fit a derivative form because of experimental strains and may be of the previous consolidation algorithm form. After experimental data from Migniot (5) who experienced consolidation in test-tubes, the following semi empirical law can be suggested :

- if $C \leqslant C_1$, flooded bottom   (slow consolidation) $\dfrac{\partial C}{\partial t} = a_1 + b_1 \sigma'$

                    uncovered bottom (fast consolidation) $\dfrac{\partial C}{\partial t} = a_2 + b_2 \sigma'$

- if $C_2 > C > C_1$, $\dfrac{\partial C}{\partial t} = \dfrac{1}{t}(a_3 + b_3 \sigma')$

                                                                           (7)

- if $C \geqslant C_2$ or $t > t_2$, $\dfrac{\partial C}{\partial t} = 0$

where σ' is the effective stress (i.e. : N.Q where N and Q are the number and the dry weight of overlying   quanta).

The first stage is related to the primary consolidation which consists of the expulsion of pore water, more or less easily according to the presence of water over the surficial sediment. The second stage (secondary consolidation) squares with a rearanging of the skeleton and induces a logarithmic increase of the concentration.

The a_i and b_i terms have been adjusted in order to restore the results of Migniot experiments :
$a_1 = 0,002$; $b_1 = 10^{-4}$; $a_2 = 0,02$; $b_2 = 7.10^{-4}$; $a_3 = 50$; $b_3 = 2,5$ (u. S.I.)
$C_1 = 230$ kg.m^{-3} ; $t_2 = 20$ hours

The disadvantage of this new soil algorithm is the computer cost when a refined vertical description is advisable.

Spatial resolution

The exchanges between water column and sediment are drastically dependant on local water height and can considerably vary in a same section so that it becomes necessary to discretize the bottom according to the water depth. Actually, for each cross-section of the estuary, five reference levels have been considered, evenly distributed between the channel level and the bank level at high spring tide (see for example the bottom "cutting" near Locquenolé on figure 2). By this way a local water height and then a bottom shear stress, a consolidation rate and a mass transfer between sediment and suspended matter can be computed at any time. Note that any transversal change in channels or mud flats cannot be simulated : such an attempt would not be consistent with the one dimensional modelling of flow.

The three dimensional discretization of the soil is consistent with any development in considering vertical exchanges between quanta inside the sediment (for instance contaminant diffusion ...) or even horizontal ones (for instance surficial mass transfert in order to model mud collapses when the bank slope is steep).

4. Results

4.1 Reference simulations and parameters

The models have been calibrated with a few parameters (friction factor for hydrodynamics, rugosity length for the bottom stress, erosion rate). This calibration has been achieved with a periodic spring tide of constant amplitude. Other parameters have been selected within common limits suggested by various authors. Thus our "reference parameters" are the following :
Friction factor (Strickler form) : 33 ; rugosity length 2.10^{-3} m
Diffusion coefficient : $K = 10 + U.R_h$ $(m^2.s^{-1})$, R_h is the hydraulic radius.
Quantum : 0,3 kg ; initial stock : 50 quanta.m^{-2} ;
Maximum computed stock : 150 quanta.m^{-2}
Settling velocity : 10^{-3} $m.s^{-1}$; $u*_d = (\tau d/\varrho)^{0,5}$: 0,01 $m.s^{-1}$
Erosion rate : $k_1 = 4.10^{-5}$ $kg.m^2 s^{-1}$; $u*e = (\tau e/\varrho)^{0,5}$: $0,26.10^{-4}$ (Cs)

In situ measurements have shown large variations of turbidity according to a realistic behaviour of the sediment. Thus a new simulation has been attempted : beginning on spring tide, the model is run during a whole spring /neap tidal cycle, until the following spring tide which related results will be discussed further.

The initial conditions are the same as for the previous simulation : nearly consolidated sediment profile with dry density varying from 380 $kg.m^{-3}$ near the surface to 500 $kg.m^{-3}$ 0,03 m below, everywhere in the estuary, even in the channel. The run begins at high tide, when suspended matter can be neglected.

4.2 Simulation analysis

The figure 4 shows the simultaneous variations of water level, depth averaged current, friction velocity, erosion and deposition rates, suspended matter and surficial sediment concentrations, and sediment stock during 25 hours, after 14 days simulation at station A.

The first result is a quasi periodicity : both figured tides are very similar and can be considered as a representative spring tide.

From the hydrodynamic point of view, the asymmetry between flood and ebb is very strong : actually, station A is in the upper part of the estuary. Secondarily we can observe a quasi-steady river flow at the end of the ebb : the fresh water discharge is only 1 $m^3.s^{-1}$ but the cross-section is so tiny at this time that velocities reach 0.5 $m.s^{-1}$. Consequently, friction velocities are relatively large because of the small depth and any fresh deposited sediment cannot stay during the ebb.

The tidal height diagram shows off the periods of uncovering benches according to their bottom level reference. Consequences on friction velocities are considerable : for instance over a mud flat which is flooded at mid-tide (level reference : 3), this friction velocity is at least twice less than in the channel.

Such differences between banks and channel can be seen on the erosion/deposition diagram. But they also depend on the sediment stock : for instance, in the channel no quantum is available during the flood period and no erosion occurs. But at the end of the flood, deposition is possible with suspended matter coming from lateral erosion. This deposition stops near high tide where no suspended matter is available. Then an erosion begins with the ebb but stops suddenly (at 14 h 30 on the figure) when the stock is over. On both sides of the estuary, processes are very different : deposition occurs at the end of the flood and erosion occurs when a strong velocity and a sufficient water height are concomitant, that is within a short time during the flood or at the beginning of the ebb.

As for the sediment stock, its behaviour seems nearly stationary and far from the initial condition which consists of a uniform stock of 15 $kg.m^{-2}$. Thus the channel has been cleaned out, except during a brief moment at high tide. Just above the channel, the first banks are slowly eroded whereas deposition is maximum at the bottom level reference 3. Lastly, the upper banks are balanced because of the lack of suspended matter when they are flooded.

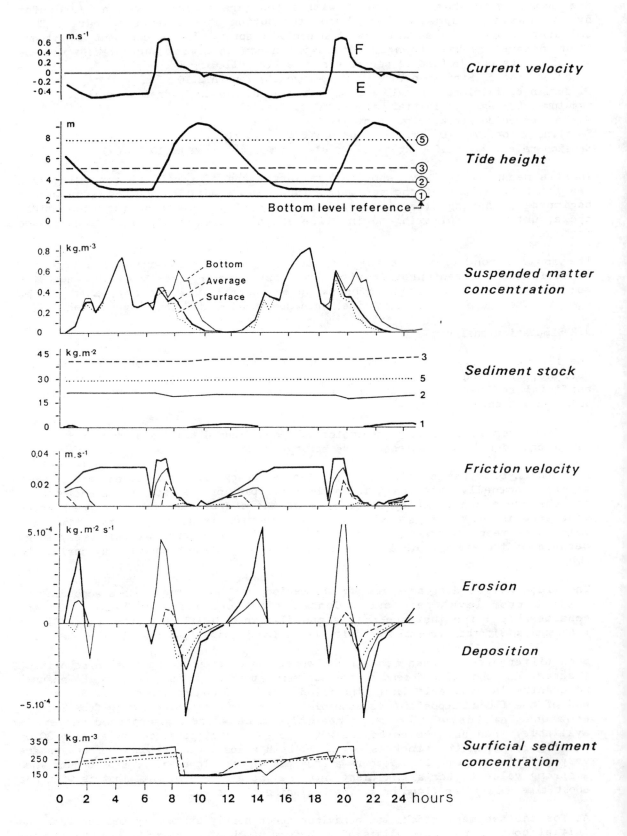

Figure 4 : Estuary of Morlaix. Computed hydrodynamic and sedimentary parameters at station A (see Fig. 1 for location). Spring tide - low river discharge.

Many features can be seen on the suspended matter diagram : first the local effect of resuspension during the flood and the ebb, with relative maxima of concentration ; secondly a concentration falling down at high tide. But the overall maximum occurs between low water and the flood turn : cross-sections are small and suspended matter comes from upstream erosion, according to the available sediment stock.

Lastly the vertical concentration gradient is important during the flood, just after the peak velocity. This gradient is computed for the channel water column and the surface concentration is taken into account for the lateral deposition : by this way the feeding of the upper banks is less important than in a fully vertically integrated system.

The evolution of the surficial sediment is more complicated. First the channel sediment is never consolidated, with always a small concentration. On the other hand, the first benches (bottom level reference 2) show large concentration variations.

Let us describe the processes related to these benches since the deposition time following the first flood. From 10 to 14 hours, the surficial concentration increases slowly, first because of a slow consolidation (flooded mudflat) and then because of an erosion of this fresh deposited matter. From 14 h to 14 h 30, deposition occurs when velocity decreases, just before the uncovering time : then fast consolidation sets up, until the concentration reaches 230 kg.m^{-3} (at 16 h). Then a logarithmic secondary consolidation follows until the flood turn (at 19 h). After a short and small deposition that cannot be seen on the deposition diagram), the flood velocity induces erosion which reaches previous deposited layers (at 20 – 21 h). The tidal residual effect is this small excess erosion.

Analogous interpretations could be given for the other bottom level references.

Longitudinal variations of sedimentary parameters are not presented in this paper : processes are the same along the estuary, the only change being the relative weight of local processes and transport effects, especially for the suspended matter concentration, inducing time lags on maxima values.

Model validation

The computed suspended matter concentration on figure 4 can be compared to the data of station A plotted on figure 3. The same evolution is observed, with concentration relative maxima related to peak velocities, and an absolute maximum during the end of the ebb. This surprising feature, induced by upstream erosion, shows the importance of transport effects before local ones, as explained by the model.

4.3. Bottom evolution

The computed sediment stock is presented on figure 5, after a whole spring/neap tidal cycle simulation (14 days). The upstream shoaling is obvious, especially on the sides along 2 km. Downstream, these side benches seem stable (the initial stock was 15 kg.m^{-2}). On the other hand the channel and the first lower benches are largely eroded, especially in the middle area of the estuary. The reason is less a longitudinal variation of current velocities than a simple exchange of sediment between the upper and the middle estuary. Actually, sediment is mainly eroded during the flood period, and then carried upstream, where it is deposited at high tide, everywhere along a cross-section. During the ebb, waters flow out of the lateral benches before the peak ebb current and erosion is very weak, except in the channel or just above. Then either these ebb eroded sediments are deposited farther in the channel and thus will be eroded during the next flood period, or they are ejected in the bay where they will eventually be deposited.

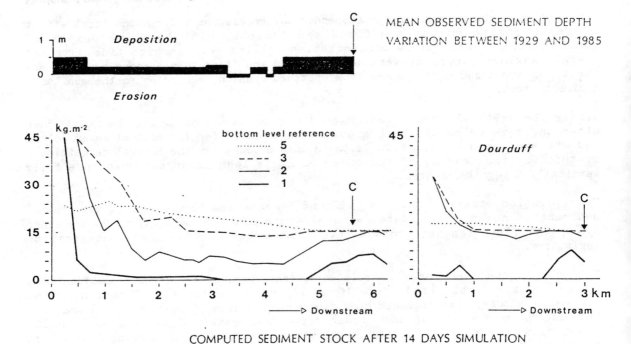

COMPUTED SEDIMENT STOCK AFTER 14 DAYS SIMULATION

Figure 5 : Estuary of Morlaix. Computed and observed bottom evolution.

So in the Morlaix estuary, the main sedimentary processes do not result from the asymmetry between ebb and flood, but rather from the morphological change of the estuary between high tide and low tide : schematically, erosion occurs at mid-tide with ebb/flood current and deposition occurs at high tide, preferentially upstream and over the lateral benches, whereas during low water, cross-sections are so small that ebb velocities remain high.

A comparison with observed bottom variations between 1929 and 1985 shows a qualitative agreement between measurements and computation (figure 5 and L'Yavanc, 3). However the real estuary seems in a steady state with a slight deposition rate upstream : this could be due to a non-modelized process that consist of steep slope collapses from lateral benches towards the channel. Besides, the simulated tidal cycle cannot be considered as a completely realistic events series : in particular a peak discharge from the river can increase the ebb erosion upstream and reduce the shoaling.

5. Discussion

Some points are worthwhile being developped from a general point of view.

5.1 Bottom friction evaluation

For smooth bottoms, a classic implicit logarithmic velocity profile has been fitted (for instance Migniot, 5) : $u = u^* [2,5 \, \mathrm{Log} \, (zu^*/\nu) + 5.5]$ where ν is the water viscosity. When applying this formula, we got a very steep velocity profile and consequently a small friction velocity, not in agreement with measured velocity gradients. Actually a rugosity length of 2.10^{-3} m (instead of 2.10^{-4} m, a common value for a mud bottom) is necessary to get realistic velocity gradients. This is likely due to large splits in the surficial sediments looking like bedforms for sandy bottoms. A special investigation should be carried out on these mud bedforms.

5.2 Spatial discretization validation

This modelling was achieved with a special care for simplifying as much as possible : thus a cross-section averaged model was choosen. However, the first tests have shown the necessity to distinguish several bottom levels, as processes are very dependent on the local water height and the uncovering time. The resulting mechanism of channel erosion and side deposition in the upper estuary warrants this lateral discretization.

Moreover, to prevent an excess deposition on the upper benches, a vertical gradient for suspended matter concentration had to be introduced. But the presently used formula (2) is fitted to a steady flow and does not take into account accelerations and decelerations, or even the near bottom location of the exchange with sediment layers. For this reason as well as for simulating a stratified flow during high fresh water discharge and neap tide, a two dimensional model with vertical calculation for current and concentration could lead to a perceptible progress.

5.3 Model sensitiveness to boundary and initial conditions

The model results depend on the downstream concentration input during the flood. In the present model the condition has a conservative form, that is the input has not to exceed the output. Naturally, a better way would consist of modelling the transport and even simulating the sedimentary processes in the bay downstream.

With regard to initial conditions, the problem is more difficult. Of course during the first simulated tides, the model will be sensitive to the initial sediment concentration profile, but after several weeks, would the results become similar ? Achieving tests on various initial conditions could give an answer and may be lead to a sedimentology time scale related to the study area.

6. Conclusions

The main objectives of this study were first to develop software for computing mud transports in coastal areas and then to improve the knowledge of sediment processes in macrotidal estuaries such as those opened to the English Channel.

Mud transport model
The model is operational, with an original algorithm for the sediment consolidation. It is suited to one dimensional flows but a lateral discretization of the bottom sediment allows to take into account the lateral benches uncovering at low tide. This software could easily be improved by running with two or three dimensional flows.

Sediment processes in the Estuary of Morlaix
The model is the best tool to show off the respective weights of transport phenomena and local erosion or deposition processes : thus it has been possible to explain the maximum turbidity at low water instead of during the maximum current, and to simulate the sediment exchange between the middle channel and the banks of the upper estuary.

But many other studies could be set by means of the model : for instance the fate of the upstream input of suspended matters, the trajectories of sediment samples during one tidal period, the impact of a high river discharge, the pore water output when the sediment consolidates (with environmental consequences), the effect of the highest spring tide and generally speaking the sensitiveness of the model to parameters and initial conditions.

7. Acknowledgements

The authors wish to thank G. De Nadaillac (Orga-Conseil, Lyon) for his work on elaborating the mud transport model. Thanks are also due to M. Breton (IFREMER) for her helping in running the model, J.P. Annézo and D. Guillerm (IFREMER) for drawing the figures, typing the manuscript and improving the general presentation.

8. References

1. Bassoullet, P., L'Yavanc, J., Guillaud, J.F., Loarer, R., Breton M. : "Etudes sedimentologique et bathymétrique de l'estuaire de Morlaix". Rapport interne IFREMER DERO-87.21-EL, 1987, pp. 1-58 (In French).

2. De Nadaillac, G. : "Transport en suspension des vases - étude bibliographique" : Rapport interne IFREMER DERO-85.03-EL, 1985, pp. 1-132. (In French).

3. L'Yavanc, J. : "Evolutions bathymétriques et morphologiques de la rivière et de la baie de Morlaix". Rapport interne IFREMER DERO-87.15-EL, 1987, pp. (1-105). (In French)

4. Mehta, A.J. : "Characterization of cohesive sediment properties and transport processes in estuaries". In Mehta, A.J. : "Estuarine Cohesive Sediment Dynamics" Springer-Verlag 14, 1984, pp. 290-325.

5. Migniot, C. and Bouloc, J. : "Erosion et sédimentation en mer- et en rivière" in Filliat, G. : "La pratique des Sols et Fondations". Paris, Editions du MONITEUR, 1981, pp. 629-731 (In French).

6. Salomon, J.C. and Breton, M. : "Modèle de circulation et de dispersion dans l'Estuaire de la Rivière de Morlaix". Rapport interne IFREMER DERO-88. EL, 1988, (à paraître). (In French).

Chapter 5
MATHEMATICAL MODELLING OF TIDAL CURRENTS AND SEDIMENT TRANSPORT RATES IN THE HUMBER ESTUARY

R A Falconer
University of Bradford, UK
P H Owens
Daresbury Laboratory, UK

SUMMARY

The paper describes a mathematical model which has been refined to predict the water elevations, the two-dimensional tide induced velocity fields and the sediment transport rates in the Humber Estuary. The water elevations, depth averaged velocities and depth mean sediment concentration distributions were evaluated within the estuary by solving numerically the time dependent non-linear equations of mass, momentum and advective-diffusion for suspended sediment, in the horizontal plane, using an alternating direction implicit finite difference scheme. An improved technique was developed for representing the drying and flooding of tidal flats, which permitted the use of a larger grid space and time step than was previously possible. For the erosion and deposition of sediment from the bed, an empirically based source-sink term was developed from three mobile bed flume studies, with the sediment model consisting of several different empirical sediment transport formulae. The model was calibrated against extensive field data for the Humber Estuary, with the predicted results from the model found to give an encouraging degree of agreement with the corresponding field measurements.

NOMENCLATURE

c	=	suspended sediment concentration at elevation z above bed
C	=	de Chezy coefficient
C	=	depth average suspended sediment concentration
C_e	=	depth average equilibrum concentration
c_{-h}	=	suspended sediment concentration near bed
C*	=	air-water interfacial resistance coefficient
f	=	Coriolis parameter
g	=	gravitational acceleration
h	=	bed elevation below datum
H	=	total depth of flow
L	=	horizontal distance of particle travel
t	=	time
u	=	horizontal velocity at elevation z above bed
U,V	=	depth average velocities in x,y directions
W_s	=	particle settling velocity
W_x, W_y	=	wind velocity components in x,y directions
x,y	=	co-ordinate directions in horizontal plane
z	=	co-ordinate direction in vertical plane
α	=	profile factor for non-uniform concentration profile
β	=	momentum correction factor
γ	=	profile factor relating near bed and average concentrations
ϵ	=	depth average eddy viscosity
η	=	water surface elevation above datum
ρ	=	density of water
ρ_a	=	density of air

INTRODUCTION

Siltation in coastal and estuarine waters is a major problem, with the economic viability of many ports, harbours and estuaries being affected by the cost of maintaining dredged navigation channels. With increasing environmental awareness, it is also important to assess the environmental impact and the effect on erosion and deposition rates of major engineering schemes, such as the proposed tidal barrages in the Severn, Mersey and Humber Estuaries. The ability to predict accurately the siltation and erosion rates resulting from such proposed coastal and estuarine engineering structures would therefore be of considerable benefit.

In the past hydraulic laboratory models with mobile beds have frequently been used to study such engineering projects, but there are problems associated with scaling the large number of independent variables affecting the sediment transport rates. In addition to the variables that must be scaled in a normal hydraulic model, suitable sediment particles must be chosen with the correct distribution of shape, size and density to represent the hydraulic properties of the prototype sediment in a particular flow. Furthermore, the physical modeller is unable to control the bed friction, which is dependent upon the nature of sediment particles and the bed forms produced by the flow.

Mathematical models are now widely used to predict tidal flows and solute transport rates in coastal and estuarine waters, although such models are not used to anywhere near the same extent in connection with sediment transport studies. However, a major obstacle to the development and application of mathematical models for sediment transport studies is the absence of a complete understanding of the physical processes involved, despite the large effort devoted to this end. It has not yet been possible to derive a generic set of equations to describe completely the processes of erosion and deposition, because of the large number of independently varying parameters affecting the sediment transport rates and the difficulty of obtaining measurements. Most of the work to-date has been concerned with the development of empirical formulae correlating the sediment flux with some hydraulic property, such as velocity or bed shear stress, of a stream under steady state conditions.

The principle objective of the study described herein was, therefore, to extend a two-dimensional depth integrated finite difference model, capable of predicting tidal flow fields, to encompass transient non-cohesive sediment transport processes. The refined model has been specifically applied to the Humber and Gironde Estuaries, with field calibration and verification data of water elevations, velocities and sediment concentration distributions for the Humber Estuary being made available by Associated British Ports Research and Consultancy Limited (formerly British Transport Docks Board).

DEPTH INTEGRATED EQUATIONS

In the mathematical model the partial differential equations governing the conservation of mass and momentum in the horizontal plane were integrated over the depth, with the momentum equations including the effects of the earth's rotation, a wind stress, bottom friction and a turbulence model. Full details of the derivation of these equations is given in Falconer (1), with the conservation equations of mass and momentum in the x and y horizontal co-ordinate directions being respectively given here for completeness:-

$$\frac{\partial \eta}{\partial t} + \frac{\partial UH}{\partial x} + \frac{\partial VH}{\partial y} = 0 \tag{1}$$

$$\frac{\partial UH}{\partial t} + \beta \left[\frac{\partial U^2 H}{\partial x} + \frac{\partial UVH}{\partial y} \right] - fVH + gH \frac{\partial \eta}{\partial x} - \frac{\rho_a \, C* \, W_x \, (W_x^2 + W_y^2)^{\frac{1}{2}}}{\rho}$$

$$+ \frac{g \, U \, (U^2 + V^2)^{\frac{1}{2}}}{c^2} - \left\{ 2\frac{\partial}{\partial x} \left[\epsilon H \frac{\partial U}{\partial x} \right] + \frac{\partial}{\partial y} \left[\epsilon H \frac{\partial U}{\partial y} \right] + \frac{\partial}{\partial x} \left[\epsilon H \frac{\partial V}{\partial y} \right] \right\} = 0 \tag{2}$$

$$\frac{\partial VH}{\partial t} + \beta \left[\frac{\partial UVH}{\partial x} + \frac{\partial V^2 H}{\partial y} \right] + fUH + gH \frac{\partial \eta}{\partial y} - \frac{\rho_a \, C* \, W_y \, (W_x^2 + W_y^2)^{\frac{1}{2}}}{\rho}$$

$$+ \frac{g \, V \, (U^2 + V^2)^{\frac{1}{2}}}{c^2} - \left\{ \frac{\partial}{\partial x} \left[\epsilon H \frac{\partial V}{\partial x} \right] + 2\frac{\partial}{\partial y} \left[\epsilon H \frac{\partial V}{\partial y} \right] + \frac{\partial}{\partial y} \left[\epsilon H \frac{\partial U}{\partial x} \right] \right\} = 0 \tag{3}$$

where η = water surface elevation above horizontal datum, t = time, x,y = cartesian co-ordinates in horizontal plane, U,V = depth average velocities in x,y directions, H = total depth of flow, β = momentum correction factor, f = Coriolis parameter, g = gravitational acceleration, ρ_a = air density, C* = air-water interfacial resistance coefficient, W_x, W_y = wind velocity components in x,y directions, ρ = water density, C = de Chezy coefficient, ϵ = depth average eddy viscosity.

For the sediment transport predictions in the model the depth integrated form of the advective-diffusion equation was used, together with source and sink terms to represent the erosion and deposition rates respectively. Neglecting the horizontal diffusion terms of the advective-diffusion equation, since these are generally small in comparison with the horizontal advection (see Galappatti and Vreugdenhil (2)), the corresponding depth integrated equation can be expressed in the following form (see Owens (3)):-

57

$$\frac{\partial HC}{\partial t} + \alpha \left[\frac{\partial UHC}{\partial x} + \frac{\partial VHC}{\partial y} \right] = \gamma \, W_s \, (C_e - C) \tag{4}$$

where C = depth average suspended sediment concentration, α = profile factor for non-uniform concentration profile – which is analogous to the momentum correction factor β, γ = profile factor defined as c_{-h}/C, W_s = particle settling velocity, and C_e = depth average equilibrium concentration determined from an appropriate sediment load formula.

In applying the mathematical model to the Humber Estuary, field measured velocity and sediment concentration distributions were integrated numerically using the trapezoidal rule to determine typical in-situ values for β, α and γ. These data were acquired by the British Transport Docks Board (4, 5) for the Humber Estuary, with measurements being taken at the three sites shown in Figure 1 and at elevations of 0.05, 0.1, 0.15, 0.3, 0.5, 1.0, 2.0, 4.0 and 8.0m above the bed and at 1.5m below the surface. The mean value for each coefficient and the corresponding standard deviation for the Humber Estuary were found to be (see Owens (3)):–

$$\beta = \frac{1}{H \, U^2} \int_{-h}^{\eta} u^2 \, dz = 1.07 \quad (S.D. = 0.09) \tag{5}$$

$$\alpha = \frac{1}{HUC} \int_{-h}^{\eta} u \, c \, dz = 0.82 \quad (S.D. = 0.14) \tag{6}$$

$$\gamma = \frac{c_{-h}}{C} \qquad\qquad = 4.07 \quad (S.D. = 2.84) \tag{7}$$

Owens(3) undertook a sensitivity analysis of these parameters for well established semi-theoretical velocity and concentration profiles and, in-line with this sensitivity analysis, the values of β and α were found to fall within a fairly tight range of values – with a low standard deviation – while the values of γ were more scattered.

Miles et al (6) carried out a similar analysis from profiles measured in the Conwy Estuary. Their findings were in broad agreement with the results obtained for the Humber Estuary, with the mean value for α being 0.84. Calculations for γ showed a considerable scatter, with a mean value of 6.2 and a standard deviation of 4.2

NUMERICAL MODEL DETAILS

The finite difference equations corresponding to the differential equations (1) to (4) were expressed in an alternating direction implicit form, with all terms being fully centred in both space and time by iteration. Full details of the finite difference code are given in Falconer (7).

Although at first sight the Humber Estuary appeared to be a relatively straightforward estuary to model mathematically, this did not prove to be the case and particular emphasis had to be placed on the development of a new flooding and drying technique in the mathematical model. The mathematical modelling difficulties experienced in representing the flooding and drying processes were not anticipated since the original model DIVAST (Depth Integrated Velocities and Solute Transport) had previously been successfully developed to model flooding and drying in Poole Harbour and Holes Bay (see Falconer (7)) where the plan-form wetted area changed considerably (up to 80%) throughout each tidal cycle. The treatment of these complex flow processes involved the adoption of a refined grid scheme and a modified flooding and drying technique, with full details being given in Falconer and Owens (8).

HUMBER ESTUARY APPLICATION

The Humber Estuary is situated on the east coast of England, providing an outlet to the North Sea for the rivers Trent and Ouse and shipping access to a number of ports, including Hull, Immingham and Grimsby. The tidal currents outside the estuary are predominantly north-south, parallel to the Yorkshire and Lincolnshire coastlines. During the flood tide the flow is southwards with a band of water, approximately 8km wide, separating from the main current and flowing into the estuary. This pattern is reversed during the ebb tide, with the seaward boundary (shown in Figure 1) parallel to, and south east of, Spurn Head acting effectively as a free streamline.

In order to drive the hydrodynamic model water elevation data recorded at Albert Dock and Bull Fort was used, with the latter elevations being extrapolated by trial and error to give the corresponding values required at the open seaward boundary. Field measurements of water elevations, velocities and sediment concentrations were available at the three survey sites, namely Halton Middle, Middle Shoal and Sunk Channel, with tidal data also being available at Immingham. Data were available for different tidal ranges at each site, with a complete tidal cycle being monitored at half hourly intervals in each case.

The modelled area was represented by a mesh of 87 x 32 grid squares, at 500m intervals, and with a time step of 180s. A three-dimensional representation of the bathymetry is shown in Figure 2. Having chosen the open boundary conditions, the model was then run to calibrate the bed roughness characteristics, with the value of Manning's number being assumed to be uniform throughout the estuary reach in the absence of suitable field data. Closest agreement between the predicted and measured water elevations was obtained for a value of n of 0.02. This value was thought to be reasonable for a sedimentary bed, with an example of the corresponding comparisons between the predicted and measured water elevations being given in Figure 3.

Several authors, including Wallis and Knight (9), have employed more complex calibration procedures, where temporal and spatial variations in the bed roughness were included. However, due to the spareness of the data and the absence of bed shear measuremesnts to confirm any calibration, the additional complexity was not justified in this case. Detailed calibration would also be of little value if it was proposed to use the model to predict the effects of a major change in the estuary, such as the proposed Humber Tidal Barrage, since the bed roughness characteristics could be markedly affected.

Full details of the water elevation and velocity field predictions, together with comprehensive comparisons with the measured field data, are given in Owens (3). Typical velocity field predictions are illustrated in Figures 4 and 5 at mean water elevation flood and ebb spring tide phases respectively. Furthermore, a typical comparison of the predicted and measured velocities at the three measuring sites is shown in Figure 6, for a spring tidal range. As can be seen from Figure 6, the agreement between the computed and measured velocities was found to be very close at all of the sites considered. The accurate numerical predictions at the survey sites suggested that the hydrodynamic model was performing satisfactorily and that predictions elsewhere in the estuary were also likely to be reliable. Furthermore, the reliability of the sediment transport model could be better assessed with errors due to the flow field predictions being minimised.

No sediment concentration measurements were available at the open boundaries. Hence, boundary conditions for the sediment transport model were prescribed by applying the time varying sediment transport formula, derived by the British Transport Docks Board (10) specifically for the lower Humber Estuary. These boundary conditions are relatively unimportant, since the model will adapt within a fairly short reach of the estuary to new conditions. An estimate of the length required to achieve this can be obtained from the horizontal distance, L, travelled by a particle settling from the surface to the bed, given by:-

$$L = HU/W_s \qquad\qquad (8)$$

Taking typical values within the Humber Estuary for H, U and W_s of 15m, 2ms^{-1} and 10mms^{-1} respectively, gives an adaption length of 3km, which is relatively small in comparison with the modelled reach shown in Figure 1.

The median sediment diameter was determined from a particle size distribution and was found to be about 0.11mm. This median diameter gave a full velocity of 9mms^{-1} and a critical shear velocity of 12mms^{-1}. In determining the equilibrium concentration, the formula of Engelund and Hansen (11) was chosen, with other relationships being programmed and considered and subsequently found to be less computationally efficient (see Owens (3)).

The model was calibrated by comparing measured and predicted sediment concentrations for the spring tide range. The resulting comparisons used for calibration are shown in Figure 7 with reasonable agreement being obtained between the measurements and the predictions. The corresponding results for the mid-tidal range are given in Owens (3) and were again shown to give reasonably encouraging agreement. The significantly lower concentrations measured and predicted during the mid-tidal range confirmed the sensitivity of the depth mean sediment concentration to relatively small changes in the velocity field, with the over prediction of the sediment concentration during the spring flood tide at Middle Shoal therefore being partly accounted for by the slight over estimation of the corresponding depth mean velocity (see Figure 6).

Examples of the shear stress and sediment concentration distributions are given for the spring tidal range in Figures 8 to 10 respectively, with the complete set of results being given in Owens (3). As expected from the depth mean velocity predictions shown in Figures 4 and 5, the flow followed the route of the main channels and the higher sediment concentrations tended to correspond to the higher bed shear stresses and hence to regions of higher velocity. The sediment concentrations were considerably greater for the spring tide than at corresponding times during the mid-tide. Weak eddies were generated in the lee of Spurn Head and this was most apparent at low water spring tide. Significant areas of the estuary were observed to become dry at low water without the generation of spurious waves – this was in contrast to the modelling of flooding and drying using the original scheme developed by Falconer (7). However, weak disturbances were apparent on the north west bank at high water during the mid-tide, with this anomaly almost certainly being caused by the flooding of a grid square and resulting in the water elevation oscillations observed at Immingham.

An interesting spurious anomaly occurring for the Humber Estuary, was the existence of an area of prounced sediment deposition located just off Spurn Head, near Bull Fort. This area of deposition was located in the centre of the main channel of the estuary and could not readily be explained from the instantaneous velocity field predictions. However, when a tidal residual velocity field prediction was reproduced for the Estuary (see Figure 11), it became clear that a tidal residual eddy appeared to exist off Spurn Head, with the eddy being centred about the area of pronounced deposition. In generating such a tidal eddy it is necessary for the inward pressure gradient and the outward centrifugal force to be in local equilibrium near the surface, although near the bed the effects of bottom friction result in a reduction of the centrifugal force and thereby give rise to a net force acting towards the centre of the eddy. This flow phenomenon can be illustrated by stirring a cup of tea and observing the net transport of the tea leaves (near the bottom of the cup) to the centre of the cup. Hence, in relation to the Humber Estuary, the sediment particles near the bed and in suspension off Spurn Head appear to experience a hydrodynamic force towards the centre of the predicted eddy shown in Figure 11, with the result that these sediment particles are subsequently more likely to be deposited when they approach the centre of the eddy, where the velocities are relatively small. This example illustrates the manner in which mathematical models can be of particular benefit in trying to understand complex flow and sediment transport rates.

Following on from the tidal residual velocities, the residual erosion rates were determined with the erosion being expressed in terms of a bed thickness. By using a simple sediment continuity equation, an erosion rate of $1gm \ m^{-2} \ s^{-1}$ was shown to be equivalent to a reduction in the bed elevation of 54mm day^{-1} (see Owens (3)). The corresponding residual erosion rates resulted in relataively large values being predicted at the seaward end of the model, with this result being attributed to the seaward open boundary condition – which was derived from the field data collected within the estuary and which may not strictly apply in the North Sea. The predictions obtained for the rest of the estuary are probably more realistic, but could not necessarily be regarded as absolute values or extrapolated to produce long term morphological predictions since no field data of erosion were available for confirmation. Also, any erosion or deposition could be attributable to the model adjusting to an imprecisely defined bathymetry. However, the model could be used to make comparisons between the effects of various proposed schemes on the local sediment flux and bed elevation variations.

CONCLUSIONS

A finite difference numerical model, capable of predicting two–dimensional depth averaged tidal flows, has been refined and extended in an attempt to represent non–cohesive sediment transport processes.

The transport of suspended sediment has been characterised by the advective–diffusion equation and effort has been directed towards the development of an appropriate source–sink term to represent erosion and deposition of bed material. The sediment transport capacity of the flow was represented by Engelund and Hansen's formula because of its computational efficiency, although other formulae were also considered.

The model was tested by simulating tidal flows and sediment fluxes in the Humber Estuary. The hydrodynamic model generated very accurate predictions of the measured water elevations and velocities at the measuring sites along the estuary. Using these flow predictions, it was possible to calibrate the sediment model to obtain good sediment concentration predictions.

In summary, the results of this study have shown that it is possible to predict sediment concentration distributions in a two–dimensional tidal flow field reasonably accurately by means of a depth integrated numerical model. However, the erosion rate predictions can only be used for comparative or qualitative studies, since no direct measurements were available for this process.

ACKNOWLEDGMENTS

This paper describes a study which was partially funded by a Science and Engineering Research Council CASE research studentship in association with the Central Electricity Generating Board Researach Laboratories. The authors wish to thank both organisations for their support and, in particular, Mr. R.W. Preston of the Central Electricity Research Laboratories for his assistance and encouragement.

This study would not have been possible without the use of field data which were made available by the Humber Estuary Research Committee of Associated British Ports (previously the British Transport Docks Board). The authors wish to express their thanks to all members of the Research Committee for making the data available and, in particular, to Mr. N.E. Denman and Mr. D. Cooper (Director of ABP Research and Consultancy Limited) for their subsequent assistance and encouragement.

During the period of this study the authors had several valuable discussions with Dr. G.V. Miles at Hydraulics Research Limited and the authors are grateful for his support and encouragement.

REFERENCES

1. Falconer, R.A.: "Mathematical Modelling of Jet-Forced Circulation in Reservoirs and Harbours", PhD Thesis, University of London, 1976, pp.237 (unpublished).

2. Galappatti, G and Vreugdenhil, C.B.: "A Depth Integrated Model for Suspended Sediment Transport", Journal of Hydraulic Research, IAHR, Vol.23, No.4, 1985, pp.359-377.

3. Owens, P.H.: "Mathematical Modelling of Sediment Transport in Estuaries", PhD Thesis, University of Birmingham, 1986, pp.220 (unpublished).

4. British Transport Docks Board: "Collection of Field Data for the Design and Operation of the Humber Tidal Model", Humber Estuary Research Committee, Report No. H1, 1974.

5. British Transport Docks Board: "Humber Estuary Sediment Flux, Part 1: Field Measurements", BTDB Research Station, Report No. R283, 1980.

6. Miles, G.V., Webb, D.G. and Ozasa, H.: "Sediment Transport Relations for Estuary Models", Marine Studies Group Meeting on Shelf and Nearshore Dynamics and Sedimentation, London, December 1980.

7. Falconer, R.A.: "A Two-Dimensional Mathematical Model Study of the Nitrate Levels in an Inland Natural Basin", International Conference on Water Quality Modelling in the Inland Natural Environment, BHRA Fluid Engineering, Bournemouth, June 1980, Paper J1, pp.325-344.

8. Falconer, R.A. and Owens, P.H.: "Numerical Simulation of Flooding and Drying in a Depth Averaged Tidal Flow Model", Proceedings of the Institution of Civil Engineers, Part 2, Vol.83, March 1987, pp.161-180.

9. Wallis, S.G. amd Knight, D.W.: "Calibration Studies Concerning a One-Dimensional Numerical Tidal Model with Particular Reference to Resistance Coefficients", Estuarine, Coastal and Shelf Science, Vol.19, 1984.

10. British Transport Docks Board; "Grimsby Middle Shoal: Bed Sample Analysis Series V", BTDB Research Station, Report No. R929, 1981.

11. Engelund, F. and Hansen, E. "A Monograph on Sediment Transport in Alluvial Streams", Teknisk Forlag, Copenhagen, 1972.

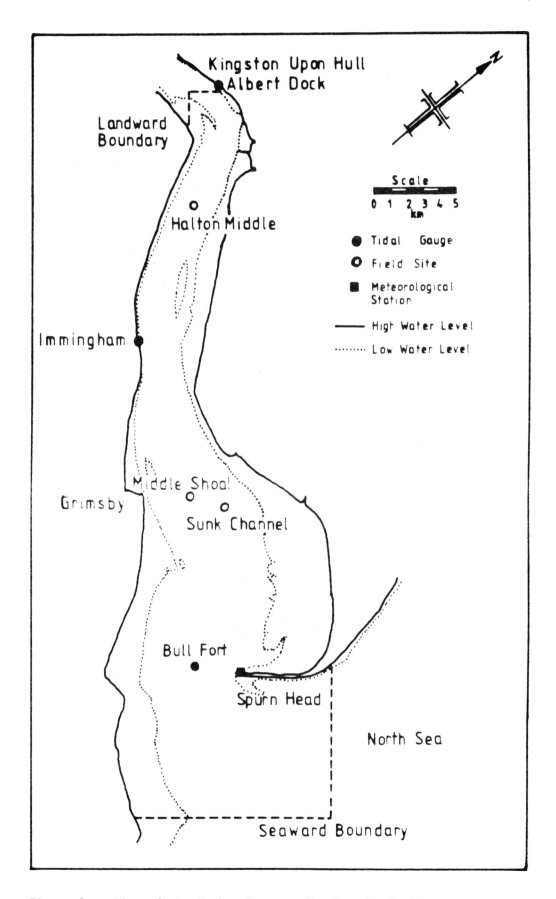

Figure 1. Plan of the Humber Estuary Showing the Field Measuring Sites

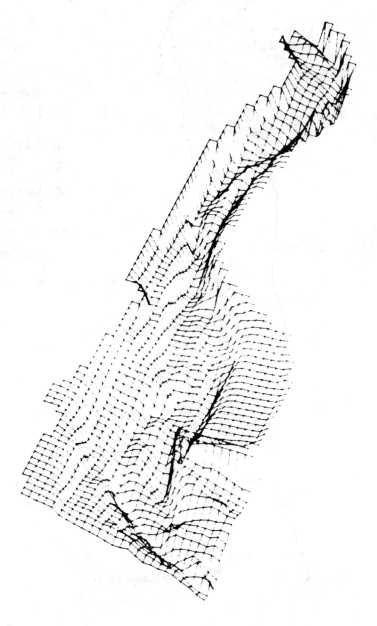

Figure 2. Isoparametric Projection of the Humber Estuary Bathymetry

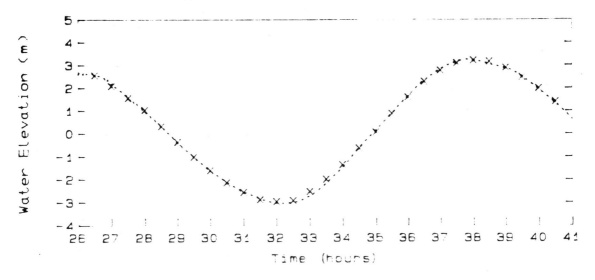

(a) Comparison of Water Elevations at Immingham

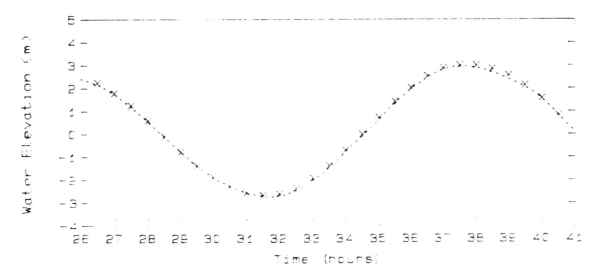

(b) Comparison of Water Elevations at Bull Fort

    X   Measured

    ·····   Computed

Figure 3. Comparisons of Computed and Measured Water Elevations at
Tidal Gauge Sites for Spring Tidal Range

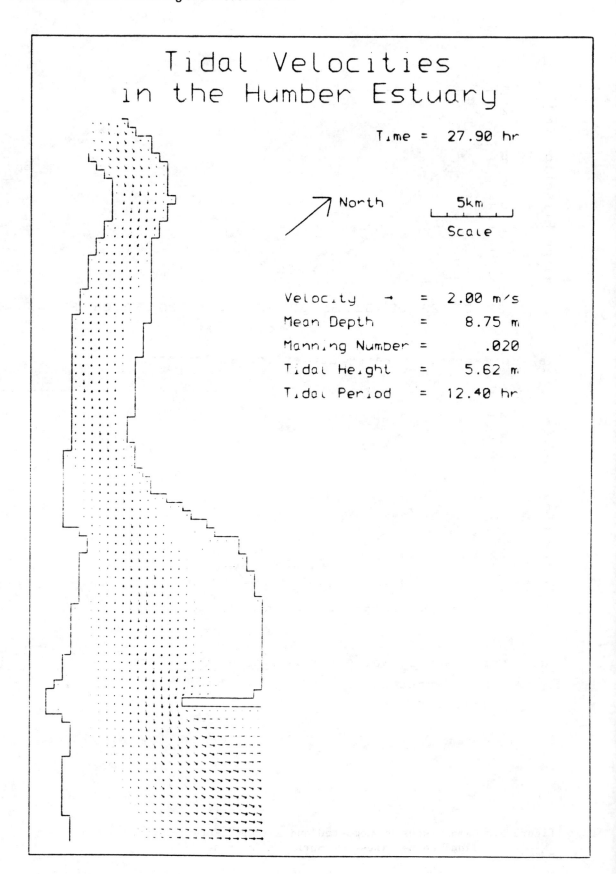

Figure 4. Computed Velocity Field at Mean Water Level Flood Spring Tide

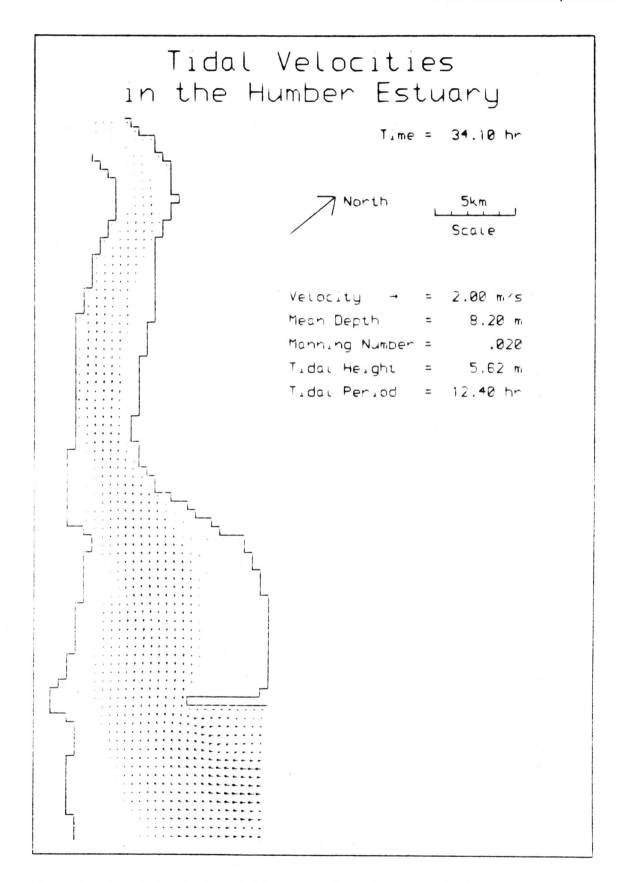

Figure 5. Computed Velocity Field at Mean Water Level Ebb Spring Tide

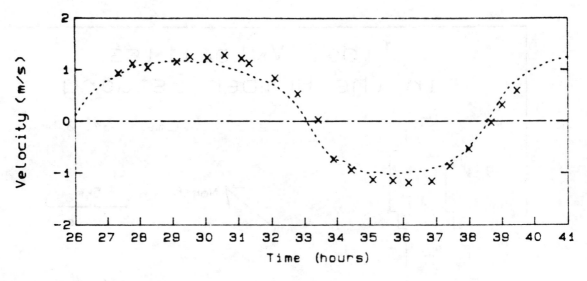

(a) Comparison of Velocities at Halton Middle

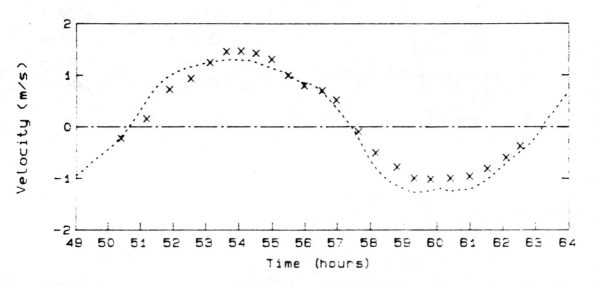

(b) Comparison of Velocities at Middle Shoal

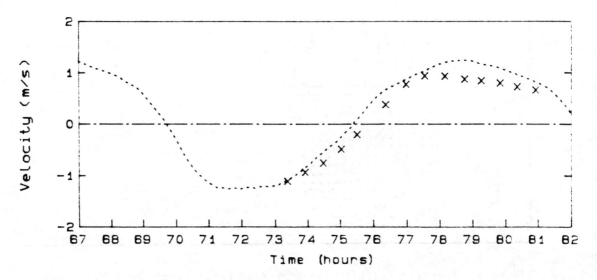

(c) Comparison of Velocities at Sunk Channel

Figure 6. Comparison of Computed and Field Measured Velocities at Measuring Sites for Spring Tidal Range

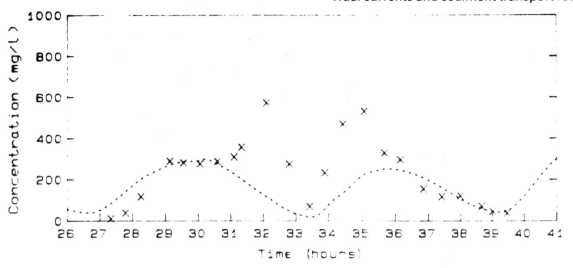

(a) Comparison of Sediment Concentrations at Halton Middle

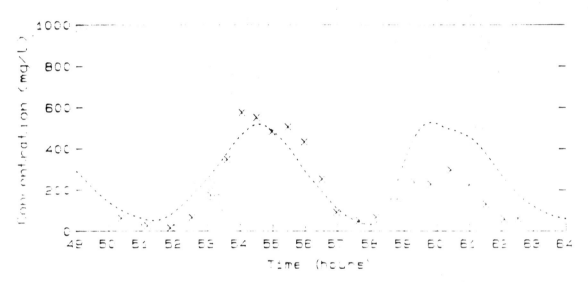

(b) Comparison of Sediment Concentrations at Middle Shoal

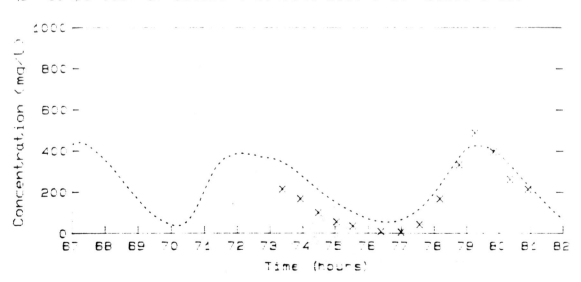

(c) Comparison of Sediment Concentrations at Sunk Channel

Figure 7. Comparison of Computed and Field Measured Sediment Concentrations
at Measuring Sites for Spring Tidal Range

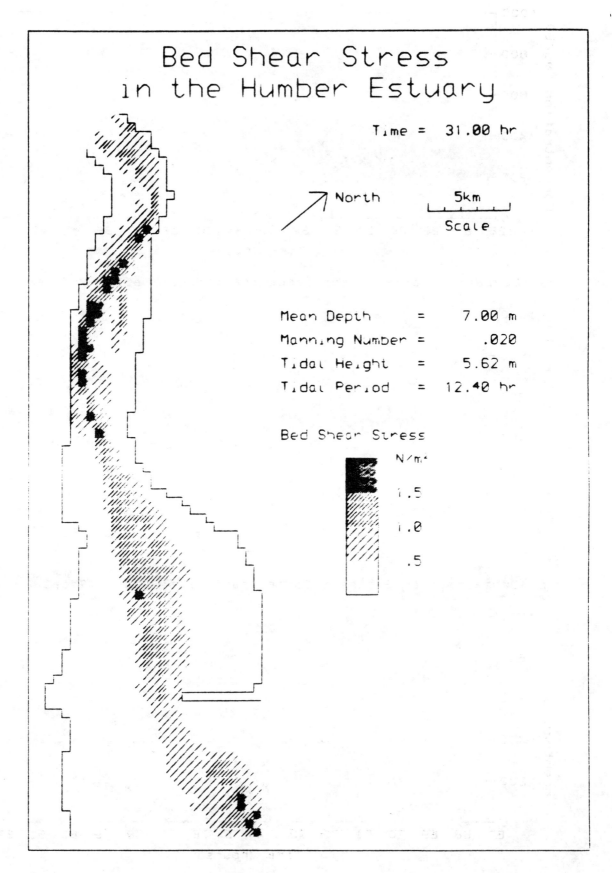

Figure 8. Computed Bed Shear Stress Distribution at Low Water Level Spring Tide

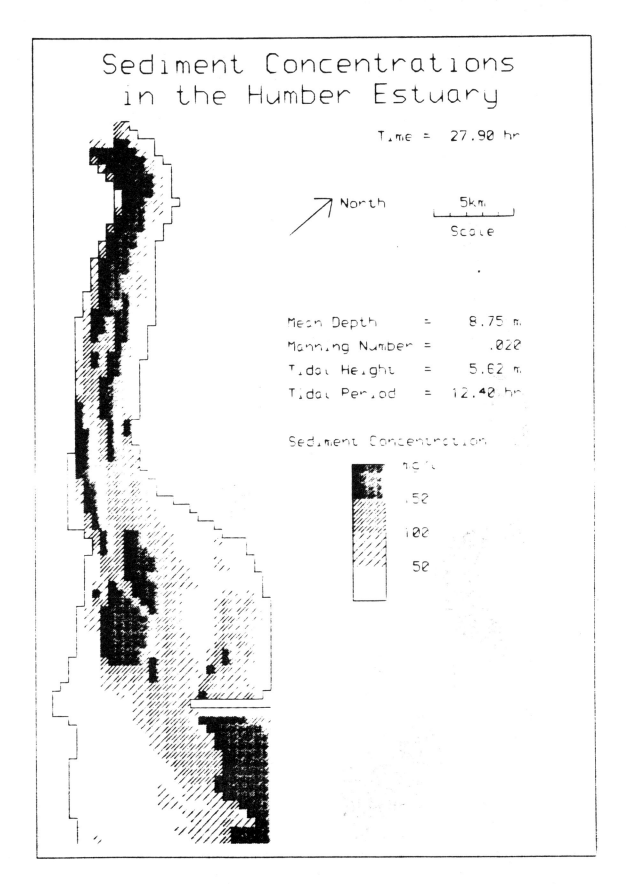

Figure 9. Computed Sediment Concentration Distribution at Mean Water Level
Ebb Spring Tide

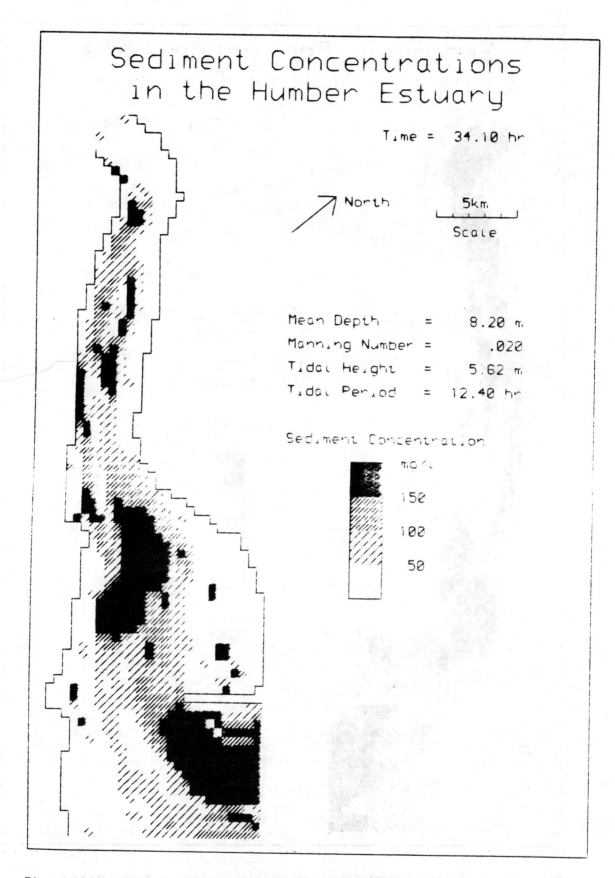

Figure 10. Computed Sediment Distribution at Mean Water Level Flood Spring Tide

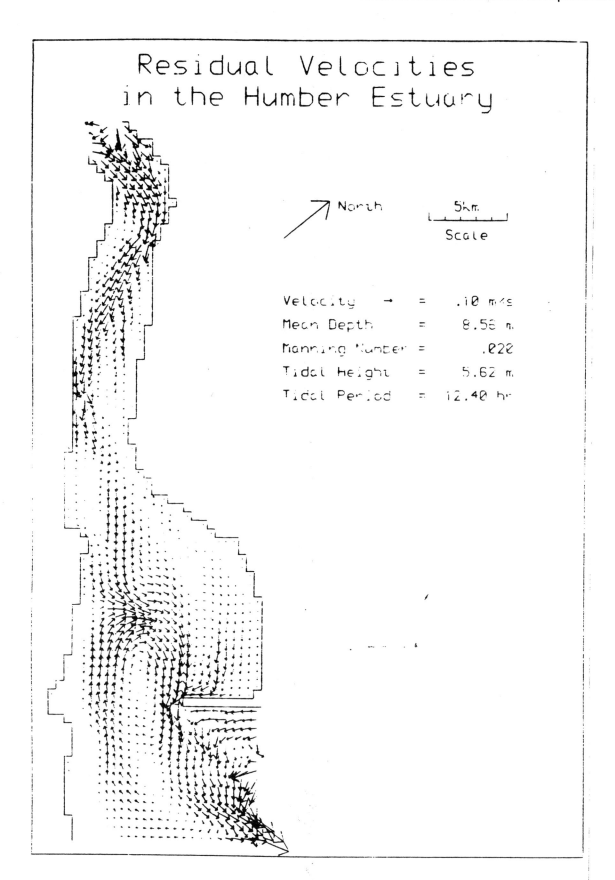

Figure 11.   Computed Tidal Residual velocity Field for Spring Tidal Range

Chapter 6
THE MEASUREMENT OF TURBULENCE IN ESTUARIES

R W Brierley, J R West and A P Parsons
University of Birmingham, UK
K O K Oduyemi
University of Cambridge, UK

Summary

The importance of turbulent transport phenomena for estuarine management models is discussed and a system for measuring turbulent fluctuations of velocity, salinity, suspended solids concentration and temperature is described. Special attention is given to the measurement of velocity which is particularly difficult.

Examples of results are given for near bed and near surface parts of the water column. It is shown that the turbulence sensors, along with dye tracing techniques and turbulent mean measurements may be used to extend our knowledge of the effects of solute and suspended solids induced vertical density gradients on vertical transport processes. Also, the relative importance of bed generated and internally generated turbulent phenomena may be investigated.

Notation

A	cross-sectional area
c	solute concentration or suspended solids concentration
D	longitudinal dispersion coefficient
D^v	vertical shear dispersion coefficient
d,h	flow depth
R	correlation coefficient
Ri	local gradient Richardson number
s	salinity
t	time
u_i	components of velocity in a Cartesian coordinate system ($u_1 = u = $ longitudinal component, $u_2 = v = $ transverse component, $u_3 = w = $ vertical component)
x_i	cartesian coordinate system ($x_1 = x = $ longitudinal direction, $x_2 = y = $ transverse direction, $x_3 = z = $ vertical direction - positive upwards)
Δt	a small time interval
ε	turbulent transport coefficient
ρ	density
τ_{zx}	shear stress on the x direction on a plane perpendicular to the z direction

Subscripts

c	solute concentration or suspended solids concentration
m	momentum
s	salinity

Superscripts

overbar	turbulent mean value
prime	a root mean square value over time Δt or a spatial or tidal variation about a spatial or tidal mean value.

Other item

< >	indicates a mean value

1. INTRODUCTION

Mathematical models of water level, salinity, suspended solids and water quality parameters for estuaries have gained a wide acceptance as useful management tools in the fields of flood control, water quality and navigation. At present the predictive capability of models is often limited by an incomplete knowledge of the mixing mechanisms that are found in unsteady stratified estuarine flows. The three dimensional turbulent mean solute mass balance equation may be given as

$$\frac{\partial \bar{c}}{\partial t} + \frac{\partial}{\partial x_i}(\bar{u}_i \bar{c}) + \frac{\partial}{\partial x_i} <u_i c>_{\Delta t} = 0 \qquad (1)$$

where c = turbulent fluctuation of solute concentration, u_i = turbulent fluctuation of the component of velocity in the x_i direction in a Cartesian coordinate system (1 = longitudinal, 2 = transverse and 3 = vertical), t = time and an overbar and $<>_{\Delta t}$ indicate turbulent mean values. It is common practice to write the turbulent flux term as

$$<u_i c>_{\Delta t} = -\varepsilon_{ci} \frac{\partial \bar{c}}{\partial x_i} \qquad (2)$$

where ε_{ci} is a turbulent diffusion coefficient for transport in the x_i direction.

The solution of the three-dimensional form of the solute mass balance equation is often too expensive and simpler two or one-dimensional spatially averaged solutions are sought. The one-dimensional equation is

$$\frac{\partial}{\partial t}(c_A A) + \frac{\partial}{\partial x}(Au_A c_A) + \frac{\partial}{\partial x}(A << uc>_{\Delta t} + u_A' c_A' >_A) = 0 \tag{3}$$

The shear induced dispersion term may be written as

$$u_A' c_A' = -D \frac{\partial c_A}{\partial x} \tag{4}$$

In the above D = longitudinal dispersion coefficient, subscript A and $< >_A$ indicate a cross-sectional mean value and subscript A' indicates a perturbation about the spatial mean. For simplicity the subscript 1 has been omitted. The dispersion coefficient may be considered to be composed of vertical and tranverse components. For example the vertical component is given by (Elder (Ref. 2))

$$D^V = \frac{-1}{d} \int_0^d u_d' \, dz \int_0^z \frac{dz}{\varepsilon_{sz}} \int_0^z u_d' \, dz \tag{5}$$

where u_d' is the deviation of turbulent mean velocity about the depth mean value u_d. Equation (5) shows that D is inversely proportional to ε_{cz} (where $z \equiv x_3$) and hence the vertical solute flux. The eddy viscosity is given by

$$\tau_{xz} = \rho \, \varepsilon_{mz} \frac{\partial \bar{u}}{\partial z} \tag{6}$$

where τ_{zx} is the shear stress in the x_1 direction on a plain perpendicular to the x_3 axis. Thus u_d is a function of ε_{mz} and hence the vertical turbulent transport of horizontal momentum ($\tau = -\rho < uw>_{\Delta t}$) where $w \equiv u_3$.

As basically similar relationships may be applied to suspended solids concentration equations it may be concluded that a clear understanding of turbulent transport phenomena for momentum, solute and suspended solids is required if the diffusion and dispersion coefficients in estuarine mathematical models are to be specified correctly.

For steady straight homogeneous open channel flow, laboratory measurements have established semi-empirical relationships for the turbulent transport parameters of eddy viscosity and diffusivity. In estuaries, the flow is unsteady due to tidal action, stratified by temperature, salinity and suspended solids and irregular in direction due to spatially and temporally varying channel geometry and bed features. As the reproduction of these conditions in laboratories is very difficult, there is a need to undertake field observations.

In shallow homogeneous laboratory flows the turbulence is generally bed generated. In two layer stratified flows turbulence is generated by breaking waves at the interface if sufficient shear is present. In the presence of naturally occuring finite vertical density gradients internal waves undergo generation, propagation and breaking to also produce internally generated turbulence. Thus estuarine flows may be expected to experience both bed generated and internally generated turbulence phenomena.

2.   MEASUREMENT TECHNIQUES

Estuarine channels are often between 5-20 m deep. A good spatial resolution over the flow depth requires measurements at about 10 points, that is 0.5 m intervals. To a first approximation the macroscale, the size of the eddies containing much of the turbulent energy, is of the same order as the distance above the bed. Thus in order to obtain a reasonable spatial resolution of turbulent eddies at 0.5 m above the bed a desirable transducer size is one tenth of 0.5 m, that is 50 mm. As the eddies will be convected past a mounted transducer, a similar temporal resolution in a flow of say 0.5 m/s requires readings at 0.1 second intervals.

Brierley et al (Ref. 1) have measured two velocity components at right-angles with 50 mm diameter disc-shaped eletromagnetic (EM) current transducers (Colnbrook Instrument Developments Ltd). Salinity was determined with inductive type conductivity cells (type 2105, Aanderaa Instruments, Norway) and suspended solids concentration was measured using a light-extinction technique (type 7000 3RP Mk II Siltmeter with S-100 or SDM 10 transducers, Partech Electronics Ltd). Using a bed-mounted mast, measurements were made at four points simultaneously within 3 metres of the bed. The data were multiplexed and transmitted to shore-based receivers, data-loggers and microcomputer. Both salinometers and siltmeters have measurement paths of less than 50 mm.

The system has recently undergone alterations to improve even further its overall effectiveness. Modifications have included circuit changes to allow the use of a single power supply, packaging of shipborne instruments into two light, splash-proof enclosures and the provision of hard copy facilities (plotter and printer) at the shore station.

Experience gained over the last five years has shown that it is flow measurement which needs the most care. Several modifications have proved necessary to enhance the performance of the commercially-available instruments so as to produce data of the required integrity. For those not familiar with EM current measurement, the technique involves a mixture of analogue and digital signal processing and it is not uncommon for 'digital spikes' to be present in the analogue outputs. If fed to any subsequent equipment having fast A/D conversion, such spikes can corrupt the data, giving misleading results which are not immediately self-evident. Once aware of the problem, it can dealt with effectively by additional screening/filtering within the EM equipment.

Recently added to the system for the 1988 programme of fieldwork is a four-sensor temperature measuring array. For measurement comparability with the other instrumentation, its design objectives included 0.1°C accuracy and a response time capable of detecting a 5°C change in not greater than 100 msecs, i.e. at 10 Hz. The latter target, coupled with the usual demands of underwater work - robustness, corrosion-resistance, sealing, etc - pointed to thermocouples as the sensors and, indeed, they have proved very successful. Type K thermocouples are employed in conjunction with a monolithic chip, type AD 595 AQ (Analog Devices Ltd). This device is a complete amplifier and cold-junction compensator and produces an output of 10 mV/°C. Thermocouple and chip are mounted on and within a bullet-shaped housing, 110 mm long by 25 mm diameter, the whole assembly being capable of submersion to 50 metres.

3.   EXAMPLES OF RESULTS

An example of turbulence data collected at 0.15 m above the bed at Calstock in the upper reaches of the Tamar estuary is shown in fig. (1). The instantaneous horizontal (cu) and vertical (cw) turbulent fluxes of suspended solids and the vertical flux of horizontal momentum (uw) are plotted over a period of 300 sec. All three terms have been non-dimensionalised by using the root mean square values (u', w', c') of the measured component terms. It may be seen that the vertical momentum fluxes are predominantly negative (w is positive in the upward direction) indicating a transport of momentum towards the bed. The vertical flux of suspended solids is upwards as is necessary to balance the gravitational settling in steady conditions. The horizontal turbulent flux of suspended solids indicates an upstream transport. All three flux records show short periods of more than average turbulent activity which illustrates the intermittent effect of the bursting process which results in turbulence generation by flow instabilities near to the bed.

Fig. (2) shows the variation of turbulent mean solute flux parameters with local solute induced density gradient ($\partial \bar{\rho}/\partial z$). The salinity (s) and velocity measurements were made at between 1.0 and 1.25 m above the bed in the Teign estuary. The vertical solute flux correlation coefficient ($R_{sw} = (\overline{sw})/s'w'$) is seen to slowly reduce in value as the density gradient increases. A similar trend also occurs if the local gradient Richardson number ($Ri = -(g\ \partial\bar{\rho}/\partial z)/(\rho\ (\partial\bar{u}/\partial z)^2)$) replaces the vertical density gradient (West and Shiono (Ref. 4)). For these data it was found (West and Shiono (Ref. 4)) that close to the bed in a solute stratified flow the velocity relative intemities $u'/\bar{u}$, $w'/\bar{u}$ were a strong function of relative depth but not substantially affected by acceleration or Richardson number effects. The intensity of the turbulent fluctuations of salinity (s') was a function of the vertical salinity gradient. A conditional sampling analysis suggested that the short time scale fluctuations tend to become more wavelike in stratified conditions with u and w being $\pi/2$ out of phase and u and s being π out of phase. This is consistent with R_{sw} decreasing and R_{su} increasing with Ri and $\partial\rho/\partial z$ (fig.(2)). The ratio ($\overline{sw}/\overline{su}$) decreases with increasing vertical density gradient emphasising the difference in behaviour of the two turbulent fluxes under stratified near bed conditions.

An example of near bed vertical turbulent flux of suspended solids data is shown an fig. (3). The vertical suspended solids correlation coefficient R_{cw} ($= \overline{cw}/c'w'$) is plotted against Richardson number with the relative depth plotted along side each data point. The density gradient in this case was generated by suspended solids with a discrete particle size of predominantly less than 63 μm. The correlation coefficient declines with increasing Richardson number and the data suggest that the decline is greater in the upper part of the flow (z/h ~ 0.4 - 0.7).

It may be anticipated that the effects of suspended solids and salinity vertical turbulent mixing processes may be different as the steepest solute gradients usually occur near to the free surface whereas the steepest suspended solids concentration gradients usually occur near to the bed. Thus it may be anticipated that suspended solids concentration will have the greater effect on bed generated turbulence for a given value of vertical density gradient.

In deeper water (>3 m deep) which is moving at the order of 1 m/s it is very difficult to deploy turbulence sensing instruments from fixed structures without substantially interferring with the flow or incurring a very substantial expenditure. Some success has been achieved in the study of vertical mixing by using a combination of techniques. Parsons et al (Ref. 3) have measured turbulent mixing in the lower Tamar estuary (~ 15 m deep) by determining the concentration distribution in a dye plume from a continuous point source at the free surface.

The dye tracing results from the Tamar (Parsons et al (Ref. 3)) yield values for the transverse turbulent mixing coefficient (ε_{sy}) of 0.05 m²/s on flood tides and 0.5 m²/s on ebb tides in the surface layers. Dye tracing results and turbulent mean salinity profile data for salinity suggested that the vertical turbulent mixing coefficient (ε_{sz}) had values of the order of 0.002 m²/s, that is an order of magnitude less than ε_{sy}. The use of an array of inductive salinometers showed that in the surface 3.5 m of the flow a complex pattern of internal waves and fronts existed during both ebb and flood tides (fig. 4). It was found that Richardson numbers were lower on the ebb tide. It is tempting to speculate that during the lower Ri conditions internal wave breaking, mixing and subsequent spreading horizontally of the zone of well mixed fluid could explain the higher transverse diffusion coefficient values on the ebb tide.

4.   CONCLUSIONS

1.   The direct measurement of turbulent fluctuations of velocity, salinity and suspended solids concentration is possible to a useful level of accuracy within about 3 m of the bed at an economic cost. With care, commercially available equipment can be adapted for these measurements.

2.   In deeper flows a combination of dye tracing, turbulent mean and turbulence sensing measurement can be used to investigate turbulent transport processes.

3.   The above techniques show considerable promise for developing further insight into turbulent transport processes which are fundamental to the improvement of the predictive capability of many estuarine management mathematical models.

REFERENCES

1.   Brierley, R.W., Shiono, K. and West, J.R. An integrated system for measuring estuarine turbulence. Proc. Int. Conf. on Measuring Techniques of Hydraulics Phenomena in Offshore, Coastal and Inland Waters. London, England, 9-11 April 1986. pp 359-366.

2.   Elder, J.W. The dispersion of marked fluid in turbulent shear flow. J.Fluid Mechanics, 5, 1959, pp 544-560.

3.   Parsons, A.P., West, J.R. and Lynn, N.M. The evaluation and representation of turbulent diffusion mechanisms in the lower Tamar estuary. Proc 5th IAHR Int. Symp. on Stochastic Hydraulics, Birmingham, England. 2-4 Aug 1988.

4.   West, J.R. and Shiono, K. Vertical turbulent mixing processes on ebb tides in partially mixed estuaries. Est. Coastal and Shelf Science, 26, 1988, pp 51-66.

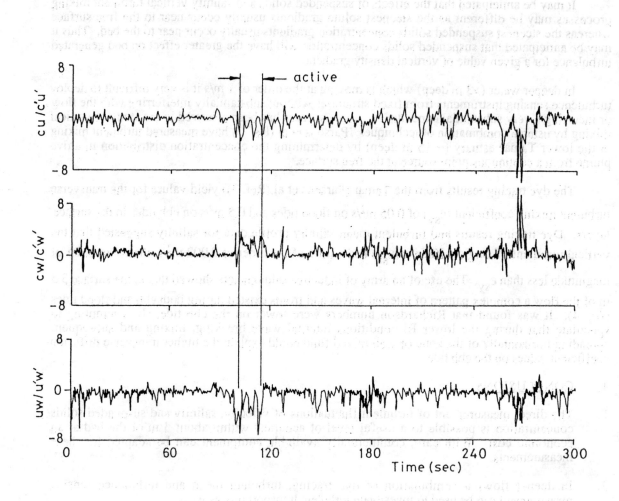

Fig. 1   Example of normalised fluxes (R. Tamar 7.7.85)

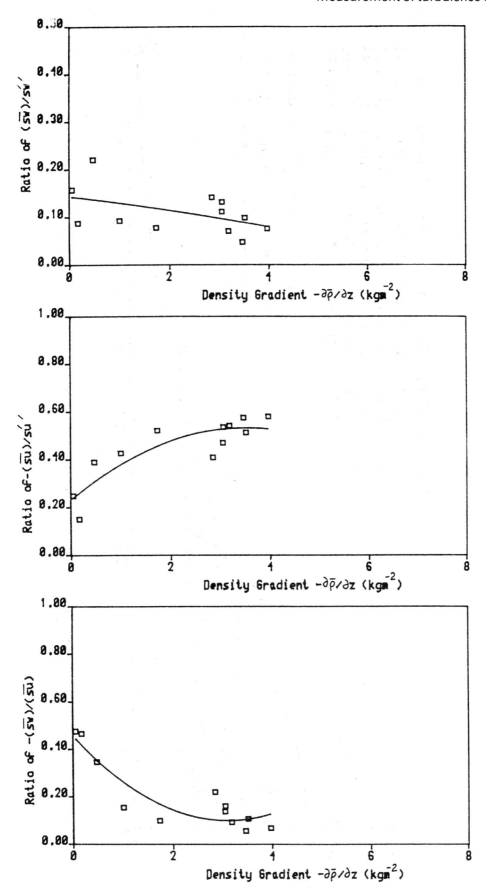

Fig. 2. Variation of Dimensionless Salinity Turbulence Parameters with Density Gradient, —— Trend Line. R. Teign 20.6.83.

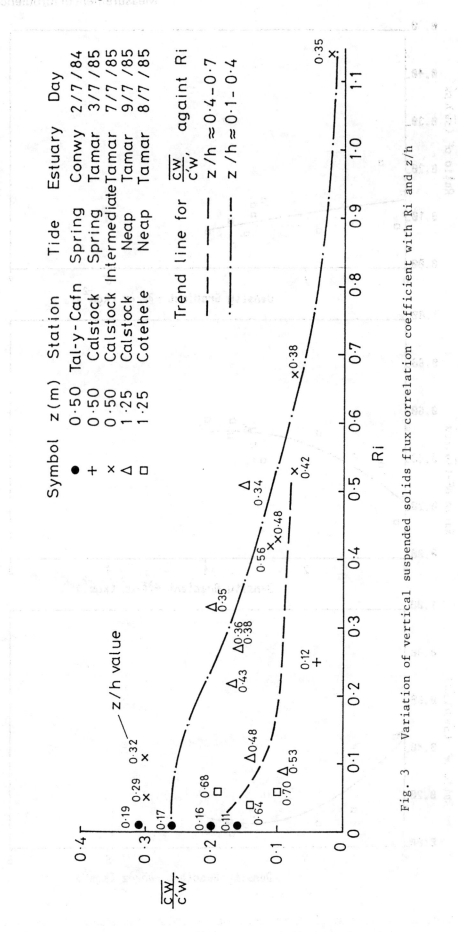

Fig. 3   Variation of vertical suspended solids flux correlation coefficient with Ri and z/h

82

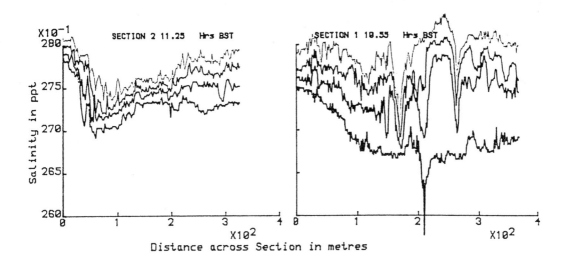

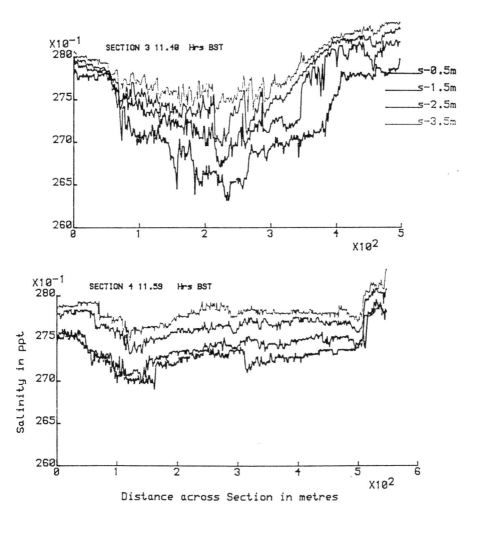

Fig. 4. Inductive Salinometer Survey 09-07-86
Ebb Tide

Chapter 7
BED QUALITY DEVELOPMENT IN THE WESTERN SCHELDT ESTUARY

L Bijlsma and F Steyaert
Ministry of Transport and Public Works
The Netherlands

SUMMARY

Increasing domestic and industrial discharges in the watershed of the river Scheldt have resulted in a rapid growth in the river's pollution load over the last decades. These events have coincided with a significant deepening of the estuary's navigation channel to accommodate larger vessels. Some 10 to 15 million m3 of spoil are now dredged up each year in order to maintain the depth of the navigation channel.

The high pollution levels in the Scheldt river system are reflected in the substantial accumulations of micropollutants in the bed sediments and microorganisms found in the many channels, intertidal zones and marshlands in this area. To aid policy makers dealing with such issues, a study was recently carried out of the transport and accumulation phenomena taking place in the various subsections of the estuary. An essential part of this analysis involved examining the silt regime in the estuary. This was particularly important since contaminants tend to become attached to the finer fractions of suspended sediments.

Differences in the proportions of the various carbon isotopes found in the organic fractions were used as a means of determining a silt balance between the fluvial and marine sediments. Sedimentation velocities were calculated by detecting the presence of radio active cesium and lead in sediment cores.

Dispersion rates, residence times and the accumulation of contaminated fluviatile silts were investigated with the help of models describing the sedimentation and entrainment processes in the different estuarial sections. This approach allowed activities such as retrospective forecasting and the supervision of the quality of bed sediments to be performed on the basis of existing data. A series of measurements of the quality of suspended material in the Scheldt and information derived from bed samples were used for this purpose.

The results presented in this study allow the impact of remedial action programmes and the effects of relocating dredging spoil to be assessed. Moreover, the models and balances derived enable a detailed analysis to be made of various time-dependent factors.

List of symbols

B	definitive storage in a given section	t/
BA, ST	quantities dredged up and dumped in a given section	t/
E, S	erosion and sedimentation of the deep bed layer in a given section	t/
U	rate of exchange of the active top layer in a given section	t/
KB	quality of the material permanently stored in a section	mg/
KSR	quality of the incoming fluvial silt in a section	mg/
KSUS	quality of the suspended material in a section	mg/
KU	quality of the active bed layer in a section	mg/
KE, KS	quality of the eroded and deposited material associated with the deep bed layer in a section	mg/
PR, PZ	percentage of fluvial and marine silt in a section	
OPP	surface area of a section	m
P	percentage of silt in the bed phase of a section	
SR	resulting fluvial transport in a cross-section	t/
SZ	resulting marine silt transport in a cross-section	t/
SG	specific gravity of dry matter	t/m
j	section number	.
i	index for incoming in the section j	.
u	index for outgoing in the section j	.

1. INTRODUCTION

One of the major issues confronting those responsible for the management of the Western Scheldt (Fig. 1) is how to maintain the bed quality in the estuary. This is complicated by the fact that micropollutants tend to become attached to the fine particles in the suspended sediment and are subsequently transported downstream, eventually accumulating in the sedimentation areas in the river. Pollutants are also transported and dispersed when materials that are removed to maintain the depth of navigation channels and harbours are dumped elsewhere.

In order to be able to assess the degree to which the bed of the estuary has become contaminated with micropollutants, a comprehensive analysis of the particle flow stream is essential. Of major importance in this context is the availability of scenarios describing the pollution loading caused by discharges and by the transport of material from maintenance work carried out on navigation channels and harbours.

The approach followed in the current study was to derive a balance for the pollution levels associated with floating matter. This work also involved modelling the accumulation taking place in the bed of the estuary. The advantage of such a method is that it also allows time-dependent effects to be assessed such as the cycle of pollution build-up in the bed followed by subsequent recovery. Similarly, the impact of transferring dredging spoil to other locations can also be determined with this approach.

2. PROCEDURE

An estimate of the bed quality in a specific area can be obtained by establishing a balance between the inflows and outflows of (polluted) river sediment. As the majority of the pollution loading normally occurs in an associated form and since the pollutants are mainly attached to the finer fractions, particular attention has to be focused on the silt balance in the river. For practical reasons, it was decided to define silt as the fraction containing particles smaller than 0,05 mm. The approach employed in the current study was to consider the problem in two parts. First, an estimate was made of the overall silt balance in the estuary, including both fluvial and marine silt. Secondly, a pollution balance was derived for the fluvial silt by equating the levels of contaminants present.

The aim of the investigation was to provide an assessment of the primary dispersal and build-up of pollutants and to determine how quickly the bed of the estuary recovers from such effects. For this purpose, the river was considered to be the only source of pollution and lateral discharges were regarded as insignificant.

The following assumptions were made in the analysis:

- the associated pollutants are essentially present in a stable form and do not undergo change;
- the hydraulic morphology of the area as well as the degree of salt intrusion remain constant;
- biological processes such as accumulation, decay, sedimentation and bioturbation can be neglected.

3. SILT MODELLING

3.1 Silt balance

A schematic representation of the silt balance at a given section of the river is shown in figure 2 (for definitions see the appended list). The diagram indicates how the inflows and outflows of fluvial and marine silt can be equated. In each section of the estuary, the silt is distributed over three compartments: the water phase (i.e. suspended material), the bed phase and the material that is stored outside the system.

The amount of silt present in the water phase can generally be neglected when it is compared with that contained in the other compartments. In contrast, substantial quantities of silt are present in the bed phase, divided over two subcompartments: an active top layer and a much thicker sedimentary layer below. The top layer is subject to intensive exchange with the water phase due to hydraulic effects, whereas the thicker underlying stratum is only influenced by major morphological changes such as shifts in the position of channels.

The storage of material outside the system normally involves sediment being removed from a particular section of the estuary. Typical examples of this are sand extraction and the sedimentation that occurs in salt marshes.

Relocation of material also occurs between the various compartments in the form of silt transport. For instance, fluvial silt flowing from upstream regions and inflows of marine silt due to dispersive processes interact directly with the water phase. The continuous exchange between the water phase and the active top layer of the bed phase is another form of silt transport. In addition, silt settles out from the water phase in the sedimentation areas associated with salt marshes. Finally, silt is transported when dredging spoil is dumped in other sections of the estuary or is extracted from the system.

In analysing the processes occurring in the estuary, it is assumed that the sedimentation and erosion rates within a given compartment remain in equilibrium. If the ratio between the amounts of fluvial and marine silt is known, a silt balance can be derived as follows (for an explanation of the symbols see appendix).

Ratio fluvial/marine silt:

$$(SRi - SRu) : (SZi - SZu) = PR : PZ \qquad\qquad (1)$$

Silt balance:

$$SRi + SZi - SRu - SZu = B + (BA - ST) \qquad\qquad (2)$$

This equation can be solved if B, BA, ST and the silt loading of the river are known.

Since the amount of materials that are dredged up and dumped in the estuary tends to vary, time series were introduced into the calculations to allow the development of the silt balance to be determined as a function of time. It was assumed that the ratio of fluvial and marine silt remains constant throughout this process.

3.2   Flow of polluted fluvial silt

The fluvial silt entering the area described in the model is assigned a certain quality rating (pollution level) which varies as a function of time. This polluted silt disperses in the system, interacting rapidly with the active bed layer. Moreover, exchanges also take place with the deeper bed layers due to morphological changes in the estuary. The processes of sedimentation and erosion which shift the position of channels in the estuary are largely responsible for this. The interaction that takes place with the deeper-lying sediment is characterized by a long geological time-scale. However, the current analysis is concerned with a much shorter time frame of the order of decades. The exchange process with the deep bed layers can therefore be seen in terms of the burial or storage of polluted material and the erosion of clean sediment (undisturbed bed phase). Figure 3 shows how the balance of polluted fluvial silt can be established in the different sections of the estuary.

The amount of material actually present in suspension in the water phase is considered to be negligible when compared to the volume of silt in the riverbed. Instantaneous mixing can be assumed to occur in the context of the time scale applicable for this study (years). Time-dependent effects are introduced into the model via the bed compartment.

In this way, assessments can be made of the flow of particulate matter as well as of the transport between the various compartments and the quantities stored there. The quality attributable to the suspended matter can be determined as follows:

$$KSUS = (KSRi*SRi + KU*U*PR + KE*E*PR)/(SRi + U*PR + E*PR). \quad (3)$$

Hence the condition of the active top layer can be expressed as:

$$dKU/dt = (KSUS*U - KU*U + KST*ST - KU*BA)/OPP*1*P*PR*SG. \quad (4)$$

The quality ratings of the other compartments and of the materials transported can be determined by drawing the following parallels:

-   the quality of material that is stored or buried is equivalent to that of the suspended material;
-   the quality of the eroded layer is equal to that of undisturbed bed material;
-   the quality of dredged material is equivalent to that of the active top layer.

4.   BOUNDARY CONDITIONS

4.1   Division of the estuary

The area under consideration in the present study comprises the full extent of the Western Scheldt estuary up to and including the mouth of the River Scheldt. In the east, this region is bordered by the Belgian part of the River Scheldt at the point at which the influence of the sea becomes negligible.

In setting up a model to describe the Scheldt estuary, it was decided to divide the area into three sections: the Belgian part (up to the Dutch border), and the eastern and western parts of the Western Scheldt (fig. 1). The borderline between the two parts of the Western Scheldt lies slightly to the west of Hansweert. The decision to divide the Scheldt estuary into three sections was based on clear differences in the geometry of these areas, on the different influences exerted by the sea and on the large variation in dredging and dumping activities carried out in the three sections.

4.2   Supply of fluvial silt

Various researchers have tried to quantify the amount of silt discharged by the river Scheldt [1, 2, 3]. The most recent estimates, which have been used as the basis for the silt balance derived in this paper, suggest that some 750,000 tonnes of fluviatile silt [3] are brought down by the Scheldt each year. In the course of time however, the volumes of silt transported will vary as a function of the discharge from the River Scheldt.

In addition to natural variations, the amount of suspended material in the river is also affected by waste water discharges in the catchment area. Untreated industrial and domestic waste increases the load of suspended material carried by the water phase. Estimates [3, 4] of the significance of such discharges are given in table 1.

It is recognized that the growth in the population coupled with rises in prosperity and industrial activity have led to increasing volumes of waste being discharged over the last decades. As a result, the turbidity of the river has risen accordingly. The introduction of specific clean-up and purification programmes will probably change this trend in the coming years and the loading of suspended matter is therefore expected to decrease.

In calculating the flow of polluted fluvial silt, it was assumed that the upstream supply of particulate matter is constant. However, the effect that purification processes will have in reducing turbidity in the river should be borne in mind when interpreting the results.

4.3 Quantities of silt in the bed phase

The quantities of silt present in the three sections of the Scheldt estuary are given in table 2. These values were calculated on the basis of surveys of the bed of the estuary [5,6] combined with the results of samples taken at the sills in the Western Scheldt.

Assessments were also made of the appropriate dimensions of the two subcompartments of the bed phase outlined in section 3. The active surface layer, which is characterized by vigorous interaction with the water phase was estimated to be 1 m thick. In contrast, the underlying layer, which is subject to much slower exchange processes, was taken to be 9 m thick. This implies that the total depth of the silt reservoir in the estuary extends 10 m into the bed sediment (see also section 4). Salient points relating to the sedimentation and resuspension of silt in the three sections of the estuary are given in table 3.

4.4 Ratio of fluvial silt to marine silt

In the 1970's, techniques were developed which allowed the relative proportions of fluvial and marine silt in an estuary to be determined. The method is based on detecting the different isotopic compositions that occur naturally in fluviatile and marine environments. The results of samples taken in the period 1975-1985 have enabled researchers to establish the relative quantities of fluvial and marine silt in the River Scheldt and the Western Scheldt estuary [7, 8]. The bed of the estuary effectively "records" the events dating back over many years.

It has been suggested that the relative proportions of fluvial and marine silts could change in the course of time. This would imply that the suspended material in the water phase does not remain in equilibrium with the bed phase. Possible shifts in this ratio can be detected by comparing the isotopic composition of the suspended material in the water phase. Although a number of such measurements have been made, no systematic shifts have been observed. The relevant data are summarized in table 4.

4.5 Exchange processes

The rate of exchange of suspended material between the water and bed phases can be estimated on the basis of the rate at which the sediment in channels and intertidal areas is transformed. In accordance with the division of the bed compartment into an active surface layer and a deeper underlying layer, two separate components of the conversion process can be identified: short-term and long-term changes. The component processes can be defined as follows:

- Short-term changes.
  Short-term changes are disturbances in the top layer of channels, shallows and edges. Such changes typically refer to annual variations in level whereby the bed either becomes deeper or shallower without being part of a long-term trend;
- Long-term changes.
  Long-term changes are associated with developments such as the erosion and deposition that occur at opposite sides of the bends in channels.

In order to determine the extent of such changes, an assessment was made of the annual variations in the bed phase. On the basis if soundings with an interval of two years, it was estimated that in this interval on average a layer of one meter was reworked. The extent of the long-term changes were calculated by summing the major sediment transfers from the channels. This allowed the sedimentation and resuspension rates between the water phase and bed compartment in the Western Scheldt to be determined. The results are shown in table 3.

4.6   Dredging and dumping statistics

There are two main reasons for carrying out dredging operations in the estuary: sand extraction and maintenance work. When sand is removed from the estuary, quantities of entrained silt leave the system at the same time. In the case of dredging operations to maintain the depths of navigational channels, a large proportion of the dredging spoil that is removed is dumped in other parts of the estuary. This material is therefore returned to the system and can subsequently be transported to another compartment. The nautical dredging spoil that is not returned to the estuary is generally used as infill material. An overview of the amounts of material involved in the dredging and dumping activities in the estuary is given in table 4.

Table 4 shows that of the 392 million m3 of spoil removed in dredging operations since 1950, 266 million m3 have been returned to the estuary. This means that 40% of the material removed from the estuary leaves the area permanently. Although the bulk of this material is sand, a certain amount of silt is also included. Figure 4 shows how the sand extraction operations, including the associated nautical dredging work, is distributed over the three sections of the estuary.

Another item which is of importance with respect to the silt balance is the transfer of dredging spoil. In the present study which is concerned with dividing the estuary into relatively large sections, estimates were restricted to the transfer of spoil between the Zeeschelde and the eastern and western parts of the Western Scheldt. Further analysis of the dumping statistics shows that there is a bias in the transfer of material between the different sections of the estuary from east to west. This is mainly the result of a lack of dumping capacity in the immediate surroundings of the dredging locations and by land suppletion on the exposed western banks. The transport data given in figure 5 show that the amounts transferred are far from negligible. In fact, the total exchange between the Zeeschelde and the eastern part of the Western Scheldt is approximately 10 million m3 whereas some 20 million m3 are transported between the eastern and western parts of the Western Scheldt.

During the period under consideration the main navigation channel in the area was deepened. Since the initial excavation work on the channel can essentially be regarded as a one-off activity it would be wrong to include the entire amount of spoil removed during this operation in the overall silt balance. However, the long-term extractions of silt as a result of maintenance work have been considered in the analysis.

4.7 Storage due to sedimentation, Land van Saeftinghe

Various sedimentation phenomena can be distinguished in the Western Scheldt which involve the permanent deposition of silt:

- sedimentation on salt marshes;
- silting up of intertidal areas;
- deposition in secondary channels.

An outline of the processes that are of major importance in determining the silt balance in the estuary was given in earlier sections. The burial and erosion processes referred to previously which are associated with large-scale changes in the bed formation are clearly related to the phenomena involved in the silting up of intertidal areas and the deposition in secondary channels. In contrast, the permanent deposition taking place in salt marshes such as the Land van Saeftinghe (figure 1) requires further discussion. This area is particularly important both because of its position in the system (following a turbidity zone) and because of the extent of the sedimentation taking place there.

In 1987, the level of the deposits in the Land van Saeftinghe was measured along a number of different lines of direction. The results obtained, when combined with level data from 1962 and the results of soundings and level measurements carried out in 1931, were used to construct deposition profiles for the years concerned. Figure 6 clearly shows the extent of the silting-up that has occurred over the last period.

Core samples were also taken in 1987 and analysed for the presence of 134Cs, 137Cs and 210Pb. The aim of these tests was to collect information about the local sedimentation rates. All the core samples examined showed evidence of a maximum concentration of 137Cs which relates to the fallout from atomic bomb tests carried out in the atmosphere [9] in 1963. A further peak was also detected which was clearly identifiable as the 1986 Chernobyl fallout. On the basis of this and other information related to the level of 210Pb (half-life 22.4 years), it was possible to determine the sedimentation rate in the area (figure 7).

An analysis of the results of level measurements and core samples showed that the rate of silting-up generally varies from between 0.01 to 0.02 m per year. The decrease in the effective volume of the Saeftinghe area calculated on the basis of level measurements is thought to proceed at the rate of 0.5×10^6 m3 per year. This reduction in volume is partly caused by sand sedimentation in the channels and partly by silt deposition. It is assumed that the reduction in volume of the channels and salt marshes is approximately equal. When the percentage of silt in the creeks, natural levees and basins [10] is taken into account, it can be concluded that 410,000 tonnes of silt per year are imported into the area. Given that the ratio of marine silt to fluvial silt is 55% : 45%, this amounts to a total import of fluvial silt of about 200,000 tonnes per year.

5.   RESULTS

The model developed to describe the silt balance in the Western Scheldt is particularly useful for reconstructing the pollution build-up in the bed of the estuary. The results of such calculations can be compared with the levels of micropollutants found in samples taken from the bed sediment. However, since the model is only intended to indicate general trends and as there are no systematic sets of data concerning the quality of bed sediments, it was decided to limit the scope of the calculations by only considering a number of metals for which most information exists.

5.1   Silt balance

The model outlined in section 3 is able to describe variations in the transport of both fluvial and marine silt with respect to time. These variations are primarily due to changes in the intensity of dredging operations in the area and to the fact that the dredged materials are no longer reintroduced into the estuary. Changes in the sedimentation rates in, for instance, salt marsh areas also affect the situation. Figure 8 shows how the calculated silt transport varies over the period 1945-1985. Although these results cannot be interpreted in a strictly quantitative manner, because of the limitations in the model referred to earlier, they do indicate general trends. In 1945, for example, little dredging work was carried out in the estuary and a large proportion of the fluvial silt brought down settled in the Zeeschelde. A significant amount of silt was also deposited on the Land van Saeftinghe. Due to the regular dredging operations carried out since 1970 and the policy of storing large amounts of spoil from the Zeeschelde on land, the silt balance have tended to shift gradually and increasing amounts of fluvial silt has settled in the Zeeschelde. This has resulted in ever smaller amounts of fluvial silt reaching the sea.

5.2   Quality of sediment in the River Scheldt

Since systematic measuring programmes for pollutants have only recently been introduced, it is not possible to provide accurate estimates of the quality of the sediment that has been brought down by the River Scheldt over the entire period covered by the present study. The first such measuring programmes which were directed to preventing micropollutants from reaching the surface waters were only set up in the 1970's [11]. Monitoring work conducted at the Dutch-Belgian border since 1976 provides a guide to the associated pollution carried in the Scheldt. In order to be able to compare these levels with those found in the bed sediment, use was made of the method described in Ref. 12. The ratio of the quality of the bed sediment to that of the suspended material in running water was estimated to be 0.7. In order to correct for the fact that pollution data were obtained at the Dutch-Belgian border rather than at the boundary of the model, a dilution factor of 1.2 was applied.

Reconstructions of the quality of the Scheldt sediment over the 1976-1985 period are given in figures 9, 10 and 11 for the metals cadmium, mercury and zinc respectively. To describe the development of pollution levels since 1945 it is has been assumed that the amount of contaminants present in the Scheldt rose significantly from 1945 onwards and reached a peak in 1960. This situation was thought to have persisted until about 1975 and was then followed by slight reductions in the case of a number of compounds.

5.3 Quality of the silt in the estuary

Surveys of the levels of the micropollutants found in the surface sediments of the estuary have been carried out previously and are reported in Refs. 13 and 14. In order to be able to compare the results of these investigations, corrections have been applied which normalize the data obtained in terms of a standard sediment. The basic characteristics of the standard sediment are that the 50% fraction of the CaCO3 free mineral particles is smaller than 16 micron and that it has an organic carbon content of 5%.

To relate the measured concentrations to the fractions mentioned above, separate regression lines have been set up for 8 individual areas. By applying these corrections to the information presented in Ref. 14 the temporal and spatial developments of the three selected metals cadmium, mercury and zinc were derived. An overview of the relevant data is given in table 5.

5.4 Calculated pollution build-up for cadmium, mercury and zinc

By applying specific boundary conditions to the model, it was possible to determine the build-up of Cd, Hg and Zn in the bed of the estuary. The results of these calculations are compared with the measured values in figures 9, 10 and 11. It can be seen that the Cd levels in the beds of sections II and III have clearly been overestimated. This can probably be explained by the fact that the solubility of associated cadmium increases with the salinity of the water phase. In contrast, the Hg levels recorded for 1975 can be seen to be higher than the calculated values. This may be attributable to the fact that the mercury emissions were larger than was originally envisaged. Finally, the predicted Zn levels were generally found to be in good agreement with the measured values. On balance, it would appear that the relative pollution trends within the estuary can be predicted satisfactorily with the model. A more detailed consideration of the results leads to the following observations:

-   the build-up of pollution in the bed sediment contained in the various sections of the estuary exhibits a lag which increases with proximity to the sea;
-   the pollution of the bed material discussed so far relates specifically to that of the 1 m thick active surface layer, since the samples of sediment that have been analysed were collected from this layer. In addition, the underlying sediment layer, which extends over a depth of 9 m, is known to have been subjected to localized pollution. However, this process progresses at a much slower rate than that of the active top layer. On the basis of the estimated long-term changes with regard to the position of the channels in the estuary it can be assumed that up to 1985 about 10% of the deep bed sediment in the estuary had been polluted.

6. **FUTURE DEVELOPMENTS**

In order to set up a policy plan for the Western Scheldt, it is essential to be able to assess how future developments could affect the quality of the sediment in the estuary. Scenarios have therefore been developed to take account of events such as:

- clean-up measures directed at minimizing the effects of discharges in the Scheldt basin. The results of this type of remedial action programme can be expressed in terms of an improvement in the quality of the silt in the Scheldt;
- the need to store considerably larger amounts of dredging spoil in the estuary than in previous periods, in order to maintain the depth of the Scheldt.

To test the validity of the various scenarios, a sensitivity study was performed using the silt quality model referred to earlier. This assessment was based on the predicted build-up of zinc in the estuary up to 1985. Three main issues were considered in the sensitivity analysis:

- the speed with which the discharge problem is dealt with;
- the storage of dredging spoil outside the system;
- the dispersal of dredging spoil from east to west.

6.1 **Time-dependent effects related to clean-up measures**

This part of the study was concerned with two specific conditions, namely maintaining the quality of the water and sediment in the Scheldt at 1985 levels and the introduction of a rapid clean-up programme to deal with discharges within a ten-year period.

For the purpose of the calculations, the extraction of dredging spoil was maintained at 1985 levels. The results obtained are given in figure 12. It can be clearly seen that the quality of the suspended material lags behind that of the sediment brought down by the river. In turn, the quality of the active bed layer lags behind that of the suspended material in a given section of the estuary. In the event that rapid progress is made with remedial action programmes, the quality of the bed sediment can be seen to follow in phase with these developments. This means that upstream clean-up operations have a direct and positive effect on bed quality. The half-life of the pollutants in the bed phase amounts to approximately 10 years. Once the various discharges have been cleaned up, the pollution levels in the bed sediment will decay to about 10% of the maximum levels recorded in 1985 after about 30 years. The following additional observations can be made in this context.

First, it is expected that the supply of fluvial silt will reduce following the introduction of clean-up measures directed at limiting domestic and industrial discharges. The use of conventional treatment plants will result in about 80% of the suspended material being extracted in the form of treated sludge. It is expected that the introduction of such measures will, on balance, reduce the silt load in the river by about 50%, which will promote recovery of the bed sediment in the estuary. These estimates are based on the assumption that the distribution of suspended material in the river corresponds with that given in table 1.

Secondly, it should be noted that, in addition to the delayed pollution effects due to resuspension in the "model" area, similar effects will also emanate from the upper part of the river. These conditions will tend to be reinforced by an increase in the oxygen content in the river where large amounts of metals have accumulated in the bed sediments in reduced forms.

6.2 Scenarios for storing dredging spoil

The frequency of dredging operations is expected to increase considerably in the coming years, after the scheduled deepening of the sills in the Western Scheldt has been completed. Since the existing dumping sites do not have sufficient capacity to handle the expected increased volumes of spoil, storage outside the system or dumping in more westerly parts of the estuary will have to be considered. To be able to assess the consequences of such actions for silt management in the estuary, a more detailed study of the silt balance was made with special reference to the particulate flow stream in section I (Zeeschelde) and section II (eastern part of the Western Scheldt).
The following cases were considered:

- current situation;
- lowering the sills and removing all the spoil from the system;
- lowering the sills combined with the transfer of all the spoil from the Zeeschelde and half the spoil from the eastern part of the Western Scheldt to a location one section further downstream.

In determining the quality of the bed sediment to be incorporated in the scenarios, the following equations were used:

- In the case of storage outside the system, allowance was made for the reduction in the river loading reaching the downstream section by taking account of the improved quality of the suspended material:

$$KSRi \ (new) = KSRi \ (existing) \ x \ \frac{SRi \ (new)}{SRi \ (existing)} \qquad (5)$$

With this approach, adjustments were therefore made to the ratio of fluvial silt to marine silt in the suspended material. This ratio will tend to shift if large amounts of silt are regularly removed from the system.
- In the case of spoil being transferred from an upstream to a downstream section, allowance was made for the increase in the loading of polluted fluvial silt by taking account of the reduction in the quality of the active bed layer:

$$KU \ j-1 \ (new) = \frac{V* + Kj-1 \ (existing) * OPP * 1 * PR * SG}{ST + OPP * 1 * PR * SG} \qquad (6)$$

$$V* = ST * KST * \frac{PRj}{PRj-1} \qquad (7)$$

With this approach, adjustments were therefore made to the ratio of fluvial and marine silt present in the active bed layer. This ratio will shift if large amounts of spoil dredged up for maintenance purposes are systematically dumped downstream.

The results of the calculations performed with the three scenarios are given in figures 13, 14 and 15. The scenario involving the extraction of dredging spoil and subsequent storage in a depot appears to be highly attractive. This type of approach becomes even more effective as the percentages of fluvial silt increase. The amounts of silt transported between section I (Zeeschelde) and section II (eastern part of the Western Scheldt), for instance, are reduced to about one third of the loading presently experienced.

In contrast, a marked increase in the silt loading is envisaged in the scenario involving dumping in a more westerly section. With this scenario, the silt transport is even expected to rise above current levels. Furthermore, it can be concluded that the dredging scenarios for section I (Zeeschelde) markedly affect the particle flow stream in the rest of the estuary. This effect is significantly greater than the impact of the dredging scenarios for section II (eastern part of the Western Scheldt), which is explainable by the relatively low proportion of silt in the sills of the eastern part of the Western Scheldt.

It can be clearly seen that the dumping scenario would have much more serious implications for the bed quality of the eastern part of the Western Scheldt than the storage scenario. In view of the amounts of materials that would be dumped under such a scenario and the quantities of silt present in the active top layer in the eastern part of the Western Scheldt, it is expected that, within about 10 years, the quality of the bed sediment in this area would approach that of the Zeeschelde. In comparison, the dumping of materials removed from the eastern part to the western part of the Western Scheldt will have much less effect on the quality of the bed sediment in this area. These effects will be mainly confined to localized deterioration at the dumping sites and their immediate surroundings, such as the salt marshes and mudflats.

6.3   Long-term effects on the silt balance

In predicting the outcome of the various scenarios it was assumed that a number of boundary conditions would remain constant. However, as was indicated in section 4.2, it is expected that the silt loading in the Scheldt will decline as a result of an increase in clean-up activities in Belgium. This would mean that the total amount of fluvial silt brought down into the estuary would be drastically reduced.

Furthermore, it is likely that less silt will be deposited on the Land van Saeftinghe over the coming years. The decrease in the storage capacity of the Land van Saeftinghe basin below the +3 m and +4 m A.O.D levels is illustrated in figure 16. The number of tides that exceed these levels are 125 and 1.5 per year respectively. If the capacity of the basin continues to decline at its present rate this would mean that the entire area below + 3m A.O.D. would be filled in by the year 2010. As a consequence of the reduction in the storage capacity of the Land van Saeftinghe in the coming years more silt will be transferred downstream.

7.   DISCUSSION

The following assumptions were made in setting up the silt balance:

1. It was assumed that the hydraulic morphology of the area and the ratio of saltwater to freshwater in the estuary did not change significantly during the period considered. However, it is recognized that the geometry of the estuary and hence the

transport processes have changed due to increased dredging activities. The finer fraction in the upper part of the estuary is primarily affected by longitudinal vertical circulation, which is caused by variations in density due to saltwater/freshwater gradients. If the salinity in the estuary has not changed significantly as a result of the differences in geometry, it can be assumed that the longitudinal circulation currents and therefore the silt transport have not undergone a fundamental change. In the lower part of the estuary the tidal assymetry is little affected by the deepening of the main channel. It therefore seems realistic to assume that the ratio of fluvial and marine silt in the bed of the estuary has not been subject to major shifts over the years.

2. In deriving a silt balance for the Western Scheldt, it was assumed that the sedimentation and erosion processes taking place within the individual sections of the estuary always remain in equilibrium. In view of the size of the sections used in the model, these effects are expected to balance out. It therefore seems reasonable to adopt such an approach at this stage of the analysis.

3. It was assumed that the silt loading in the Western Scheldt is principally determined by the River Scheldt. Lateral discharges have therefore been neglected. An analysis of anthropogenic influences and the concentration distribution in the estuary has shown that this approach is correct.

4. The associated pollution carried by the silt fraction was assumed to be stable and not undergo change. This basically conservative assumption is, however, not always valid for a number of micropollutants.

5. Biological processes such as accumulation, decay, sedimentation and bioturbation were not taken into account in the analysis nor were interactions occurring in the bed phase. The morphological processes of sedimentation and resuspension and the transfer of silt as a result of dredging and dumping operations were assumed to be of much greater significance.

From the above observations it is clear that the emperical box approach adopted in the present study enables specific trends to be distinguished. This method can therefore be used to assess developments in the bed quality in the Western Scheldt in respect of various clean-up scenarios and/or dredging/dumping strategies. However, it will only be possible to generate exact quantitative descriptions if a multidimensional transport model is used. This would allow residual transport effects due to differences in density and the mixing processes of freshwater and saltwater to be included, as well as the residual transport associated with tidal gullies and ebb deposition. Furthermore, a model of this type could be used to assess other geometries and systematic changes in the freshwater discharge. An description of the dynamic characteristics of various pollutants would be of importance in this context.

8. CONCLUSIONS

- A model describing the balance of silt flows in the Western Scheldt estuary has been developed which allows global predictions to be made of the build-up of micropollutants in bed sediments. Calculations are made on the basis of estimates of the relative proportions of fluvial and marine silt and the quantities of micropollutants attached to suspended material. In addition, a knowledge of dredging and dumping activities and the various sedimentation and erosion processes is also required.

- If the policy regarding the quality of the silt brought down by the river remains unchanged, the build-up of pollution in the bed sediment is expected to continue. The 1 m thick top layer is known to react quickly to changes in the quality of the suspended material. The degree of exchange that has already taken place with the polluted fluvial silt is up to 70% (in the west) and up to 95% (in the east). Moreover, the underlying sediment layer (1-10 m) has been subjected to local pollution extending over some 10% of the region.

- Clean-up programmes aimed at minimizing the effects of discharges will immediately improve the quality of the active bed layer in the whole estuary by reducing the pollution loading carried by the fluvial silt. The half-life associated with these pollution levels in respect of exchange processes is about 10 years.

- A considerable reduction in the flow of polluted fluvial silt can be achieved by extracting silt along with dredging spoil. Such an approach would be particularly effective in the Zeeschelde where the percentage of fluvial silt in the bed is highest.

- Transferring dredging spoil from east to west could have significant implications in terms of the release of polluted fluvial silt. The consequences would be most serious if spoil from the Zeeschelde is transferred. This would result in the bed quality in the eastern part of the Western Scheldt deteriorating rapidly.

REFERENCES:

1. TERWINDT, J.H.J. 1977. Mud in the Dutch Delta area, Geologie en Mijnbouw, Volume 56 (3).

2. WOLLAST, R. and J.J. PETERS 1978. Biochemical properties of an estuarine system, the river Scheldt, Biochemistry of estuarine sediments, UNESCO, Paris.

3. WOLLAST, R. and A. MARIJNS 1981. Evaluation des contributions des diffrentes sources de matire en suspension l'envasement de l'Escant, Rapport final au Ministre de la Sant Publique, 152 pp.

4. D'HONT, P. and T.G. JACQUES 1982. Suspended matter in the Scheldt, Ministerie van Volksgezondheid, Water no. 4. (in Dutch only)

5. BASTIN, A. 1974. Regional sedimentology and morphology of the southern Northsea and the Western Scheldt estuary. Doctoraatsthesis, Kath. Univ. Leuven, 91 pp. (in Dutch only)

6. WARTEL, S. 1977. Composition, transport and origin of sediments in the Scheldt Estuary, Geologie en Mijnbouw, 56 pp. 219-233.

7. SALOMONS, W., M. DE BRUIN, R.P.W. DUIN and W.G. MOOK 1978. Mixing of marine and fluvial sediments in estuaries. 18th Coastal Engineering Conference, Aug 27-Sept. 1978, Hamburg.

8. SALOMONS, W. and W.G. MOOK 1981. Field observations of the isotopic composition of particulate organic carbon in the southern North Sea and adjacent estuaries, Marine Geology, 41 (1981) pp. 11-20, Elsevier Scientific Publishing Company, Amsterdam, the Netherlands.

9. BALZER. Diagenesis and exchange processes at the benthic boundary. In: lecture notes on coastal and estuarine studies seawater-sediment interactions on coastal waters. J. Rumohr, Springer Verlag, Berlin/Heidelberg, 1987 pp. 116-119.

10. REYNDERS, J.J. 1985. Environmental aspects of the intertidal zones of the Western Scheldt., Inst. voor Aardwetenschappen, Rijksuniversiteit, Utrecht. (in Dutch only)

11. RIJKSWATERSTAAT, Rijksinstituut voor zuivering van afvalwater 1982. The waterquality of the Western Scheldt in the period 1964-1981, Rapport no. 82.063. (in Dutch only)

12. RIJKSWATERSTAAT DGM 1986, NOB interimreport of the workinggroup standardisation, March 1986. (in Dutch only)

13. SALOMONS, W. and W.D. EYSINK 1981, Pathways of mud and particulate trace metals from rivers to the North Sea, Spec. publication Int. Ass. Sediment, 1981, 5, pp. 429-450.

14. RIJKSWATERSTAAT DGW 1986, Micropollutants in the sediments of the Western Scheldt 1974-1985, notitie GWAO-86.537 (in Dutch on:

Table 1: Breakdown of the silt loading in the Scheldt basin

natural transport	0.27×10^6 ton/year
domestic	0.19×10^6 ton/year
industrial	0.29×10^6 ton/year
total silt loading	0.75×10^6 ton/year

Table 2: Quantities of silt in the different sections of the estuary

	silt percentage	area km^2	silt storage 0-10 m 10^6 ton	marine silt percentage	fluvial silt percentage
compart. I Zeeschelde	40%	5	42.5	20%	80%
compart. II eastern part 1)	4.5%	60	45.9	55%	45%
compart. III western part	4.6%	174	133	90%	10%

1) exclusive marsh land Saeftinghe

Table 3: Exchange of silt by sedimentation and resuspension in 10^6 tons/ year (silt percentages in bed from table 2)

	active top layer	deep layer	total
compart. I Zeeschelde	1.2	0.1	1.3
compart. II eastern part 1)	1.4	0.1	1.5
compart. III western part	4.9	0.2	5.1

Table 4: Amounts of silt in dredging material in the area since 1950 in
10^6 tons/year (silt percentages included in dredging and dumping
from table 2)

	comp. I Zeeschelde	comp. II eastern part	comp. III western part	total
included in dredging	1.0	0.3	0.1	1.3
included in withdrawal of sand	0.1	0.1	0.1	0.1
included in dumping	0.4	0.2	0.1	0.7

Table 5: Overview of quantities of heavy metals found in sediments

Mean values concentration heavy metals mg/kg	50%	16 µm							
		As	Cd	Cr	Cu	Hg	Ni	Pb	Zn	number of samples
comp. I	1974	84	37,2	515	195	3,75	71	260	1530	?
	1979	87	27,1	245	138	2,95	49	225	800	47
	1984	36	16,8	300	145	1,75	54	205	740	9
	1985	–	18,2	80	102	–	31	135	525	9
comp. II	1959	61	6,4	180	76	1,85	26	125	500	10
	1979	30	3,9	125	36	0,95	29	75	310	21
	1982	16	2,5	65	35	0,95	18	65	230	4
	1984	35	5,0	155	54	0,95	34	90	330	36
	1985	29	2,9	45	57	0,70	29	75	295	20
	1986	10	4,9	45	57	0,70	29	85	340	6
comp. III	1960	22	1,2	115	33	1,05	22	80	260	9
	1974	18	1,4	100	29	0,65	24	65	185	14
	1979	25	1,8	100	26	0,65	26	55	190	62
	1981	14	1,9	20	23	0,40	14	55	140	5
	1982	11	1,2	45	20	0,55	14	45	145	10
	1984	19	1,4	100	23	0,40	25	50	160	27
	1985	17	1,3	40	45	0,10	20	50	225	42
marsh land Saeftinghe	1971	55	8,9	215	60	2,15	30	120	440	15
	1974	47	7,5	170	79	1,85	34	115	390	30
natural background values		10	0,5	40	15	0,10	14	40	120	

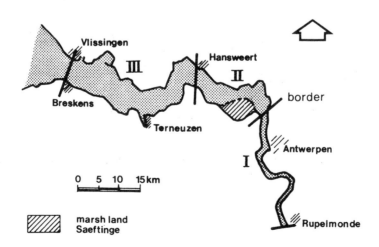

Fig. 1. Division of the Western Scheldt into three
sections for modelling purposes.

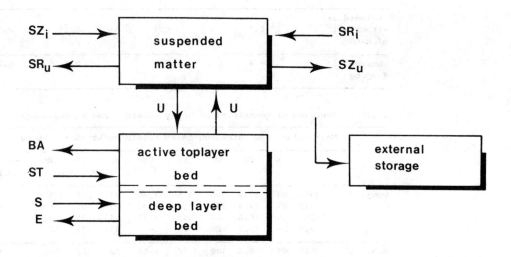

Fig. 2. Compartments and material transfer in the silt balance.

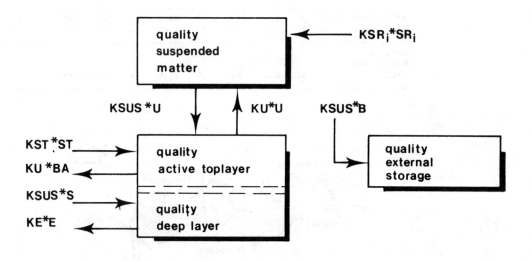

Fig. 3. Compartments and material transfer associated with contaminated silt.

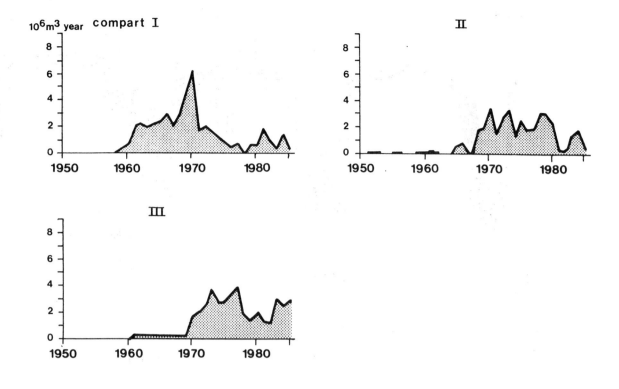

Fig. 4.   Quantities of soil discharged   per section.

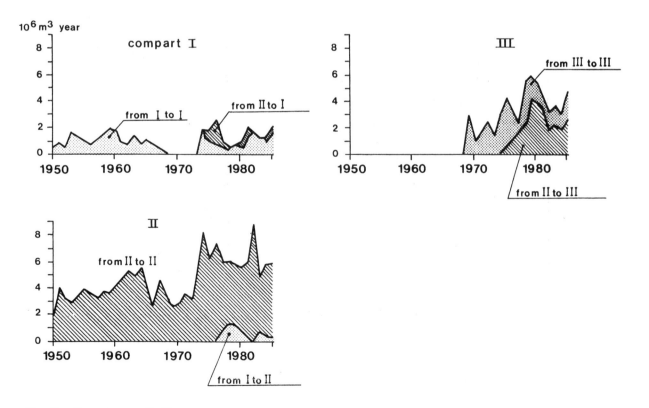

Fig. 5.   Dumping activities.

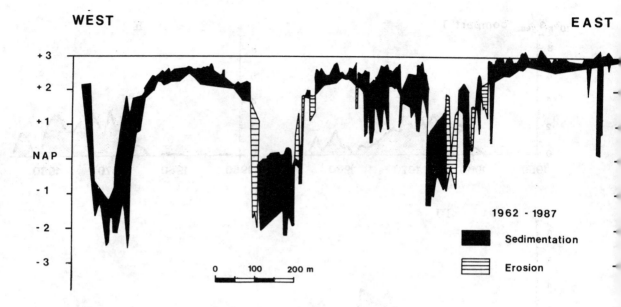

Fig. 6. Profiles indicating the height of Saeftinghe 1962, 1987.

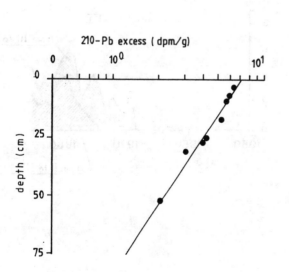

Fig. 7a. ^{210}PB content in Saeftinghe core sample.

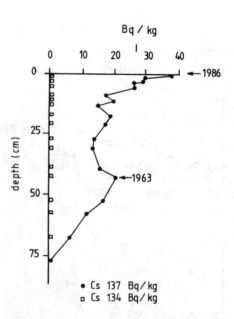

Fig. 7b. ^{137}Cs content in Saefinghe core samp[le]

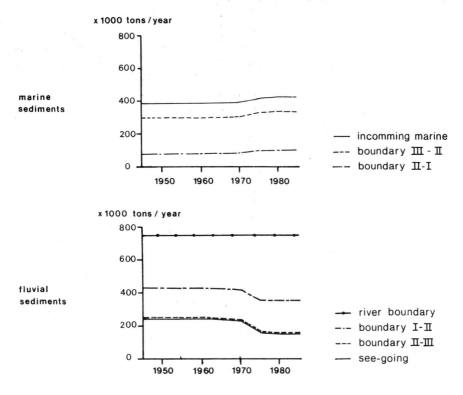

Fig. 8. Calculated silt transport in the estuary.

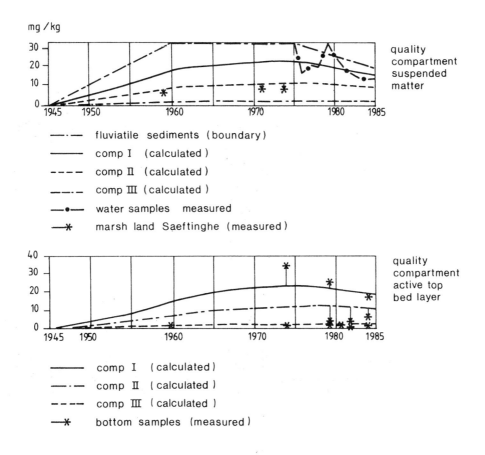

Fig. 9. Reconstruction of the pollution build-up with cadmium.

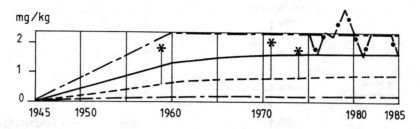

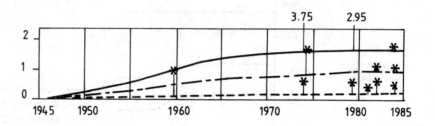

Fig. 10. Reconstruction of the pollution build-up with mercury.
(for explanation of the curves see Fig. 9.)

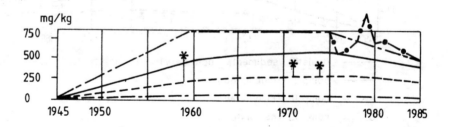

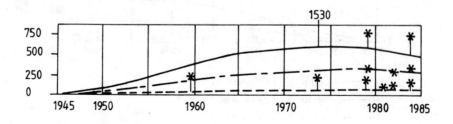

Fig. 11. Reconstruction of the pollution build-up with zinc.
(for explanation of the curves see Fig. 9.)

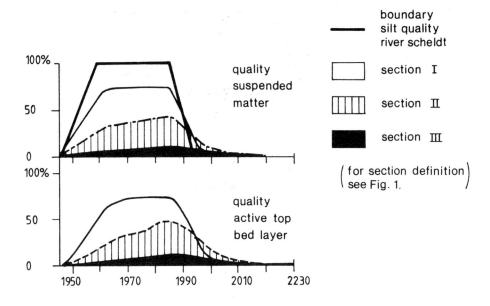

Fig. 12. Effect that clean–up measures for discharges have as
a function of time.
(100% is the maximum contamination of the fluvial silt)

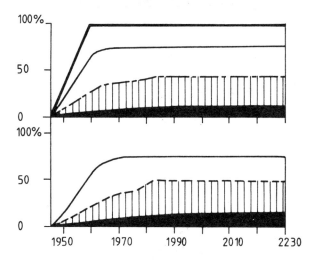

Fig. 13. Effect that continuing the present policy with regard
to the dumping of spoil has, as a function of time.
(for explanation of the curves see Fig. 12.)

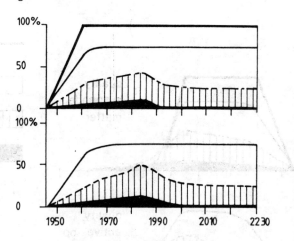

Fig. 14. Effect that storing dredginging spoil outside
the system has, as a function of time.
(for explanation of the curves see Fig. 12.)

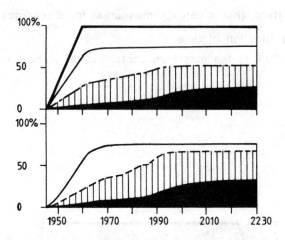

Fig. 15. Effect that dumping dredging spoil downstream
has, as a function of time.
(for explanation of the curves see Fig.12.)

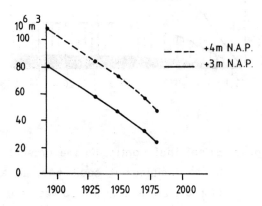

Fig. 16. Reduction in the basin storage at
Land van Saeftinghe.

108

Chapter 8
MEASUREMENTS OF REAERATION USING A FLOATING SOLUBLE SOLID

A James and J R Bicudo
University of Newcastle upon Tyne, UK

Summary

The rate of reaeration is a critical factor in the oxygen balance of most streams
and estuaries and there has been an extensive search for a simple and reliable
method of measurement. All the existing techniques such as distributed equilibrium or
the sophisticated hydrocarbon gas and radioactive tracer methods suffer from some
limitation.

A new technique using a floating soluble solid has been investigated under
laboratory and field conditions, and shows considerable promise. The laboratory
studies have examined the combined and individual effects of velocity, depth and
roughness on the reaeration (hydrocarbon desorption), and dissolution rates.
Factorial and surface response analysis have been performed over the entire set of
data in order to establish a region of the factor space in which the relationships
between the processes are satisfied. An overall correlation coefficient of about
0.85 has been obtained between the reaeration and dissolution processes (and the
shape of response surfaces are very similar). Field trials in long concrete
channels have confirmed the potential of the method when compared with the gas
tracer technique.

I. Introduction

The rate of reaeration is a critical factor in the oxygen balance of most natural
waters, so that poor estimation can seriously affect decisions on the need for
wastewater treatment. Empirical methods have been developed for the estimation of
the rate of reaeration but because of the importance of reaeration, there has been
an extensive search for a simple and reliable method of measurement. This paper
reviews the earlier techniques of disturbed equilibrium, radiotracer and hydrocarbon
gas tracers and compares them with a new technique using a floating soluble solid.

II. Reaeration Measurement by Established Techniques

Since these techniques have been previously described, this review is limited to a
brief outline of each method together with some notes of the drawbacks.

a. Disturbed Equilibrium Technique - this involves reducing DO levels in streams
    by dosing with Sodium sulphite and measuring the rate of recovery of DO
    downstream.

    It is limited to streams with no temporal variation of DO due to natural causes
    and to stretches of streams without sources and sinks of DO.

The main errors are associated with poor mixing of the sulphite and continued oxidation in the reaeration zone. (For further details see Edwards et al (1)).

b. Radiotracer Technique – this involves dosing the stream with a mixture of Krypton 85, Tritium and some fluorescent dye. The change in concentration of the Tritium is a measure of the dispersion rate and the difference between the rates of change of Krypton and Tritium is a measure of the rate of surface exchange. The fluorescent dye provides a visual indication for sampling.

This method is accepted as the most accurate technique available but is limited in application by the expense of the tracers and monitoring equipment and the limitations on the use of radio-isotopes (for details see Tsivoglou (2)).

c. Modified Gas Tracer Technique – this involves dosing the water with a suitable tracer gas such as Propane, Ethylene, Methyl Chloride and determining the rate of surface exchange by the decrease in concentration of the gas downstream.

There are some practical problems in dispersing the gas in wide streams (Yotsukura et al (3)) and for large flows the technique would be expensive. It also requires sophisticated equipment for measurement.

One general limitation of the above methods is that they rely upon unidirectional flow and therefore cannot be applied to estuaries, lakes or in the sea.

III. Solid-Dissolution Technique – Theoretical Background

The physical analogy between the rate of dissolution of a floating solid and the rate of reaeration is illustrated graphically in Fig. 1 and may be represented mathematically by:

$$\frac{d\ DO}{dt} = K_2 (C_s - C) \tag{1}$$

and
$$\frac{dS}{dt} = K_s (C_s - C) \tag{2}$$

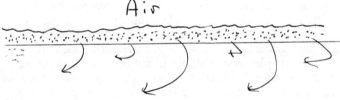

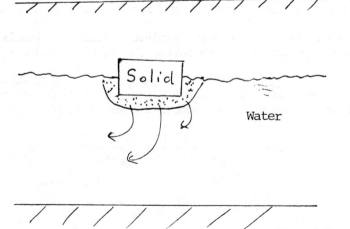

Fig. 1 Analogy between reaeration and solid dissolution

where

DO = concentration of dissolved oxygen

$\dfrac{dDO}{dt}$ = rate of reaeration

$\dfrac{ds}{dt}$ = rate of dissolution of floating solid

K_2 = reaeration coefficient

K_s = dissolution coefficient

C, C_s = actual concentration and saturation concentration

The theoretical background to the dissolution model was first developed by Hixson & Crowell (4) and then extended by Hixson & Baum (5) in their studies of mass transfer coefficients in liquid solid agitation systems. According to them, the velocity of solids dissolution is viewed as a coefficient which is the resultant rate of the combined processes of transformation of the solid into its dissolved and distributed products without any assumption regarding the mechanism by which the process takes place.

From the assumptions that (1) the process of dissolution takes place normal to the surface and the effect of agitation of liquid against all parts of the surface is essentially the same; (2) the change in concentration is so small that the average concentration driving force can be expressed by the difference between the solids weights; the solids dissolution process may be simply described by:

$$W_o - W_t = \rho . A_s . V_s . t \qquad (3)$$

where

W_o and W_t = the solids weights at the beginning and at time t,

A_s = the surface area of the solid given by ΠR^2,

V_s = the velocity of solids dissolution,

ρ = the solid density, and

t = the time of exposure.

Other studies on solid/liquid mass transfer systems (e.g. Johnson & Huang (6)) led to the conclusion that the surface renewal theory developed by Danckwerts (7) for gas/liquid systems was also applicable to solid/liquid systems.

It was experimentally observed by Bicudo (8) that the physical processes of both oxygen absorption and benzoic acid dissolution are governed by the same diffusion processes so that physical mass transport determines the overall reaction rate. It was also demonstrated that the main resistance to both processes is in the liquid phase and that agitation has a considerable effect on the overall mass transfer coefficient.

Laboratory results revealed a good agreement between both K_2 and K_L with V_s (refer to Figs. 2 and 3), statistically confirmed by a correlation coefficient of 0.71 and 0.80 respectively.

The relationship between the reaeration and velocity of benzoic acid dissolution was found to be a function of the level of turbulence within the system and was mathematically expressed by:

$$\dfrac{K_2}{V_s} = 1.534 \times 10^{-3} \left(\dfrac{Re}{H}\right)^{1.15} \qquad (4)$$

where

K_2 = the reaeration rate coefficient in hours^{-1},

V_s = the velocity of solids dissolution in cm min^{-1},

Re = the Reynolds number and

H = the average depth of flow in m.

The practical application of such a relationship for the determination of reaeration rates in small streams is discussed in the next section.

IV Comparative Studies - Methodology

A comparison was made between reaeration rates and determined by 3 different techniques :

a. Disturbed equilibrium
b. Gas tracer
c. Floating solid

The radiotracer was excluded from the comparison as the experiments were carried out in channels which formed part of a public water supply system.

Two series of experiments were performed in uniform concrete channels approximately 2.50 m wide using sections of 1000 m in the Whittle Dene channel and 630 m in the Hallington channel. (The channels form part of the water collection system for Newcastle & Gateshead Water Company).

Altogether seventeen tests were carried out using the following procedure:

1. Water velocity was measured by dilution guaging using sodium chloride measured by conductivity and also by current meter.

2. The length of channel necessary for complete mixing from a single mid-point injection was determined from the relationship

$$L_m = 0.1 \frac{UW^2}{E_z} \tag{5}$$

where

E_z is the lateral mixing coefficient = $0.2 \, H \, U_*$ $(m^2 sec^{-1})$

U_* is the shear velocity =$\sqrt{gHS}$ $(m.sec^{-1})$
U = velocity $(m \, sec^{-1})$
W = width (m)
g = gravitational constant $(m \, sec^{-2})$
S = slope

3. Time of travel studies were carried out using instantaneous injections of Sodium chloride.

4. The reaeration coefficient was measured by the disturbed equilibrium technique as described by Gameson & Truesdale (9). The Sodium sulphite and Cobalt-Chloride dosing solution (about 450 1) was fed into the stream over a 20-30 min period. The reaeration coefficient was calculated using

$$K_2 = \frac{1}{t_p} \, \ln \left(\frac{N_1 \, Q_1}{N_2 \, Q_2} \right) \tag{6}$$

where

t_p = time of travel of peak concentration

N_1 & N_2 = areas under the $(C - C')$ time curves

Q_1 & Q_2 = channel discharges

subscripts refer to upstream and downstreatm sampling points.

5. The reaeration coefficent was measured by the steady-state propane gas tracer method as described by Yotsukura et al(10). The propane gas was injected into the stream through a porous permeable plastic. The desorption rate was measured at the downstream stations by the method described by Holley & Yotsukura(11) using

$$\frac{(CQ)_1}{(CQ)_2} = \exp \left[- K_p \, (\bar{t}_2 - \bar{t}_1) \right] \tag{7}$$

where   C is the steady state propane concentration

$\bar{t}$ is the travel time determined by centroids of the salt tracer concentration curves and the subscripts refer to upstream and downstream sampling sites, respectively.

The reaeration coefficient is then computed from the following equation: (Rainwater & Holley (12))

$$K_2 = \frac{K_p}{1.36} \tag{8}$$

where K_2 = the reaeration coefficient, and

K_p = the propane gas desorption coefficient.

6. The solids dissolution coefficient was measured by the technique adapted from Giansanti and Giorgetti (13). A set of benzoic acid discs were labelled and weighed in an analytical balance to the nearest 1/1000g and were then placed in the stream at the upstream sampling station and collected afterwards at the downstream sampling station. The solids were left drying overnight at a constant temperature (20°C) and weighed again the following day. The velocity of dissolution coefficient was determined from equation (3) by least squares analysis and K_2 was estimated from equation (4).

V. Comparative Studies - Results & Discussion

The results obtained from the first series of field experiments at Whittle Dene are summarised in Table 1. The most significant correlations among the hydraulic variables and rate coefficients are given in Table 2. It can be seen from Table 2 that, although depth and velocity did not vary significantly between tests, the K_2 value measured by the disturbed equilibrium technique did show a considerable variation and could not be used as a reliable guide to the rate of reaeration.

Date	U m sec^{-1}	H m	Q m^3 sec^{-1}	*$K_{2,20}$ obs hour^{-1}	V_s obs cm min^{-1} x10^{-6}	$K_{2,20}$ pred(1) hour^{-1}	$K_{2,20}$ pred(2) hour^{-1}
10.9.86	0.71	0.78	1.38	1.0941	96.0	0.1473	2.225
16.9.86	0.88	0.88	2.16	0.2547	103.0	0.2420	3.271
18.9.86	0.83	0.87	1.81	0.1329	98.0	0.1602	2.585
25.9.86	0.82	0.92	1.89	5.3800	106.0	0.1305	2.685
30.9.86	0.79	0.87	1.72	0.0212	100.0	0.1403	2.492
13.1086	0.83	0.82	1.70	-	102.0	0.1923	2.766
14.10.86	0.84	0.81	1.70	-	100.0	0.2063	2.764
15.10.86	0.89	0.80	1.78	-	101.0	0.2505	3.001
16.10.86	0.82	0.83	1.70	-	101.0	0.1793	2.724

*reaeration rate coefficient measured by disturbed equilibrium method;

$K_{2,20}$ pred (1) — reaeration rate coefficient predicted by Churchill et al(14) equation at 20°C;

$K_{2,20}$ pred (2) — reaeration rate coefficient predicted by the K_2/V_s relationship (equation 4) at 20°C.

Table 1   Results of field measurements of reaeration rates in the channels at Whittle Dene

Variable	Velocity	Depth	Discharge	K_2 obs	K_2 pred(1)	K_2 pred(2)
Depth	0.324					
Discharge	0.893	0.705				
K obs	− 0.134	0.459	0.066			
K pred	0.781	− 0.332	0.429	− 0.440		
K pred	0.979	0.228	0.835	− 0.040	0.829	
V_s	0.534	0.551	0.690	0.670	0.132	0.596

Table 2   Correlation coefficients obtained from Whittle Dene surveys

The rate of reaeration was estimated using the equation by Churchill et al (14):

$$K_2 = 5.02 \ V^{0.969} \ H^{-1.673} \qquad\qquad (9)$$

where K_2 = reaeration rate coefficient (day^{-1})
        V = average stream velocity (m.sec^{-1})
        H = average stream depth (m)

This gave good correlation with the results of K_2 predicted from the rate of solids dissolution but the two differed by a factor of almost 20 times.

It was concluded from these preliminary tests that neither the empirical equation by Churchill nor the disturbed equilibrium techniques provided a sufficiently reliable datum against which to judge the accuracy of the solids dissolution tests. It appeared that failure to comply exactly with the main assumptions of the disturbed equilibrium technique plus the introduction of procedural and random errors (as listed by Hovis et al (15) had led to the acquisition of inaccurate data.

A second series of field experiments was carried out in channels at Hallington where a greater range of flows and velocities could be explored. The results are summarised in Table 3.

Date	U m sec^{-1}	H m	Q m^3 sec^{-1}	$K_{2,20}$ obs hour^{-1}	$K_{2,20}$ pred(1) hour^{-1}	$K_{2,20}$ pred(2) hour^{-1}	V_s cm min^{-1} x 10^{-6}
9.6.87	0.65	0.33	0.52	1.7919*	1.4662	2.095	75.0
11.6.87	0.63	0.33	0.51	0.6410*	1.4331	1.913	71.0
18.6.87	0.64	0.31	0.49	1.1328*	1.6173	2.171	78.0
18.6.87	0.63	0.30	0.47	1.8660*	1.6933	2.011	73.0
26.6.87	0.58	0.28	0.43	1.0324*	1.7986	1.881	74.0
19.7.87	0.50	0.23	0.25	1.7189	2,2772	1.536	69.0
31.7.87	0.75	0.45	0.78	2.7960	0.9459	2.905	96.0
6.8.87	0.75	0.43	0.73	2.0721	1.0242	2.393	78.0

Table 3   Results from field experiments at the Hallington channel.

In Table 3, the reaeration coefficient K_2 pred(1) was calculated using the equation by Owens et al (16):

$$K_2 = 6.92 \, V^{0.73} \, H^{-1.75} \qquad\qquad (10)$$

where K_2, V and H are defined as before. In the column K_2obs, the results marked with a single asterisk were measured by the disturbed equilibrium method and the remaining results were obtained by the gas tracer technique.

The results in the column K_2 pred(2) were calculated from the rate of dissolution of benzoic acid.

Correlations between the hydraulic variables and reaeration coefficients are summarised in Table 4.

Variable	Velocity	Depth	Discharge	K_2obs	K_2 pred(1)	K_2 pred(2)
Depth	0.848					
Discharge	0.987	0.806				
K_2 obs	0.634	0.432	0.661			
K_2 pred	− 0.981	− 0.877	− 0.987	− 0.462		
K_2 pred	0.921	0.837	0.942	0.664	− 0.900	
V_s	0.740	0.747	0.786	0.699	− 0.729	0.94

Table 4   Correlation coefficients from experiments in the channel at Hallington

From the coefficients in Table 4 it is clear that a considerable improvement was obtained in the field data resulting in good correlation between velocity, depth and discharge.

With the exception of the second test (11.6.87), the reaeration rates calculated from the disturbed equilibrium appear to be in reasonable agreement with those computed from Owens equation and the hydraulic characteristics of the channel.   But by contrast the K_2 values calculated from the gas tracer technique (19.7.87, 31.7.87 and 6.8.87) are in total disagreement with the disturbed equilibrium values.

In the absence of any reliable datum it remains difficult to assess the validity and accuracy of the solids dissolution technique from the field data alone.

Data from laboratory studies (Bicudo(8))were therefore combined with the field data as shown in Fig. 4.   This shows that the correlation between (K_2/V_s) and (R_e/H) observed in the laboratory data gives good predictions when applied to the field data where K_2 was estimated by solids dissolution or the gas tracer technique.   The remaining field data where K_2 was estimated by the disturbed equilibrium technique, do not fit with this correlation, which casts further doubts on the accuracy of the disturbed equilibrium technique.

In a further phase of data analysis the results from Whittle Dene and Hallington were combined to give the correlations summarised in Table 5.

The most important conclusion from Table 5 is the strong correlation between the rate of solids dissolution and the 3 hydraulic parameters.   This is further confirmed in Fig.5 which shows K_2 predicted from solids dissolution plotted against energy dissipation as expressed by SU (where S = slope and U = velocity),

Variable	Velocity	Depth	Discharge	K_2 obs	K_2 pred(1)	K_2 pred(2)
Depth	0.863					
Discharge	0.921	0.984				
K_2 obs	− 0.026	− 0.010	− 0.034			
K_2 pred(1)	− 0.888	− 0.966	− 0.948	− 0.002		
K_2 pred(2)	0.958	0.763	0.826	0.089	− 0.809	
V_s	0.984	0.947	0.944	0.142	− 0.945	0.873

Table 5   Correlation coefficients from combined field data

giving a correlation coefficient of 0.955.   The equation for this linear relationship is

$$K_2 = 0.537 \ SU \tag{11}$$

where 0.537 is the value of the escape coefficient (Tsivoglou & Wallace, (17)). The excape coefficient expresses the relationship between gas transfer, surface replacement and turbulence and hence is unique for each channel. The value of 0.537 is higher than those quoted as typical for various streams in the U.S.A. (between 0.13 to 0.25) but fits within the range reported by Tsivoglou and Neal(18) who observed values as high as 0.49 and 0.59.

VI. Conclusions

It was concluded that the rate of solids dissolution can be used as an alternative method for measuring reaeration in streams. Although not as accurate as the gas tracer technique it is more reliable than disturbed equilibrium or the empirical equations such as Owens et al. which are commonly employed.

The floating soluble solid technique has significant advantages in cheapness, simplicity and the ability to function in lakes, estuaries and seas where the other techniques cannot be employed.

Acknowledgements

The work was in part supported by a research scholarship from Conselho Nacional de Desenvolvimento Cientisico e Tecnologico.

VII. References

1. EDWARDS, R.W., OWENS, M. and GIBBS, J.W.(1961). "Estimates of Surface Aeration in Two Streams". Institute of Water Engineers Journal. Vol.15, No. 5, pp.395-405.

2. TSIVOGLOU, E.C. (1967) "Tracer Measurement of Stream Reaeration". Report to the Federal Water Pollution Control Administration, U.S. Dept. of Interior, Washington D.C., 89 p.

3. YOTSUKURA, N., STEDFAST, D.A. and JIRKA, G.H. (1984). "Assessment of Steady State Propane Gas Tracer Method for Determining Reaeration Coefficients, Chenango River, New York". U.S. Geological Survey Water Resources Investigations, 84-4368.

4. HIXSON, A.W. and CROWELL, J.H. (1931). "Dependence of Reaction Velocity upon Surface and Agitation, I. Theoretical Considerations". Industrial and Engineering Chemistry. Vol. 23, No. 8, pp. 923-931.

5. HIXSON, A.W. and BAUM, S.J. (1941). "Mass Transfer Coefficients in Liquid-Solid Agitation Systems". Industrial and Engineering Chemistry. Vol. 33, No.4, pp. 478-485.

6. JOHNSON, A.I. and HUANG, C.J. (1956). "Mass Transfer in an Agitated Vessel". American Institution of Chemical Engineers Journal. Vol.2, No.3, pp.412-419.

7. DANCKWERTS, P.V. (1951). "Significance of Liquid-Film Coefficients in Gas Absorption". Industrial and Engineering Chemistry. Vol. 63, No.6, pp.1460-1467.

8. BICUDO, J.R. (1988). "The Measurement of Reaeration in Streams". Thesis presented to the University of Newcastle upon Tyne in partial fulfilment of the requirements for the degree of Doctor of Philosophy

9. GAMESON, A.L.H. and TRUESDALE, G.A. (1959). "Some Oxygen Studies in Streams". Journal of the Institution of Water Engineers. Vol. 13, No.2, pp. 175-187.

10. YOTSUKURA, N., STEDFAST, D.A., DRAPER, R.E. and BRUTSAERT, W.H. (1983). "Assessment of Steady-State Propane Gas Tracer Method for Reaeration Coefficients, Cowaselon Creek, New York". U.S. Geological Survey Water Resources Investigations, 83-4183.

11. HOLLEY, E.R. and YOTSUKURA, N. (1984). "Field Techniques for Reaeration Measurements in Rivers. In Gas Transfer at Water Surfaces. D. Reidel Publishing Company, Holland, pp. 381-401.

12. RAINWATER, K.A. and HOLLEY, E.R. (1984). "Technical Report on Laboratory Studies of the Hydrocarbon Gas Tracer Technique for Reaeration Measurement". Report CRWR 189, Center for Research in Water Resources, Univ. of Texas at Austin.

13. GIANSANTI, A.E. and GIORGETTI, M.F. (1986). "Contribuicao para a Determinacao do Coeficiente de Reoxigenacao Superficial em Corpos D'agua - II Presented at the 2nd Latin American Congress of Heat and Mass Transfer held in Sao Paulo, Brazil, 12-15 May.

14. CHURCHILL, M.A., ELMORE,H.L. and BUCKINGHAM, R.A. (1962). "The Prediction of Stream Reaeration Rates". Journal of the Sanitary Engineering Division, ASCE, Vol. 88, No. SA-4, pp. 1-46.

15. HOVIS, J.S., WHITTEMORE, R.C.,BROWN, L.C. and McKEOWN, J.J. (1982). "An Assessment of the Measurement Uncertainty in the Estimation of Stream Reaeration Coefficients Using Direct Tracer Techniques". Proceedings of the Stormwater and Water Quality Management Modelling Users Group Meeting, 25-26 March. U.S.EPA 600/g-82-015, pp. 36-53.

16. OWENS, M., EDWARDS, R.W. and GIBBS, J.W. (1964). "Some Reaeration Studies in Streams". International Journal of Air and Water Pollution, Pergamon Press, Great Britain. Vol. 8, pp. 469-486.

17. TSIVOGLOU, E.C. and WALLACE, J.R. (1972). "Hydraulic Properties Related to Stream Reaeration". Report No. EPA-R3-72012, Georgia Institute of Technology, Atlanta, 324 p.

18. TSIVOGLOU, E.C. and NEAL, L.A. (1976). "Tracer Measurement of Reaeration III, Predicting the Reaeration Capacity of Inland Streams". Journal of Water Pollution Control Federation. Vol. 48, No. 12, pp. 2669-2689.

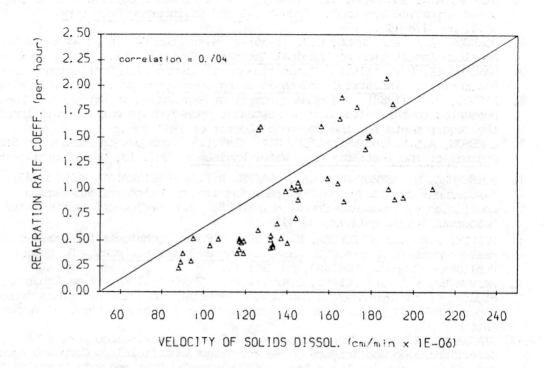

Fig.2   Reaeration rate coefficient versus solids
dissolution correlation

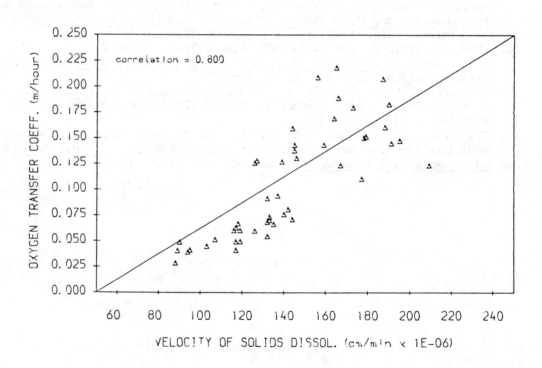

Fig. 3   Oxygen transfer coefficient versus
solids dissolution correlation

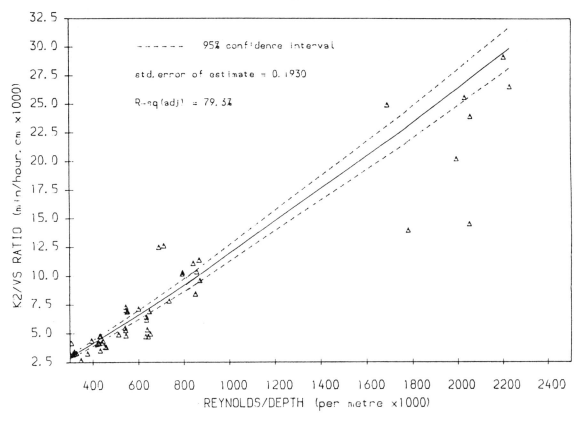

Fig. 4   Laboratory and field tests combined data

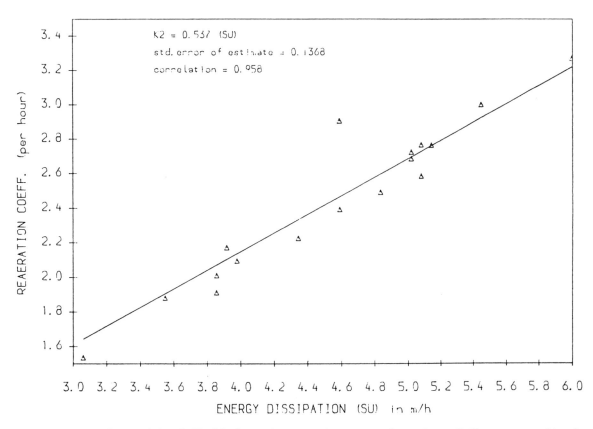

Fig. 5   Combined field data (reaeration as a function of the energy dissipated)

Chapter 9
MODELLING OF THE DISPERSION OF RADIOACTIVITY IN COASTAL WATERS

J MacKenzie, M J Egan, M E D Bains and D A Wickenden
UKAEA, UK

Summary

This paper describes a numerical model which has been developed to simulate the dispersion of radionuclides following the discharge of liquid effluent to coastal waters. The general aim in developing the model was to be able to provide more detailed information on spatial and temporal variations in concentration levels than can be obtained from the compartment-type models commonly used in radiological assessments. The basic model framework involves the solution of the two-dimensional depth-averaged advection-diffusion equation on a finite-difference grid. Tidally-varying current velocities and water elevations are input to the model, allowing the movement of the contaminant plume to be followed during successive tidal cycles. Decay processes and partitioning of the radionuclides between seawater and sediment may also be taken into account. In one application, the model has been used to predict radionuclide concentrations in Weymouth Bay following discharge from the UKAEA's establishment at Winfrith, Dorset. A comparison of the model results with measured radionuclide concentrations in seawater is presented.

1. Introduction

The marine environment has received artificial radionuclides from many sources - initially as fallout from the atmospheric testing of nuclear weapons, while subsequent, more localised inputs have occurred as the result of the disposal of low and intermediate level wastes into the deep oceans and the controlled discharge of low level liquid waste from nuclear installations into coastal waters.

Discharges to coastal waters in the UK are bound within authorised limits, which are in turn determined such that radiation exposure of the public is well below the limits recommended by the International Commission on Radiological Protection (ICRP). The ICRP has developed the concepts of individual and collective dose-equivalents to assess the health detriment from irradiation to both individuals and populations. In order to predict individual and collective doses arising from discharges to the marine environment, it is necessary to possess the capability to assess both the dispersion of radioactivity in seawater and the subsequent return of activity to man through internal and external exposure pathways.

Traditionally, mixed-box models have generally been used for radiological assessments. For example, the UKAEA's collective dose model COLDOS[1], which is based on methods developed by the National Radiological Protection Board (NRPB) and the Ministry of Agriculture, Fisheries and Food (MAFF)[2,3], consists of a number of large inter-connected compartments covering the seas of Northern Europe and ocean waters. Each compartment is assumed to be homogeneously mixed and transfer rates simulate the movement of water, and hence radionuclides, between compartments. This is an appropriate method for collective dose calculations given that the large populations and long time scales being considered necessarily involve some degree of averaging. Similarly, for individual doses, the initial dispersion of discharged activity can be estimated using a single-box model. Here, the compartment is again assumed to be well-mixed, so that a steady-state concentration may be calculated on the basis of a suitable water volume and turnover rate within the box. Simple models of this type are commonly used for assessment of the radiological effects of routine discharges. The appropriate integration period for comparison with recommended dose limits is at least one year and it is therefore not generally considered necessary to take account of short term fluctuations in discharge rates or seawater concentration fields. The uptake of activity, particularly by marine organisms, will also tend to have an averaging effect upon radiation exposure rates.

Nevertheless, compartment models such as these provide little insight into temporal and spatial variations in the radionuclide concentrations in seawater. They yield only limited information on concentration levels in the vicinity of the release and none on concentrations outside the box area. A need has therefore been identified for a more sophisticated model which would be able to simulate the progress of the contaminant plume and thereby provide a more detailed picture of concentration distributions. The model could then be used in situations where it is required to demonstrate more precisely the dilution of the discharge due to dispersive processes – for example, at distances outwi h the immediate vicinity of the release location, where it is not practicable to set up detailed monitoring programmes, or where concentration levels are so low as to be below detection limits. A model of this type, once validated, would also be useful in predicting the effects of pulsed releases, in providing input to monitoring programmes or in interpreting the results of monitoring.

In the following sections we describe the model which has been developed in response to these requirements and present the results of comparisons between model predictions and field observations taken in Weymouth Bay, Dorset.

2.   Description of model

Since the model was intended, in the first instance, for application to the UKAEA's two coastal sites, it was decided that a two-dimensional approach was required in order to obtain an adequate representation of the bay areas involved. This, together with the desire to be able to follow the movement of the plume over relatively short timescales, prompted the development of a two-dimensional, depth-averaged, tidal model based on the solution of the advection-diffusion equation for constituent transport, together with appropriate loss terms as follows:

$$\frac{\partial(hC)}{\partial t} = Qh - \frac{\partial}{\partial x}(huC) - \frac{\partial}{\partial y}(hvC) + \frac{\partial}{\partial x}\left(hk_x\frac{\partial C}{\partial x}\right) + \frac{\partial}{\partial y}\left(hk_y\frac{\partial C}{\partial y}\right) - \lambda hC - \lambda_s hC \tag{1}$$

where $C(x,y,t)$ is the radionuclide concentration in seawater; $Q(x,y,t)$ is the source discharge rate; $h(x,y,t)$ is the water depth; $u(x,y,t)$ and $v(x,y,t)$ are the depth-mean current velocities in the x and y directions respectively; k_x and k_y are diffusion coefficients in the x and y directions; λ is the radioactive decay constant and $\lambda_s(x,y,t)$ is the rate of loss of the radionuclide to seabed sediments.

Note that all the variables above, with the exception of k_x, k_y and λ, are functions of time, t and the longitudinal and lateral co-ordinate directions, x and y. The above equation is solved for $C(x,y,t)$ using the FACSIMILE package (4) which allows for the discretisation of the spatial derivatives on a finite difference grid. It is assumed that the water body is well-mixed in the vertical direction. The model has been designed so that all site specific information is required as input data, thereby retaining a large amount of flexibility in the model itself.

This allows the model to be applied readily to different locations.

Some of the important model details are described briefly below.

i) Depth and velocity fields

The water depth h(x,y,t) represents the total depth of the water body; ie

$$h = d + z \qquad\qquad (2)$$

where d(x,y) is the depth below mean water level and z(x,y,t) is the surface water elevation above mean water level. The advective velocity terms u and v represent the tidally varying flow velocities. It is assumed that the depth and velocity fields are known and that they satisfy the continuity equation. This implies that they need to be obtained from a flow simulation model or from a very complete set of observations made either in the field or using a physical model. To date, the model has been set up for the UKAEA establishments at Dounreay in the north of Scotland and Winfrith on the south-east Dorset coast. For both these cases, the tidal flow data was provided by Hydraulics Research of Wallingford from hydrodynamic models of the surrounding regions.

ii) Diffusion coefficients

The diffusive terms in equation (1) represent the small scale mixing processes within the contaminant plume generated by turbulence. Due to lack of data, the effective diffusion coefficients k_x and k_y are taken to be constant in space although there is provision in the model for the coefficients to vary with x and y.

In the application to plume dispersal simulations described below the diffusion coefficients were taken from measurements made during Exercise Mermaid[6].

iii) Loss terms, λ and λ_s

In addition to the transport processes, the model takes into account the loss of radionuclides from the system due to radioactive decay and uptake by seabed sediments. At present, the only mechanism included in the model for uptake of activity by the seabed is particle scavenging - that is the adsorption of activity onto suspended sediment in the water phase, followed by the gravitational settling of the material in suspension to form bottom sediments. Neither loss term is likely to be significant over the timescales involved but is included for completeness.

iv) Initial and boundary conditions

The initial value for the concentration C(x,y,t) at t = 0 needs to be specified for each cell in the model grid. In practice, effluent from a pipeline is discharged into a single cell which is assumed to be instantly homogeneously mixed. C(x,y,t=0) is set to zero in the rest of the grid. Where the model boundary is occupied by land cells there is no flow across the boundary; otherwise activity is allowed to disperse out of the grid by both advection and diffusion depending on concentrations and flow velocities inside the boundary. It is assumed that, once lost, there is no return of activity into the grid.

2.1 Importance of seabed interactions

The term for radionuclide concentration in the water column in equation (1) includes both radionuclides in solution and those attached to suspended sediment. Sediment in suspension is assumed to move with the seawater and partitioning of activity between particulate and seawater is determined by the sediment distribution coefficient (Kd) and the suspended sediment load. Wherever possible site specific distribution coefficients are used. At present, it is considered that the assumption to transport the seawater and suspended sediment together is not unreasonable given that, for the areas being considered, suspended sediment loads are low and the material is not, in general, of a cohesive nature. Also the approximation involved in this assumption is likely to be less than that involved in the use of equilibrium distribution coefficients. However, the need for a dynamic model of the interaction between the water column and the seabed is currently being investigated. Among the options being considered is the replacement of equilibrium distribution coefficients with reversible reaction rates and the separate treatment of radionuclides in solution and those associated with particulate.

In the long term, uptake of radionuclides by bottom sediments can, in certain situations, lead to the accumulation of activity in the seabed with the possibility of subsequent remobilisation of that activity back into the water column. Remobilisation may arise due to a variety of mechanisms. For example, particle scavenging of radionuclides can result in a concentration gradient between interstitial water in the seabed and the overlying water, which will cause molecular diffusion of soluble species into the water column. Physical disturbance of seabed sediments due to bottom currents, storms and fishing activities, can also facilitate the return of soluble species to the water column through the direct exchange of interstitial and overlying waters. Such disturbances may further bring the sediment particles into direct contact with the overlying water allowing for desorption of radionuclides attached to the particles. An additional significant source of particle mixing and porewater exchange may be biological mixing where the sediment is disturbed by benthic organisms burrowing and feeding in the seabed.

However, if such remobilisation does occur, it is likely to do so slowly and over a long time period. This phenomenon is therefore of interest primarily from the point of view of collective dose rather than individual dose, since remobilisation is unlikely to lead to concentrations in seawater greater than those which were responsible for the initial contamination of the sediments. The long term processes have therefore not been included in the model described here, which is concerned with much shorter timescales.

3. Weymouth Bay Application

3.1 Location and input data

The UKAEA's establishment at Winfrith discharges liquid radioactive effluent to deep water in Weymouth Bay via a pipeline which runs two miles out to sea from Arish Mell. The model described above has been set up to simulate the dispersion of activity from the Winfrith pipeline within Weymouth Bay. The model covers an area of approximately, 40 km x 16 km extending from Portland Harbour to St. Alban's Head. The depth and velocity data were provided by Hydraulics Research from a hydrodynamic model of a larger region which included the Isle of Wight and the mouth of the Solent.[5] The full region covered by the hydrodynamic model is shown in Fig 1. Hydrodynamic data were obtained for representative spring and neap tide conditions. There are complex tidal conditions existing in this area which is known for the occurrence of double high and low waters. The area depicted by the smaller rectangle in Fig. 1 is that covered by the dispersion model. In their model, Hydraulics Research used a homogeneous grid 400m x 400m and the same grid size was retained in the dispersion model. Examples of the output produced by the model are presented in Fig. 2 which shows the changing contours at different points of the tidal cycle for a hypothetical release of 10^5 Bqs^{-1} of radioactivity from the Winfrith pipeline, lasting for around 3 hours. The first two figures show the situation for a west-going tide while in the latter two figures the tide has changed direction and the plume is being swept towards St Alban's Head.

3.2 Plume trajectories - validation of hydrodynamic data

Before the Winfrith pipeline was laid, a study - known as Exercise Mermaid(6) - was carried out to investigate the dispersion characteristics of the water body around Arish Mell and to determine the optimum length for the pipeline. As part of that study a number of short period discharges of a tracer dye were made from different points around Arish Mell. Each discharge lasted for 20 minutes during which time 1 ton of dye was discharged onto the sea bottom. From analysis of the results of sampling made during the experiments, the track of the centre of the dye patch was determined in each case. Figs 3 and 4 show the comparisons between the Mermaid observations and model simulations performed for two of these discharges.

Both releases were from the same position, which was that eventually chosen for the end of the pipeline, but were made at different points in the tidal cycle. Fig 3 shows the results for a release made on a west-going tide. The boxes represent a section of the grid which has been superimposed on a map of the area. Here, the solid line is the track of the centre of the dye patch determined from Exercise Mermaid while the dotted line represents the model predictions. The points along the curves are at half-hourly intervals after the discharge. It can be seen from the size of the boxes that the model resolution is limited, with the position of the maximum concentration within the cell being determined from adjacent cells. However, even allowing for this it can be seen that there is reasonable agreement between the two trajectories.

Figure 4 shows a similar comparison, for a release made towards the end of an east-going tide. Again, the solid line represents the track of the dye patch determined from field observations while the dotted line is the trajectory predicted by the model. Both simulations were carried out for a conservative pollutant and assumed neap tide conditions since these were commensurate with the maximum tidal speeds quoted for the experiments. However, the model only represents a typical neap tide and not the exact conditions prevailing on that day. Furthermore, local winds might also have an effect on dispersion and this is not included in the model. The effect of wind may be more important in the comparison of Fig. 4, since there was a moderately strong east wind on that day (around force 4), whereas in the previous case the wind was light (force 1) from a north westerly direction. Nevertheless, these results seem to indicate that the model's predictions of speed and direction of travel are reasonable.

3.3   Simulation of plume dispersal

The above comparison was made with trajectories determined from experiments with dyes carried out before the pipeline was laid. Since then, there has of course been monitoring of activity levels in Weymouth Bay resulting from the discharges from the Winfrith pipeline. In particular, in August 1987 during the annual shutdown of the steam generating heavy water reactor, some measurements of seawater concentrations were made in certain parts of the bay. These measurements were obtained from work funded by the Department of the Environment. They were taken as part of an ongoing monitoring programme and so were not designed to form the basis for a validation exercise. Nevertheless, the opportunity was used to run the model to simulate the discharge from the pipeline for the days previous to the monitoring survey.

During the survey, seawater samples were collected from a number of locations in the bay; the collection points are shown in Fig. 5. The seawater was collected in 25 1 drums from a height of 3m above the seabed using an onboard pump system. All the samples were collected on one day. For stations 1-4 two sets of measurements were made; the first taken before the arrival of the plume carrying that day's discharge while the second was taken during the passage of the plume. This was done in an attempt to obtain some measure of the background concentration. For the remaining stations, 5-7, only one sample was taken. Radiochemical analysis of the seawater samples was subsequently carried out at Winfrith from which the levels of Cobalt-60 present were determined. Samples were also taken from the effluent tank on the day of the survey and on the three preceding days.

From analysis of the effluent samples, information on the total amount of Cobalt-60 discharged on each of the four days prior to the monitoring survey was obtained. We used this information to simulate the release of Cobalt-60 from the pipeline outfall over these four days, according to the discharge pattern shown in Fig. 6. Information on the exact timing and duration of the release was only available for the day of the survey so the previous releases were assumed to be of the same duration and to begin at 24 hour intervals. We have further assumed a uniform discharge each day, since the tanks are stirred just prior to discharging and may therefore be expected to be reasonably well-mixed.

The resulting model predictions of seawater concentrations for stations 1-4 are presented here since these are the points at which measurements were taken before and after that day's discharge was made. Figure 7 shows the predictions of concentration against time for station 4, which is nearest to the release point.

125

The solid line shows the predictions for the four day simulation and it can be seen that there is a pronounced peak in the concentration about two or three hours after the discharge each day with a secondary peak corresponding to the returning tide. The dotted line in Fig. 6 shows the simulation for the final day's release only. The two sets of results for that day are very similar, suggesting that for this station, which is close to the pipeline, there is only a limited memory of previous releases. This is further reflected in the field measurements (depicted by the two crosses in the figure) where there is a substantial difference in the measurements taken before and after the final day's discharge.

Fig. 8 shows the predictions for station 3 which is further away from the pipeline and likely to be away from the main path of the plume. Here, in contrast to the previous case, there is a build-up of concentration over successive days. There is little evidence of the first day's release after which concentration levels exhibit an increasing trend with each passage of the plume over the location. The predictions also show a marked difference between the maximum and minimum concentrations obtained and this is again reflected in the field measurements which show a similar difference for before and after the passage of the plume. Clearly the absolute values are greater than those predicted by the model but, given that the increasing concentrations suggest that we would need to model for longer than four days to get a proper estimate of the background concentration, we would expect the model to underestimate in this case. During the monitoring programme, the intention was to time the measurements such that the first measurement would provide an indication of background concentration (that is, the minimum value), while the second was planned to coincide with the peak concentration resulting from that day's discharge. The model predictions are, in fact, consistent with this picture.
Fig. 9 shows the predictions for Station 2 which is again to the west of the pipeline but further inshore than the previous location. Again there is a gradual build-up in concentration over the four days, reflecting the memory of previous days' discharges. As in Fig. 7, the dotted line represents the simulation for the final day's release only, but this time there is a noticeable difference between the two simulations. This is in contrast to the results for Station 4, which was closer to the pipeline, indicating that the four day prediction for Station 2 contains a substantial element of background concentration. Although the second field measurement was meant to reflect the maximum concentration at each location, timing of the model predictions suggest that this has not been achieved in this case. However, these predictions are consistent with the measurements themselves which are both of a similar magnitude.

The final set of results are for Station 1 and are presented in Fig. 10. This station, which is outside Portland Harbour, is now some distance away from the main trajectory in an area where the flow velocities are somewhat weaker. Therefore it would be expected that activity reaching this area would do so predominantly due to diffusive processes. This is shown in the results in that there is very little activity from the first two days discharge after which the concentration is increasing continuously over the succeeding days. It would therefore be expected that this location would retain a memory of previous days' discharges, including those taking place some days before the modelling was started, and that the contribution from the final day's discharge would only be a small fraction of the activity present. For this reason the predictions suggest that there is a considerable background concentration in this area and that four days' modelling is not sufficient to predict concentrations on the day the measurements were taken. The measured values are consistent with this picture, since it would be expected that the model would underpredict the absolute concentrations given the lack of time allowed in the simulations for the background concentration to build up. However, the measurements exhibit a greater variation between the concentrations observed before and after the passage of the plume than would be expected at this location, supposing that dispersion here is primarily determined by diffusive processes. This may be due to the measured values reflecting local fluctuations at the edge of the plume, not allowed for in the model predictions. An alternative explanation is that the plume trajectory did on this occasion extend further into the bay area above Portland Harbour than the predictions suggest - possibly due to a particularly strong tide or as a result of wind-induced currents. To determine whether this is a transient or more persistent phenomenon will require more detailed investigation than the limited data available for this comparison allows.

4. Conclusions and Discussion

A generalised 2-dimensional model for bay areas is being developed and has been applied to the sea area around the UKAEA's Winfrith pipeline. Model predictions of plume trajectories in Weymouth Bay have been compared with those obtained from dye experiments and the results suggest that the velocities used in the model are quite realistic in representing the speed and direction of flow. We have also taken the opportunity to simulate an actual discharge from the pipeline; the trend of the model predictions are in general accord with measurements taken at the same time although, since this was not a proper validation exercise, it is difficult to draw any quantitative conclusions of model validity from the comparison.

The major problem appears to be in determining the background concentration since the results show that concentration levels are increasing over successive days' discharges. Thus, in order to be reasonably confident of obtaining the correct background, it would be necessary to model the daily discharges over a longer period than the four days chosen here. In fact, more information on Cobalt-60 discharges during that month has recently been made available and it is hoped to re-run the simulations in the near future. Another difficulty is that the nature of the model – the fact that it calculates concentrations on a grid, over discrete time intervals – means that there is a certain amount of temporal and spatial averaging. The observations, on the other hand, reflect spot measurements which may result from small-scale fluctuations in concentration which would not be represented in the model. There was also some uncertainty in the exact time and duration of the releases for the first three days' discharges which could again have an effect on the results.

However, Winfrith are continuing with their programme of monitoring surveys in Weymouth Bay and it is hoped to be able to put together a proper validation study in the future. It is also hoped to use the model to look at different areas in this region – for example some monitoring has also been carried out in Poole Harbour and it would be interesting to look at predictions for a region containing this area. Finally, it is also intended to include a dosimetric model in the code whereby individual doses to critical population groups may be estimated.

5. Acknowledgements

The authors would like to thank Mr K Stammers, AEEW for his help and advice during this project.

6. References

1. MacKenzie, J. and Nicholson, S. : "COLDOS – A computer code for the estimation of collective doses from radioactive releases to the sea". SRD R389 (1987).

2. Clark, M. J., Grimwood, P. J. and Camplin, W. C. : "A model to calculate exposure from radioactive discharges into the coastal waters of Northern Europe". NRPB R109 (1980).

3. Camplin, W. C., Durance, J. A. and Jeffries, D. F. : "A marine compartment model for collective dose assessment of liquid radioactive effluents". Sizewell Inquiry Series No. 4, MAFF, Directorate of Fisheries Research (1982).

4. Curtis, A. R. and Sweetenham, W. P. : "FACSIMILE/CHEKMAT Users Manual" AERE R12805 (1987).

5. Hydraulics Research, Wallingford: "SE Dorset Water Services. Mathematical simulations of tidal currents between Portland Bill and St Catherine's Point." Report no. EX-1474 (1986).

6. Bowles, P., Burns, R. H., Hudswell, F. and Whipple, R. T. P. : "Exercise Mermaid". AERE E/R 2625 (1958).

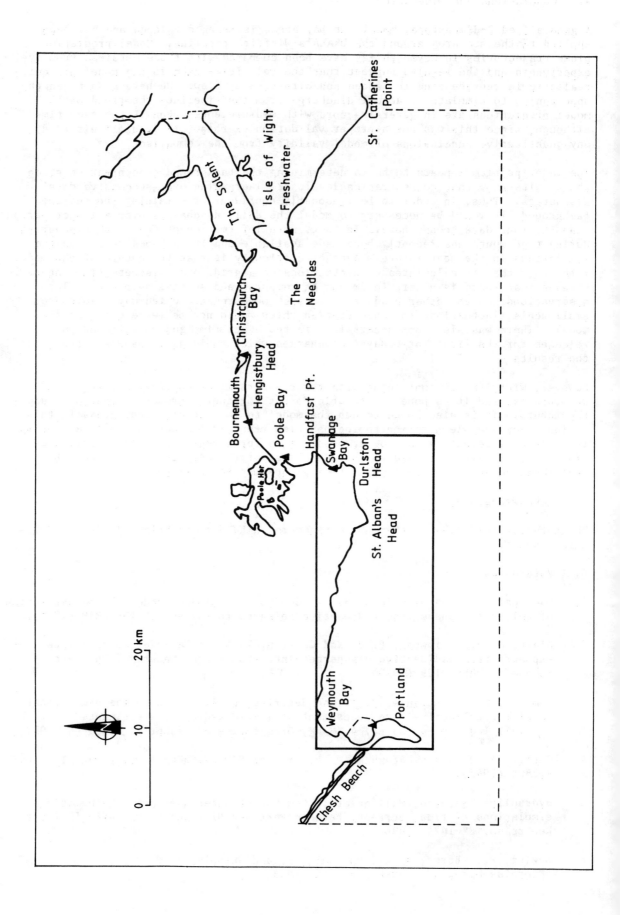

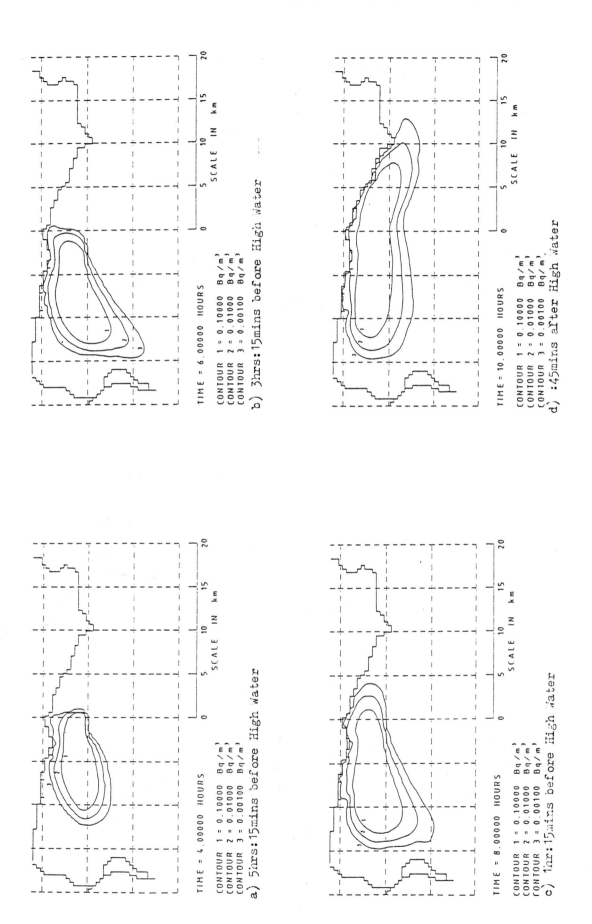

Figure 2: Effluent concentrations for discharge on spring tide.

129

Advances in water modelling and measurement

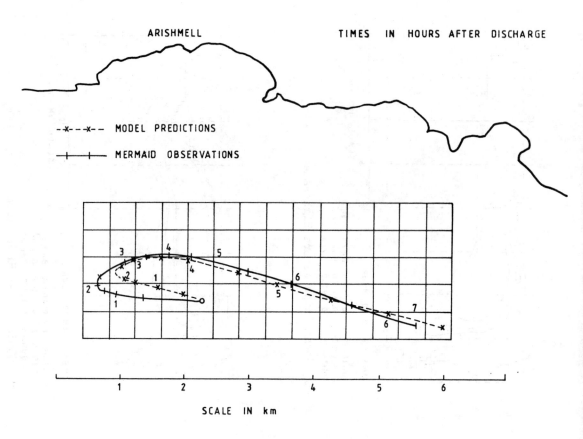

Figure 3:   Track of dye patch released on west-going tide.

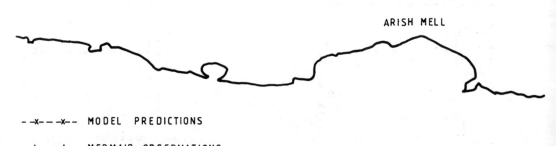

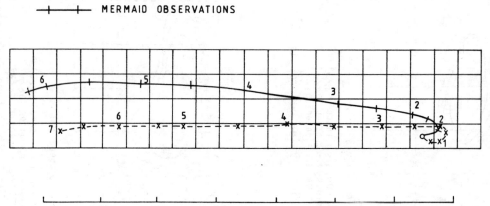

Figure 4:   Track of dye patch released on east-going tide.

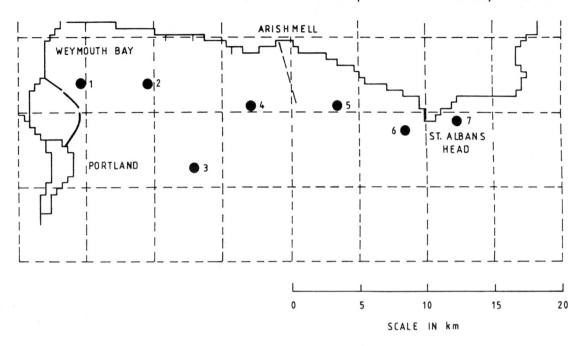

Figure 5: Survey locations.

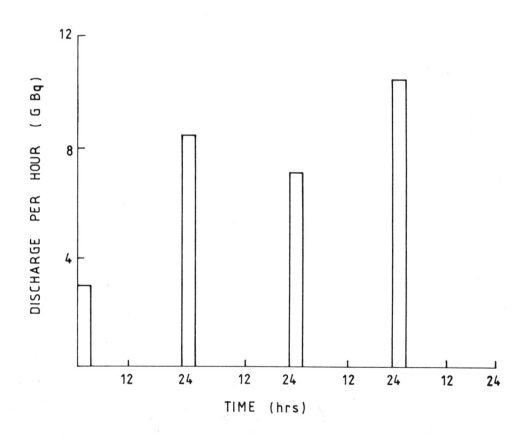

Figure 6: Discharge pattern.

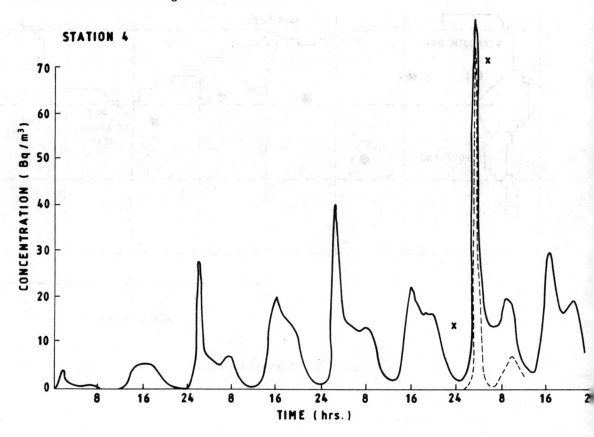

Figure 7: Model predictions for station 4.

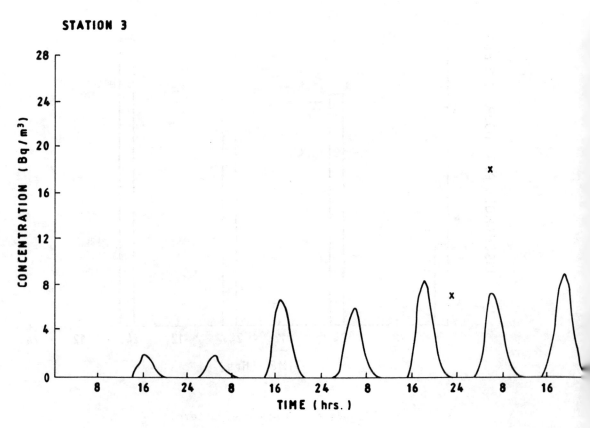

Figure 8: Model predictions for station 3.

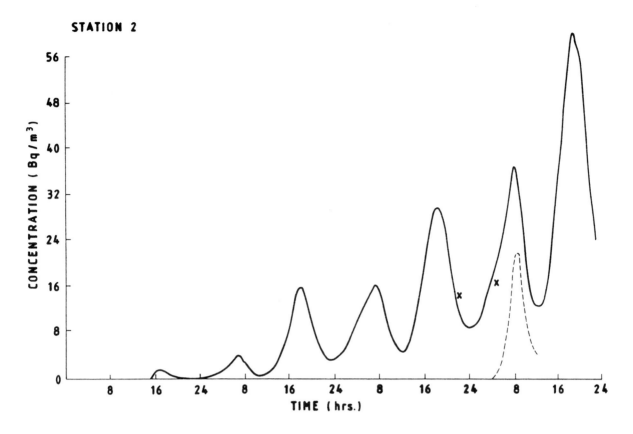

Figure 9: Model predictions for station 2.

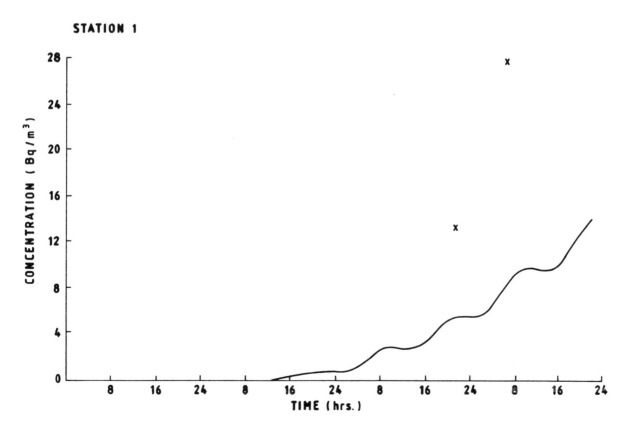

Figure 10: Model predictions for station 1.

Chapter 10
THE THREE-DIMENSIONAL PROGRAMME TRISULA WITH CURVILINEAR ORTHOGONAL COORDINATES

P van der Kuur, A Roelfzema and G K Verboom
Delft Hydraulics, The Netherlands

THE MATHEMATICAL MODEL

The TRISULA programme system for three-dimensional flow simulations is based on a hydrostatic multi-level approach. The basic mathematical description of TRISULA includes water elevation, flow velocity in three dimensions and density. Density is a function of salinity and temperature, and is coupled dynamically with the flow equations to simulate density-induced motions. This results in a complete set of non-linear equations for free surface flow and conservation of heat, salt and, in addition, dissolved pollutant constituents.

The TRISULA programme has a wide area of applications, ranging from wind generated currents in lakes and tidal flow in coastal seas to salt intrusion in estuaries, far field cooling water dispersion and large scale pollutant dispersion.

The approach used to model the water flow is based on the "shallow water assumption". Under this assumption the vertical momentum equation is reduced to the hydrostatic pressure relation. The vertical water motion is derived from the horizontal flow field, using the continuity equation.

The shallow water assumption imposes two restrictions on the applicability of the TRISULA programme:
- vertical accelerations due to buoyancy effects cannot be taken into account properly;
- sudden variations in bottom topography cannot be reproduced properly.

The fluid is taken as incompressible and density variations are neglected in the equations of motion except in the buoyancy terms. (Boussinesq approximation).

The vertical turbulent viscosity and diffusion are presently modelled by flow dependent coefficients, which are computed according to a turbulence closure technique using turbulence intensity and mixing length (k-l model) taking into account buoyancy effects.

Wind drag on the water surface can be included. This enables to study the effects of wind-induced currents and the deflection of surface plumes by wind. At the surface and bottom quadratic friction laws are applied.

NUMERICAL ASPECTS

In the numerical method used in TRISULA, the area under consideration is covered by two-dimensional horizontal curvilinear orthogonal grids of variable size. This enables more accurate schematization of complicated irregular geometries and increases the computational efficiency.

According to the multi-level concept, the vertical grid consists of layers determined by a fixed number of permeable interfaces. As a result of the use of the so-called "sigma transformation" in the vertical, the number of layers is constant over the entire computational field. The vertical grid may have a non-equidistant distribution, which allows a more detailed reproduction of features in zones of particular interest, e.g. in the top and bottom layer.

The momentum and transport equations are vertically integrated over the layer thickness, which results in 2DH-like equations. The coupling of the layers is by vertical advection and diffusion (turbulent shear stresses and fluxes of scalar quantities). The equations are solved by an implicit finite difference method (ADI). This method was chosen because it allows large time steps without negative effects on the computational stability. Grid staggering (in the horizontal mesh) is employed which gives a satisfactory degree of accuracy without excessive computer costs.

BOUNDARY CONDITIONS

A distinction is made between vertical and lateral boundary conditions.

Vertical boundary conditions:
- wind shear stress at the surface;
- free-slip at the bottom by means of Chezy-type formula.

Lateral boundary conditions:
- closed boundaries: zero normal velocities, combination of free-/no-slip;
- open boundaries: water levels or normal velocities and discharges, are to be prescribed through time-dependent data or Fourier components (frequencies, amplitudes and phases);
- recovery of salinity, heat, etc. at flow-reverse conditions (from out- to inflow) assured by a "Thatcher-Harleman approach".

INPUT DATA

The following input data are required:
- Bottom topography and boundary outline.
- Conditions at open boundaries: normal velocity or free-surface motion in a time-series or as Fourier components; time-dependent concentrations.
- Drag coefficients for wind stress and bottom stress (Chezy).
- Parameters related to the horizontal and vertical turbulent exchange of momentum and water quality variables.
- Time-dependent discharges and/or withdrawals in one or more grid cells.

OUTPUT DATA

Output data include the following options:
- Time histories of all variables in selected checkpoints.
- Velocity (vector)fields and isolines of free surface.
- Concentration fields (isolines).

SPECIAL FACILITIES

The TRISULA system includes a number of special facilities, of which the most important are:

- simulation of drying and flooding of intertidal flats;
- simulation of the discharge of heat and effluents and the intake of cooling water at any location and any depth in the computational field;
- representation of thin dams like groynes and breakwaters;

- a series of pre- and postprocessing options for the analysis and presentation of input and results;
- modular set up for the implementation of turbulence models, including damping functions to incorporate the effect of vertical stratification.

APPLICATIONS

The TRISULA programme is now applied on different cases, both for two-dimensional, vertically averaged condition as well as for three dimensional, multi-layer conditions.

- An example of the "degenerated" two dimensional vertically averaged applications is the "Manukau Harbour Dispersion Modelling".
  Manukau Harbour (Auckland, New Zealand) is an area with many shoreline irregularities and tidal tributaries. Within the harbour a pronounced system of relatively deep channels and large shallow mud flats exists. These mud flats cover about 40 to 50 percent of the harbour area. The estuary is connected with the Tasman Sea by a well defined inlet channel.
  These conditions led to the adoption of a curvilinear grid approach for the numerical model study for this area, providing an optimum balance between accuracy and cost of the geometrical representation and the hydrodynamical and water quality modelling.
  The model study was performed by using the two-dimensional flow version and, based on its results, the water quality model DELWAQ, both with curvilinear orthogonal coordinates.
  For the generation of the curvilinear grid a grid generation programme was available, satisfying conditions of orthogonality and grid size variations.
  To allocate depth values for each grid point, an automated interpolation technique was used.
  The coupling of the DELWAQ model with the flow model, which provides the details of hydrodynamics and geometry was realized by an interface module.

  Model simulations included effluent discharges at three locations, for conservative and non-conservative constituents (faecal coliforms, with different die-off times), and for conditions without and with wind.

- An example of the three dimensional application is a cooling water recirculation study for the Senoko Incineration Plant in Singapore. This application, based on rectangular coordinates, preceeded the curvilinear orthogonal grid version of the present TRISULA programme.

  The thermal recirculation was studied by means of a mathematical model of the relevant part of the Johor Strait, with computational elements of 200 x 200 m^2. The currents induced by the tide, river inflow and wind were computed on a depth-averaged basis.

  The transport and dispersion of heat discharged from the various sources was computed with a three-dimensional model, using realistic vertical velocity distributions.
  Vertical salinity gradients and submerged cooling water withdrawals were included in the model.

- A case-study is now performed on three-dimensional applications with curvilinear orthogonal coordinates.
  The area under consideration is a combination of the English Channel and the North Sea, between about 49° and 59° North latitude. For this area a model with curvilinear co-ordinates is developed with grid sizes varying between about 14 and 18 kilometers, which results in about 2,200 grid points.
  Based on this grid, tidal flows are calculated without and with wind by using wind conditions from 1953 and 1983 storms. These calculations are performed with one layer, representing a vertically averaged approach and with up to seven layers. The effect of wind on vertical velocity profiles will be compared with existing data and with the vertically averaged approach.
  Subsequently the DELWAQ programme will be coupled to the flow model, also

with a one layer and a multi-layer approach to calculate the transport of conservative constituents.
The result of this study will become available summer 1988.

REFERENCES

1. Leendertse, J.J. et al,
   A three-dimensional model for estuaries and coastal seas.
   Volumes I up to VI, The Rand Corporation, (Santa Monica), 1973-1979

2. Stelling, G.S.,
   On the construction of computational methods for shallow water flow problems.
   Rijkswaterstaat communications, No. 35,
   The Hague, Rijkswaterstaat, 1984

3. Verboom, G.K., Slob, A.,
   Weakly-reflective boundary conditions for two-dimensional shallow water flow problems.
   5th Int. Conf. on Finite Elements in Water Resources, June 1984
   Vermont, also Adv. Water Resources, Vol. 7, December 1984
   DELFT HYDRAULICS publication no. 322, June 1984

4. Blumberg, A.F., Mellor, G.L.,
   Modelling vertical and horizontal diffusivities with the sigma coordinate system.
   Monthly Weather Review, Vol. 113, No. 8, August 1985

5. Wybenga, J.H.A.,
   Determination of flow patterns in rivers with curvilinear coordinates,
   XXI Congress, IAHR, Melbourne, Australia, August 1985, also DELFT HYDRAULICS Publication No. 352, October 1985

6. Blumberg, A.F., Mellor, G.L.,
   A description of a three-dimensional coastal ocean circulation model.
   Three-dimensional Shelf Models,
   Coastal and Estuarine Sciences, 5,
   American Geophysical Union, 1986

7. Verboom, G.K., Segal, A.,
   Weakly reflective boundary conditions for shallow water equations.
   25th Meeting Dutch Working group on Numerical Flow Simulations, Delft, October 1986

8. Willemse, J.B.T.M., Stelling, G.S. and Verboom, G.K.,
   Solving the shallow water equations with an orthogonal coordinate transformation.
   Delft Hydraulics Communication, No. 356, januari 1986

9. Leendertse, J.J.,
   A three-dimensional alternating direction implicit model with iterative
   Fourth order disipative non-linear advection terms.
   WD - 3333 - NETH,
   The Netherlands Rijkswaterstaat, January 1987

10. Nihoul, J.C.J., Jamart, B.MN. (editors),
    Three-dimensional models of marine and estuarine dynamics.
    Elseviers Science Publishers B.V., 1987

11. Roelfzema, A., Perrels, P.A.J., Scholten, W.N.G., van der Wekken, A.,
    Developing a mathematical model system for the Rhine-Meuse Estuary.
    Int.Symp. New Technology in Model Testing in Hydraulic Research, Pune , India, September 1987, also DELFT HYDRAULICS publication No. 384, November 1987.

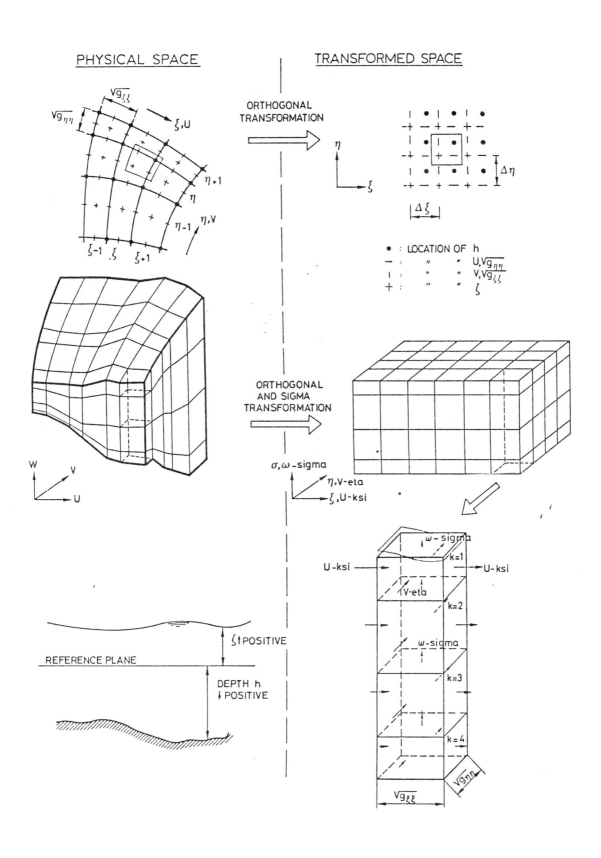

PHYSICAL SPACE

TRANSFORMED SPACE

ORTHOGONAL TRANSFORMATION

ORTHOGONAL AND SIGMA TRANSFORMATION

● : LOCATION OF h
— :   "   " $U, \sqrt{g_{\eta\eta}}$
| :   "   " $V, \sqrt{g_{\zeta\zeta}}$
+ :   "   " ζ

REFERENCE PLANE

ζ ↑ POSITIVE

DEPTH h ↓ POSITIVE

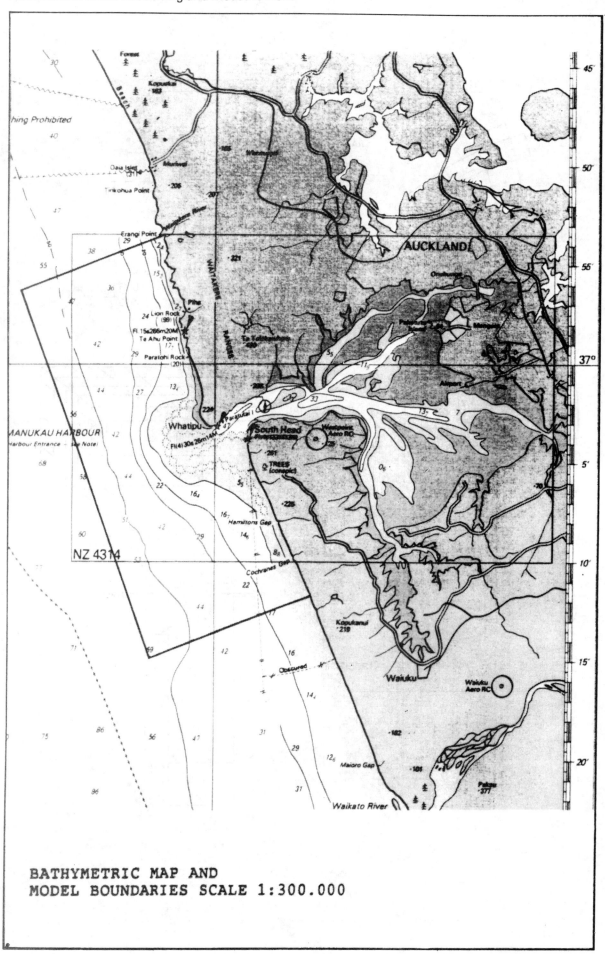

BATHYMETRIC MAP AND
MODEL BOUNDARIES SCALE 1:300.000

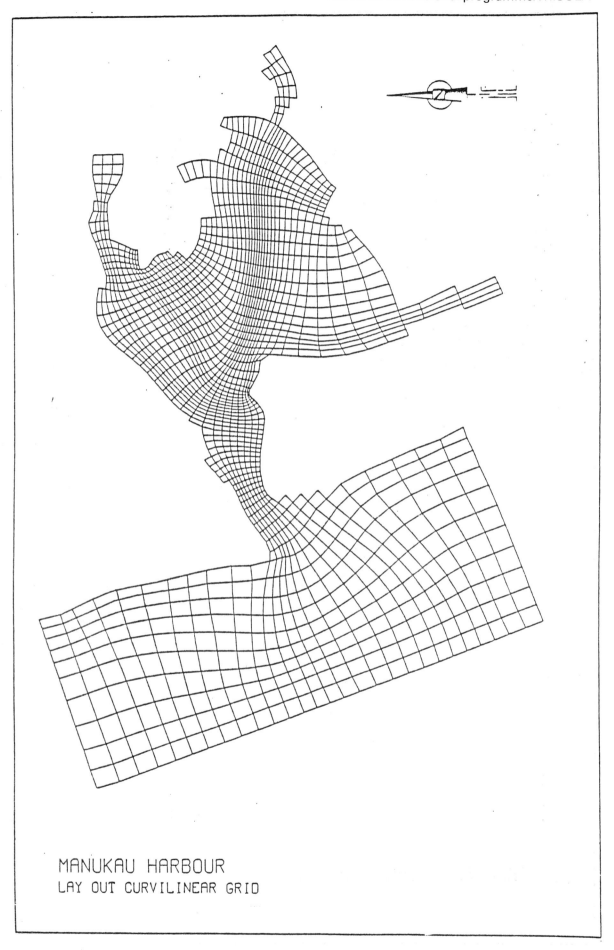

MANUKAU HARBOUR
LAY OUT CURVILINEAR GRID

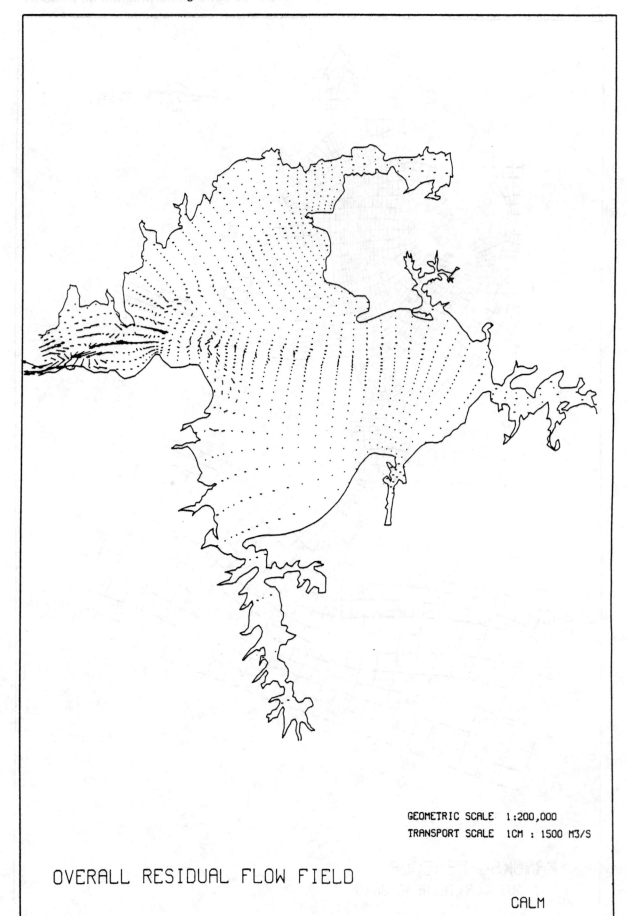

GEOMETRIC SCALE 1:200,000
TRANSPORT SCALE 1CM : 1500 M3/S

OVERALL RESIDUAL FLOW FIELD

CALM

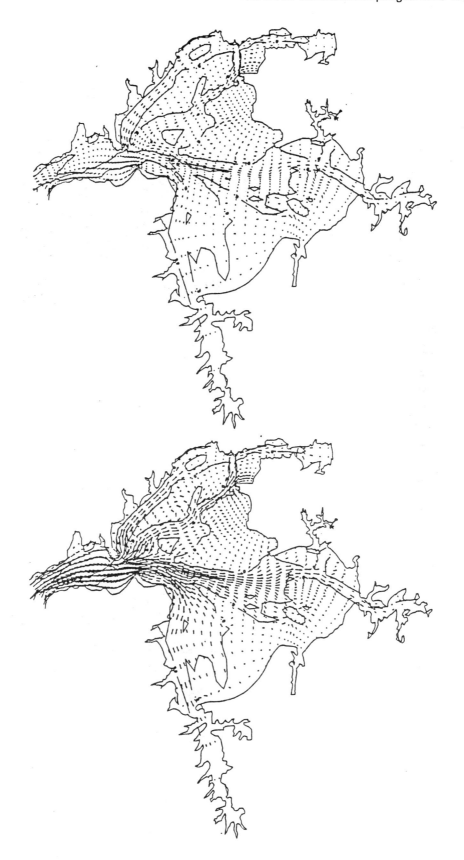

SCALE 1:300,000

VELOCITY PATTERN AT LOW TIDE AND RISING TIDE

CALM   VELOCITY SCALE : 1 M/S = 0.5 CM

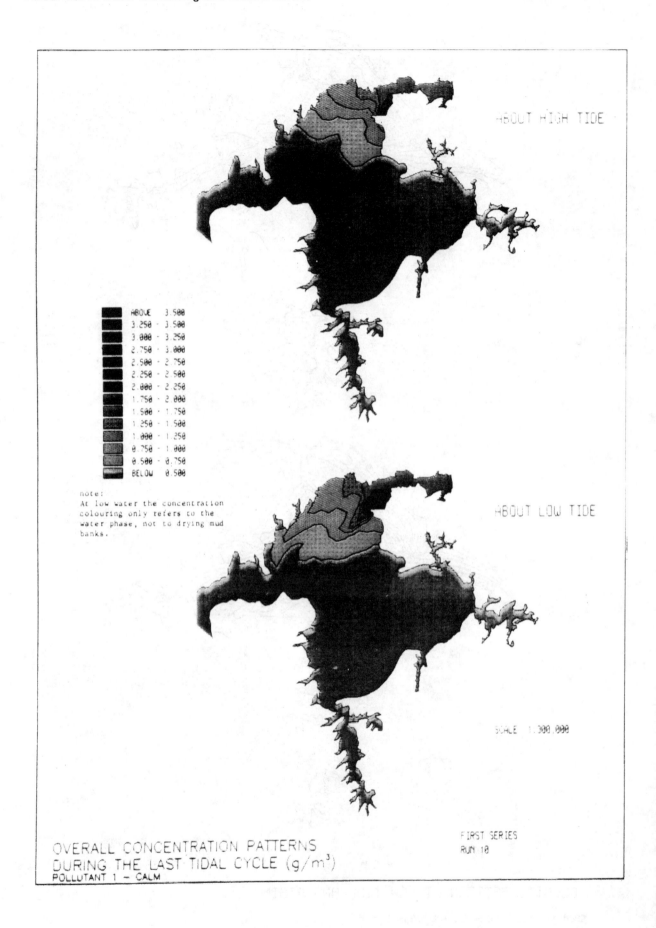

ABOUT HIGH TIDE

ABOVE   3.500
3.250 - 3.500
3.000 - 3.250
2.750 - 3.000
2.500 - 2.750
2.250 - 2.500
2.000 - 2.250
1.750 - 2.000
1.500 - 1.750
1.250 - 1.500
1.000 - 1.250
0.750 - 1.000
0.500 - 0.750
BELOW 0.500

note:
At low water the concentration
colouring only refers to the
water phase, not to drying mud
banks.

ABOUT LOW TIDE

SCALE 1.000.000

FIRST SERIES
RUN 18

OVERALL CONCENTRATION PATTERNS
DURING THE LAST TIDAL CYCLE (g/m³)
POLLUTANT 1 — CALM

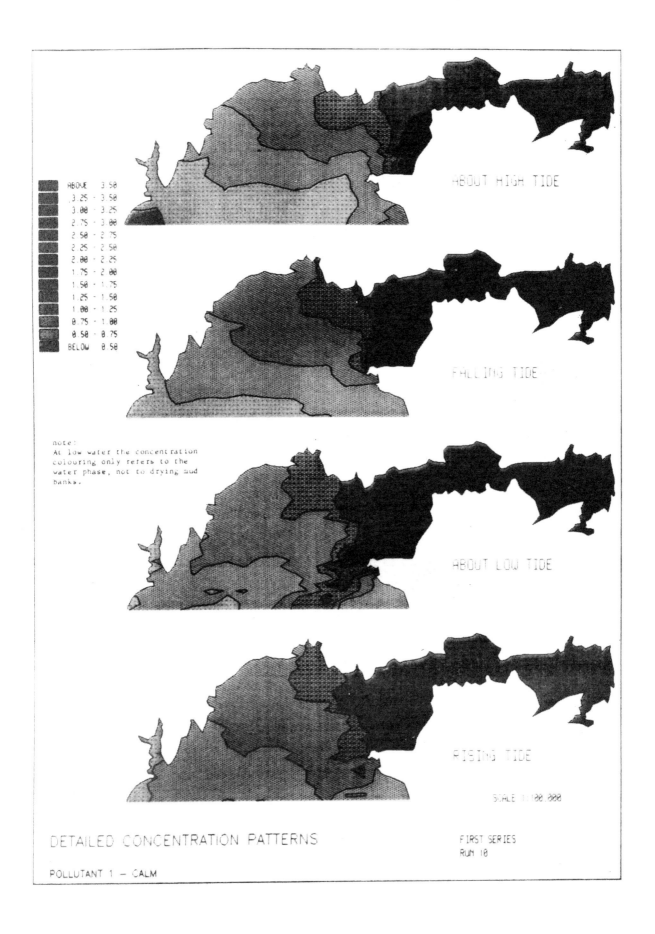

ABOVE   3.50
3.25 - 3.50
3.00 - 3.25
2.75 - 3.00
2.50 - 2.75
2.25 - 2.50
2.00 - 2.25
1.75 - 2.00
1.50 - 1.75
1.25 - 1.50
1.00 - 1.25
0.75 - 1.00
0.50 - 0.75
BELOW 0.50

ABOUT HIGH TIDE

FALLING TIDE

note:
At low water the concentration
colouring only refers to the
water phase, not to drying mud
banks.

ABOUT LOW TIDE

RISING TIDE

SCALE 1:100.000

DETAILED CONCENTRATION PATTERNS

FIRST SERIES
RUN 10

POLLUTANT 1 — CALM

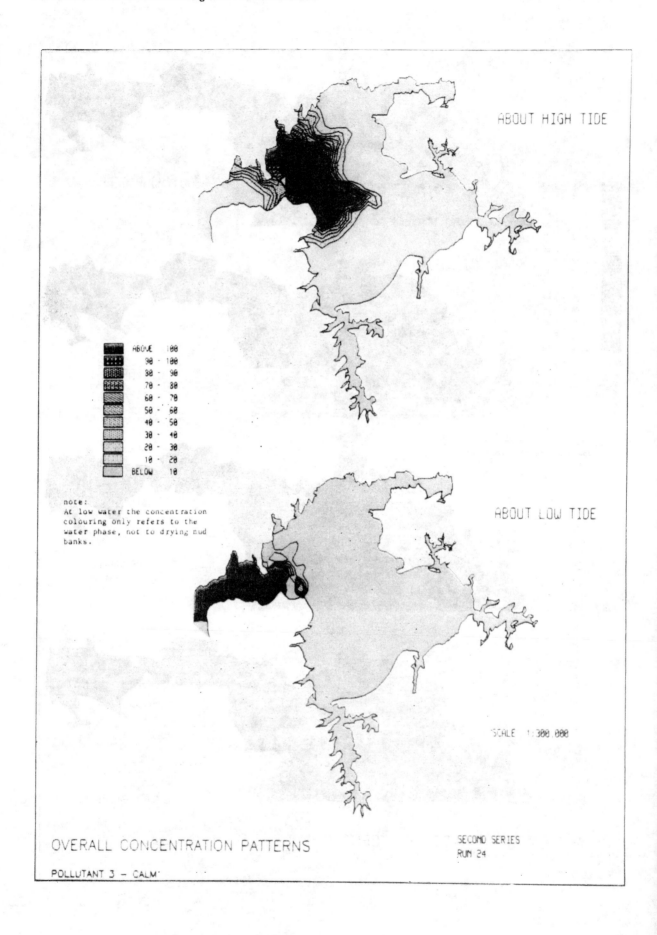

ABOUT HIGH TIDE

ABOVE 100
90 - 100
80 - 90
70 - 80
60 - 70
50 - 60
40 - 50
30 - 40
20 - 30
10 - 20
BELOW 10

note:
At low water the concentration
colouring only refers to the
water phase, not to drying mud
banks.

ABOUT LOW TIDE

SCALE 1:300,000

OVERALL CONCENTRATION PATTERNS

SECOND SERIES
RUN 24

POLLUTANT 3 - CALM

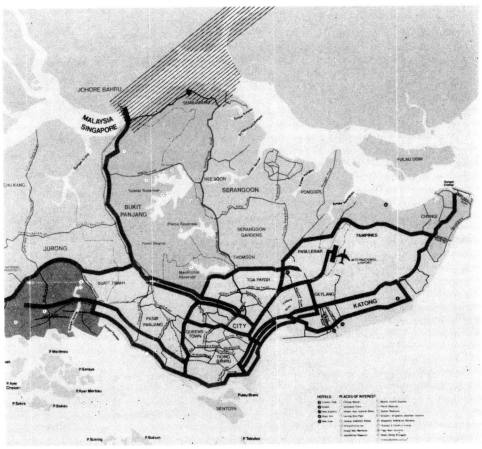

Singapore Island with the proposed site location.

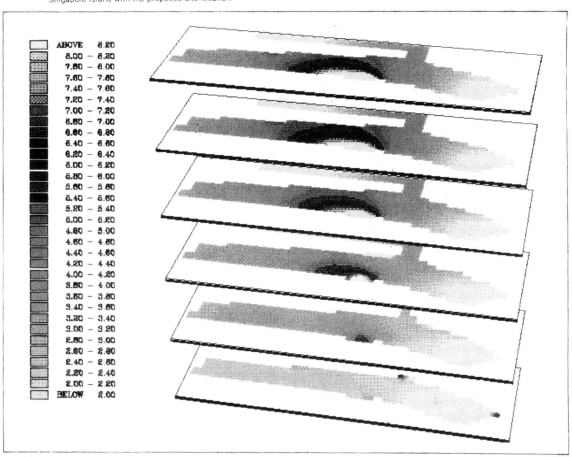

Excess temperature images around high water slack at 0.5 m, 1.5 m, 2.5 m, 3.5 m, 5 m and approx. 9 m depth.

Chapter 11
CONTROLLING AND MONITORING TIDAL WATER QUALITY – THE WAY AHEAD

V A Cooper, D Johnson and D T E Hunt
Water Research Centre, UK

Summary

The United Kingdom is heavily dependent on its tidal waters for the disposal of industrial effluents and sewage sludge, and for the marine treatment of sewage from coastal communities using sea outfalls, but also uses them for recreation and amenity, and for fisheries and shellfisheries. The development and application of suitable procedures to ensure effective management of coastal water quality is, therefore, essential.

This short paper describes a new strategy for the control and monitoring of industrial discharges and discusses the sound design and monitoring of sewage outfalls – activities which both involve dispersion modelling techniques – and examines some novel techniques for assessing the environmental impact of marine disposal and treatment operations.

1.   Introduction

The United Kingdom employs its tidal waters for a variety of potentially-conflicting uses: recreation and amenity, fisheries and shellfisheries, sewage sludge and industrial waste disposal, and marine sewage treatment via sea outfalls from coastal communities.

The consequent demand for effective management of the quality of our estuarine and coastal waters creates a need for the development and application of appropriate procedures for controlling and monitoring the environmental impact of waste disposal and of marine sewage treatment. Such control and monitoring is necessary to address public concerns about their acceptability as uses of our marine environment, and to satisfy the requirements of legislation – such as the European Community (EC) Dangerous Substances Directive (1) and its daughter Directives, the UK Control of Pollution Act – COPA (2), the EC Bathing Waters Directive (3) and the EC Shellfish Directive (4).

2.   The control and monitoring of industrial discharges

2.1   General considerations

The UK has adopted the Environmental Quality Objective/Environmental Quality Standard (EQO/EQS) approach for the control of industrial discharges. In this approach, a Use-related EQO (eg "The protection of marine life") is identified for the receiving waters and a corresponding EQS, set in terms of the permitted concentration of the contaminant in the receiving water, then provides the basis for pollution control. Gardner and Mance (5) have discussed the basis of the EQO/EQS approach in detail; they describe how proposed EQSs are derived from the results of (preferably, chronic) toxicity tests, and further evaluated by comparison with chemical and biological field data.

For all discharges whose concentration of the contaminant exceeds the EQS, there will be a region in the receiving water body in which the EQS cannot be met. The Mixing Zone (MZ) is simply that part of the receiving water body in which exceeding the EQS is acceptable to the regulatory authority; a better, if more cumbersome, name would be "Zone of Permitted EQS Transcursion".

It follows that, if the EQS is set as an annual average, the MZ must be defined as the area outside which the annual average concentration must be below it - see Figure 1. Moreover, a Mixing Zone can exist only in an area where the discharge affects the defined Environmental Quality Objective - for only in that area will there be a valid Environmental Quality Standard to define its boundary.

2.2   Problems of application to tidal waters

In a river, physical mixing of a discharge with the main flow is usually complete within a hundred channel-widths of the outfall, and it is therefore appropriate to require compliance with the EQS at that point, since little further dilution will normally occur until the next tributary junction. Thus, the zones of mixing and of permitted EQS transcursion coincide (explaining the use of the term "Mixing Zone" for the latter).

In estuarine and coastal waters, however, the concept of "complete mixing" is meaningless, and the allowable location and size of the zone of permitted EQS transcursion must be determined on other grounds, although the term Mixing Zone is still used to describe it.

Application of the EQO/EQS approach in tidal waters is also less straightforward than in rivers because regular monitoring at the boundary of the MZ to assess compliance (see Figure 1) is wholly unrealistic. As Table 1 shows, the uncertainties in resulting estimates of the mean concentration, relative to the EQS, are enormous even when a hundred samples are taken during the year, because tidal movement of the contaminant plume causes the concentration at the boundary to fluctuate wildly. Such uncertainties would preclude rational assessment of compliance and could easily lead, by pure chance, to "false alarm" calls for action or to unjustified complacency.

Problems such as these led the House of Commons' Environment Committee (6) to conclude that the EQO/EQS concept, as applied to tidal waters, was seriously flawed. However, as the following sections indicate, a valid and practical approach for the application of the concept in tidal waters can be developed, using dispersion modelling techniques.

2.3   The recommended approach

WRc has drawn up a strategy for applying the EQO/EQS approach in tidal waters - in association with the Water Utilities, the Department of the Environment, the Ministry of Agriculture Fisheries and Food and the Department of Agriculture for Scotland and the Department of the Environment (Northern Ireland). This strategy, shortly to be published by the Water Authorities Association in a successor booklet to its existing guidance document on Mixing Zones (7), is outlined in Figure 2 and described in more detail below.

2.3.1 Establishing Mixing Zones and Discharge Consent Conditions

Mixing Zones will need to be set only for the small number of "significant" discharges having an important potential impact on tidal waters (7). What constitutes an acceptable Mixing Zone for a significant discharge will be a local decision, inevitably involving a degree of judgement, made within broad national guidelines and under the principle of "Minimisation At Reasonable Cost (MARC)".

In applying MARC, the regulatory authority will strike a balance between the desire to minimise the discharge and its Mixing Zone, by attention to outfall design and/or effluent treatment, and the costs of doing so. It will need to consider the nature of the contaminant and the possible impact of the discharge upon the receiving waters, and have information on the costs associated with Mixing Zones of different sizes.

On the environmental side, there will be a need to consider the overall impact, as well as the size, of the Mixing Zone. It should not:

- have a marked deleterious effect upon the overall ecology of the receiving water body;
- cause an effect which would be aesthetically objectionable;
- have an adverse effect on the general and visible amenity of the area;
- give rise to a public health hazard;
- impinge on a shoreline where there is normal access by the public;
- impinge on a Site of Special Scientific Interest;
- occupy more than one third of an estuary's width at any point (to allow passage of fish and other mobile organisms);
- interfere with any other recognised Uses of the receiving water.

The level, extent and possible build-up of potentially toxic substances in aquatic or terrestrial organisms must be considered, as must also the possibility of "overlapping" pollution caused by adjacent discharges and the possibility of future requests to discharge to the same water body. Moreover, any area of acute toxicity to typical species adjacent to the discharge must be minimised within the constraint of "reasonable cost", by using the best available diffuser technology and by circumscribing the concentration, as well as the load, of the contaminant in the discharge, when the Consent Conditions are set.

It can be seen that the EQO/EQS/MARC approach does not sanction increasing discharges to tidal water bodies until their diluting and assimilative capacity is almost exhausted. Rather, it combines responsible use of that capacity with a clear requirement to minimise, within reasonable costs, the environmental consequences of discharges.

As shown in Figure 2, a dispersion model of the discharge, taking into account the general background levels of the contaminant in the receiving water body, will normally be required to identify the best option for outfall location and discharge level. From the average load of contaminant in the discharge estimated to be consistent with the chosen Mixing Zone, the Discharge Consent Condition (DCC) will be set in terms of both a daily average and a daily maximum load. The DCC will also need to specify a concentration limit to minimise any region of acute toxicity.

In view of the large uncertainties in model predictions, substantial "safety factors" will need to be incorporated, and model calibration carried out, when the definitive DCC is set. Calibration may be effected by determining the contaminant concentration along transects across the plume of the discharge, or by using a tracer as a surrogate for the contaminant. The proposed DCC may then be amended to allow for any consistent bias in the model predictions. The calibration will need to be repeated only when a change in hydrographic conditions (eg from civil engineering work) is likely seriously to alter the Mixing Zone.

2.3.2 Monitoring

Because there is no value in undertaking routine monitoring at the Mixing Zone boundary with the expectation of making an annual assessment of compliance with the EQS (see Table 1), day-to-day pollution control must involve monitoring the concentration of the contaminant in the discharge itself, to check compliance with the Discharge Consent Condition.

However, environmental monitoring is also essential to ensure that diffuse sources, or inadequate flushing of the consented discharge, do not increase the general levels of the contaminant in the area, above those which prevailed when the MZ and DCC were determined. Such monitoring must be performed in the "far field", distant from the relevant Mixing Zone. The concentrations involved will be much lower than the EQS and difficult and costly to determine accurately.

2.3.3 Future developments

This approach to pollution control, using Environmental Quality Standards for particular contaminants, is unsuitable for controlling discharges of complex industrial effluents of unknown and/or widely varying composition. Such effluents commonly arise from the batch production of synthetic organic chemicals - eg pharmaceuticals and pesticides.

In the future, therefore, we envisage that discharge control procedures based on direct assessment of toxicity will be used, certainly for complex and variable effluents but also, perhaps, for simpler effluents as well. As with chemical monitoring, routine discharge control could not be achieved by application of the chosen toxicity test to samples of the receiving water. Rather, such tests would be applied routinely to samples of the effluent itself, diluted with appropriate volumes of sea water; the DCC and the conditions of testing would need to be carefully drawn up to avoid ambiguity, but this is not seen as a major problem. Again, modelling could be used to help establish the Mixing Zone and Discharge Consent Conditions, by predicting the dilution and dispersion of the effluent in the receiving water body.

Of course, one could not expect to perform such routine tests on a range of representative species. Instead, one or two of the most sensitive organisms would be used - or the simplest toxicity test applied, having first established the relationship between its response and those of the representative species in general. A simple test using a non-representative species could even be used, provided that such a relationship had been demonstrated.

3.   The design and monitoring of sewage outfalls

The use of sea outfalls to treat sewage from coastal communities in the UK is of long standing, but the short outfalls of the Victorian period are not compatible with modern water quality standards, nor with our society's environmental expectations. The development and application of new procedures for the design and performance monitoring of sea outfalls seeks to ensure that modern standards are met, and current expectations fulfilled.

3.1   Computer modelling

Before undertaking the considerable capital investment required for a new long sea outfall, it is essential to ensure that the proposed location will provide sufficient dilution, dispersion and time for decay of micro-organisms so that the required standards can be met at the relevant use areas (such as bathing waters and shellfisheries). The most effective way of predicting these aspects of outfall performance is by using computer models, validated by field experiments.

WRc has developed a suite of mathematical models to simulate quantitatively the dispersion of sewage and other potentially polluting materials discharged to tidal waters. A hydrodynamic model is used to predict the pattern of tidal currents and water levels over representative tidal cycles. These predictions are validated using field data derived from tide gauging, current metering and float or drogue tracking. The hydrodynamic model interfaces with a Lagrangian, random walk dispersion model which takes account of wind-induced movements by superimposing a parabolic vertical profile on the depth-averaged profile. This model is validated by tracer studies, usually undertaken using bacterial spores (Bacillus subtilis) and/or Rhodamine dye.

After running the dispersion model over several tidal cycles, estimates of concentrations in small cells are stored at hourly intervals over a whole day. Differing bacterial decay rates and polluting loads can be accomodated, and the output is in the form of animated colour graphic displays showing the evolution of bacterial levels through time, for any chosen combination of tidal state, wind stress and decay rate.

In the past, modelling to predict the dispersion of sewage has been based on average flow conditions. The modelling of storm overflows has been largely ignored, yet these often present severe problems, preventing compliance with EC bathing water standards, because they discharge through old, short outfalls. To improve our ability to design for storm conditions, WRc is linking the dispersion model with models to predict the volumes and time-evolution of storm flows in sewerage systems. A further addition to the suite of models predicts the quality of sewage, in terms of suspended solids and biochemical oxygen demand, and work is in progress to incorporate bacterial loading as well. Current WRc research aims to extend this integrated approach to include river quality and diffuse sources, such as agricultural run-off, in order to predict water quality in estuaries and coastal seas.

Models such as those described are invaluable design tools - capable of providing an overview of the combined effects of a variety of physical and biological processes, rather than an exact simulation of events on any particular day.

3.2   Environmental monitoring of sewage outfalls

WRc - on behalf of the UK Water Utilities - is currently involved in two case studies with the aims of:

-   assessing the impact of modern, well-designed outfalls;
-   evaluating conventional and new techniques for environmental impact assessment.

Both are situated near holiday resorts - Weymouth (20,000 m3/d DWF) and Tenby (9,000 m3/d DWF) - and are well-designed, modern outfalls discharging into water of sufficient depth and current to ensure adequate dilution and dispersion. The research includes:

-   tracer studies to validate dispersion models and identify sewage-contaminated sediments;
-   benthic macrofaunal investigations;
-   virological investigations in water and sediment;
-   novel techniques for the assessment of biological effects.

3.2.1   Tracer studies

In addition to tracer work to validate dispersion models, the contamination of sediments by sewage can be traced using faecal bacteria and the faecal sterol, coprostanol. The bacterial groups monitored are Thermo-Tolerant Coliforms (TTC) and Faecal Streptococci (FS), both of which can be enumerated rapidly and cheaply. Although their mortality rates are variable, and they are not necessarily source-specific, they can give an indication of the area of immediate impact of an outfall. Coprostanol is specific to mammalian faeces, and is not quickly degraded in the environment; it has been reported (8, 9, 10) to be a reliable tracer of sewage contamination in sediments.

To illustrate the results of these tracer studies, Figure 3 shows the concentrations of coprostanol along 100 m transects on bearings of 130° and 335° from the Weymouth outfall. There is clear evidence of a localised impact, with maximum concentrations within 50 m of the outfall on the 130° bearing; additionally, a significant relationship ($p < 0.02$) was found between coprostanol concentrations and TTC levels. It is also interesting to note that the tidal current flows for 9 hours on the 130° bearing, but for only 3 hours on the 335° bearing - on which evidence of sewage contamination was lower.

3.2.2   Benthic macrofaunal ecology and sediment quality

Examination of the benthic macrofauna in the vicinity of an outfall is a well-established procedure for assessing environmental impact, but the interpretation of its results requires considerable care. The nature of the sediment itself has a profound influence upon the macrofaunal ecology, and information on the latter must be supplemented by data on the particle size distribution and organic carbon content. It is also useful to obtain data on contaminant concentrations (eg heavy metals, persistent organic compounds) in the fine fraction (<60 µm) when undertaking benthic surveys.

Multivariate statistical techniques can be used to supplement informed interpretation of the benthic macrofaunal data, and to analyse any patterns found in the light of data on sediment quality (11). The effect of sediment quality upon benthic ecology, and the often transient nature of sediment distributions, make it essential to carry out baseline benthic surveys prior to outfall construction and commissioning, in order that the effects of natural variations in sediment type are not confounded with polluting effects.

3.2.3 Virological investigations

Viruses in seawater are an increasing public health concern. Rotaviruses, which were dicovered in the 1970's, are responsible for around 50% of all clinical cases of acute gastroenteritis in children under the age of two, and can also cause severe diarrhoeal illness in adults (12).

Whilst standards and enumeration procedures for enteroviruses are established, rotaviruses have yet to be monitored in bathing waters. Methods for the enumeration of rotaviruses in environmental samples have been developed under contract to the Department of the Environment, and monitoring of bathing waters and beach sediments is in progress at the case study sites.

3.2.4 Novel biological techniques

Research on new biological techniques for assessing the environmental impact of discharges to tidal waters aims to develop tests and procedures meeting the following criteria:

- environmentally and ecologically relevant
- sensitive to contaminants
- rapid response
- simple, robust and reproducible

No single test can, of course, provide all the information required or meet all the above criteria optimally, but a number of useful approaches are being actively developed or tested, and others are at the stage where they can be routinely deployed. The techniques involved range from effects on genetic structure, through effects on juvenile organisms, to studies on the bioenergetics of adult organisms.

Pollutants can act on genetic material in two ways: by affecting the frequency of occurrence of particular genes, and by inflicting direct damage on the DNA within individual cell nuclei. The former effect is more readily observed in natural populations exposed for relatively long periods, and is therefore suitable for long-term monitoring strategies. The second type of effect, mutation, can be divided into gene mutation and chromosomal mutation or aberration. The frequency of gene mutation is low, and tends to be compensated by natural regulatory processes. Chromosomal aberration is therefore the effect exploited to develop indicators of pollution stress, and the interested reader is referred to the two chapters by Dixon in reference (13) for further details.

The next level of effect which can be exploited is embryo development. The successful development of mussel and oyster embryos to the first shelled stage can be prevented by damage at the chromosomal level, or by disturbance of the surface potentials on the cells which are differentiating as the embryo develops. Exposure of fertilised eggs to the test sample and subsequent assessment of developmental success affords a sensitive, yet robust technique which has been widely used by WRc for monitoring both effluent toxicity and effects in receiving waters. The sensitivity of this test is about an order of magnitude greater than that of any other lethal toxicity test currently in use, as illustrated in Figure 4. Application of the procedure to the sewage discharged from Weymouth and Tenby is shown in Figure 5. This work suggests that the Weymouth sewage contained a toxic component, whereas the weaker Tenby sewage did not give rise to any significant effect, compared to the control solution of distilled water in seawater (which was used to differentiate the effects of reduced salinity from those of toxicants).

The next level of testing exploits the effects of toxicants upon the growth and reproductive capacity of organisms. In freshwater, the use of reproductive capacity measurement with Daphnia magna is well-established but the approach has not been widely used in the marine sphere. Work in this area is in progress at a number of laboratories, and an application to assessment of the effects of the offshore oil industry has been reported (14).

The growth rate of individual organisms can also be used as a basis for assessing the wholesomeness of a particular environment. The measurement of the energy available for growth and reproduction, after the demands of basal metabolism have been met, provides an index known as "Scope for Growth" or "SFG" (15). Such measurements can readily be made on the common mussel Mytilus edulis (16), and the technique has been fairly widely applied – see reference 17 for a review. Figure 6 shows some results obtained by WRc from a mussel deployment in the vicinity of the Thames Estuary sewage sludge disposal ground. The lower SFG values found to the south-west of the ground are consistent with the dispersion patterns of the sludge obtained from radiotracer studies, which show the main area of sludge transport after dumping to be south-easterly from the ground, on the flooding tide.

The techniques described above all relate to water quality, and there is a pressing need for tests which can assess the effects of contamination upon benthic organisms. A bioassay employing a phoxocephalid amphipod is used in the USA, and WRc is currently developing procedures suitable for application in the UK. Pending the availability of such a test for routine use, recourse can be made to toxicity tests carried out on aqueous or organic extracts of the sediment.

Thus, Figure 3 illustrates the application of a commercially-available bioassay employing the loss of luminescence in a marine bacterium (Microtox) to organic extracts of sediments in the vicinity of sewage outfalls. The results, expressed as 15 minute EC50 values, show that the toxicity of sediment extracts increased within 50 m of the Weymouth outfall. Moreover, the Microtox toxicity was significantly correlated with both TTC and coprostanol levels (p<0.01 and 0.05, respectively). Again, no such relationships were observed at Tenby, a smaller outfall which has only been operating since 1985.

4.   Conclusions

4.1   Industrial discharges

Despite a number of problems, the EQO/EQS approach can provide a rational basis for the control of discharges to tidal waters if dispersion modelling is employed to determine the discharge level and outfall configuration consistent with an acceptable Mixing Zone. In establishing the Mixing Zone and Discharge Consent Condition, both the local environmental circumstances and the costs to the discharger must be considered and weighed, under the principle of Minimisation At Reasonable Cost (MARC). In this way, the advantages of the EQO/EQS approach can be combined with the desire to minimise, without incurring unjustifiable costs, the environmental impact of discharges.

Day-to-day pollution control cannot be effected by analysis of the receiving tidal waters, and must be based on discharge monitoring, in the usual manner. Environmental monitoring is, however, essential to ensure that the contaminant concentration in those waters is not increasing, and to check for undesirable effects on the biota.

In future, Discharge Consent Conditions – particularly for complex and/or variable effluents, but possibly for others as well – may be established directly in terms of toxicity. Such an approach can be made entirely consistent with the EQO/EQS principle of permitting usage of the diluting and dispersive capacity of receiving waters, by employing mathematical modelling to predict that capacity.

4.2 Sewage outfalls

The use of long sea outfalls represents a cost-effective and environmentally sound option for the treatment of sewage from coastal towns, provided that they are properly designed. Mathematical modelling of the hydrodynamic regime and of the dispersive character of the receiving waters is essential to effective outfall design.

A number of techniques, both novel and conventional, are available for assessing the environmental impact of outfalls. Such techniques, applied to two modern and well-designed systems, have revealed that any effects are highly localised. It is recommended that a baseline environmental survey should be undertaken prior to outfall construction/commissioning, and further surveys undertaken after commissioning and at intervals thereafter. As a minimum, such surveys should include bacterial, particle size, organic carbon and coprostanol analysis of sediments, and benthic macrofaunal work. Additional work, if costs allow, would involve diver surveys and underwater film/videotape to assess localised impact, sublethal toxicity tests on sediment extracts, assessment of contaminant uptake by benthic biota and sediment analysis for heavy metals, persistent organics and viruses.

4.3 New techniques for assessing environmental impact

A wide range of new techniques are being developed for assessing the impact on the marine environment of industrial discharges, sewage sludge disposal activities and sewage outfalls. These techniques complement and extend existing chemical and biological measurements and their use is likely to expand considerably, as a direct means of judging the well-being of our tidal waters.

In the longer term, our goal must be to combine the power of mathematical techniques to model contaminant dispersion with our growing knowledge of the effects of the contaminants upon marine organisms - and, ultimately, whole ecosystems - so that an ability to predict environmental effects directly can be brought to bear on the management of coastal and estuarine water quality.

5. References

1. Council of European Communities. Directive 76/464/EEC on the Discharge of Dangerous Substances. Official Journal of the European Communities. No. L54. 1976.
2. Control of Pollution Act. HMSO. 1974.
3. Council of European Communities. Directive 76/160/EEC Concerning the Quality of Bathing Water. Official Journal of the European Communities. No. L31/1. 1976.
4. Council of European Communities. Directive 79/923/EEC On the Quality Required of Shellfish Waters. Official Journal of the European Communities. No. L281/4. 1979.
5. Gardner, J. and Mance, G.: "Proposed Environmental Quality Standards for List II Substances in Water. Introduction." Technical Report TR 206, Medmenham, UK, WRc, 1984.
6. House of Commons Environment Committee, 1987. Third Report. Pollution of Rivers and Estuaries. HMSO.
7. Water Authorities Association: "Mixing Zones. Guidelines for Definition and Monitoring", London, WAA, 1986.
8. Walker, R. W., Wun, L. K., Litsky, W. and Dutka, B. U.: "Coprostanol as an indicator of faecal pollution". Critical Reviews in Environmental Control, 12, 1982, pp.91-112
9. McCalley, D. V., Cooke, M. and Nickless, G.: "Coprostanol in Severn Estuary sediments". Bulletin of Environmental Contamination and Toxicology, 25, 1980, pp.374-381.
10. Brown, R. C. and Wade, T. L.: "Sedimentary coprostanol and hydrocarbon distribution adjacent to a sewage outfall". Water Research, 18, (5), 1984, pp.621-632.

11. Roddie, B. D.: "Statistical analysis for marine survey data" Report ER 1260-M, Medmenham, UK, WRc, 1986.
12. Blacklow, N. R. and Cukor, G. C.: "Viral gastroenteritis", New England Journal of Medicine, 304, 1981, pp.397-406.
13. Bayne, B. L., Brown, D. A., Burns, K., Dixon, D. R., Ivanovici, A., Livingstone, D. R., Lowe, D. M., Moore, M. N., Stebbing, A. R. D. and Widdows, J.: "The Effects of Stress and Pollution on Marine Animals", New York, Praeger, 1985.
14. Girling, A. E. and Streatfield, C. M.: "An assessment of the acute and chronic toxicity of production water from a North Sea oil platform based upon laboratory bioassays with a calanoid copepod - Acartia tonsa (Dana)", Presented at the International Conference on Environmental Protection of the North Sea, London, 1987.
15. Widdows, J.: "Physiological Responses to Pollution." Marine Pollution Bulletin, 16, 1985, pp.129-134.
16. Bayne, B. L.: "Marine Mussels: Their Ecology and Physiology." Cambridge, Cambridge University Press, 1976.
17. Lack, T. J. and Widdows, J.: "Physiological and cellular responses of animals to environmental stress - case studies." in Kullenberg, G., editor, "The Role of the Oceans as a Waste Disposal Option." Dordrecht, D. Reidel Publishing Company, 1986, pp.647-655.

Table 1: Uncertainties in the Estimates of Annual Mean Concentrations on the Mixing Zone Boundary*

| Number of samples | 90% Probability intervals (as percentage of EQS) | |
	Discharge 1	Discharge 2
10	+/ 157	+/- 235
50	+/- 70	+/- 105
100	+/- 50	+/- 74

* Simulations, using a contaminant dispersion model, for two representative hypothetical discharges

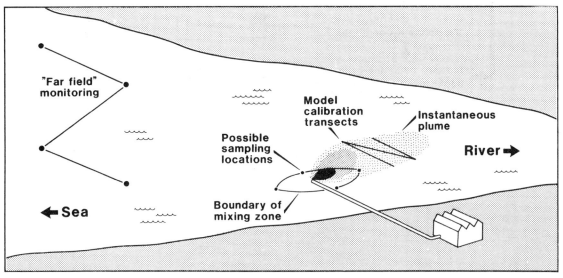

Figure 1.   The concept of the Mixing Zone, as applied to the discharge to tidal waters of a substance controlled by an EQS set as an annual average

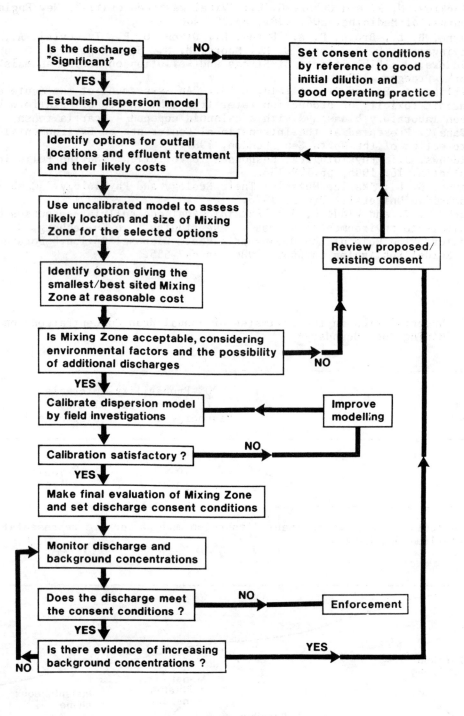

Figure 2. The recommended strategy for applying the EQO/EQS approach to industrial discharges to UK tidal waters

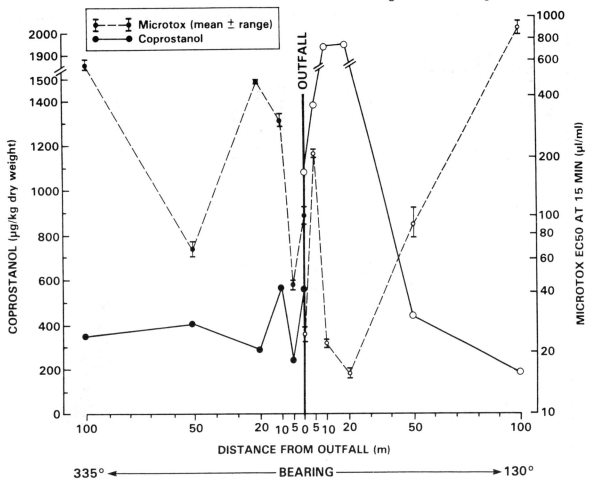

Figure 3. Levels of coprostanol near the Weymouth outfall and an example of the application of the Microtox bioassay to organic extracts of sediments near sewage outfalls.

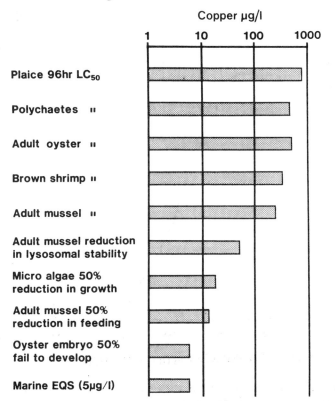

Figure 4. The sensitivity to copper of various toxicity and sub-lethal stress tests

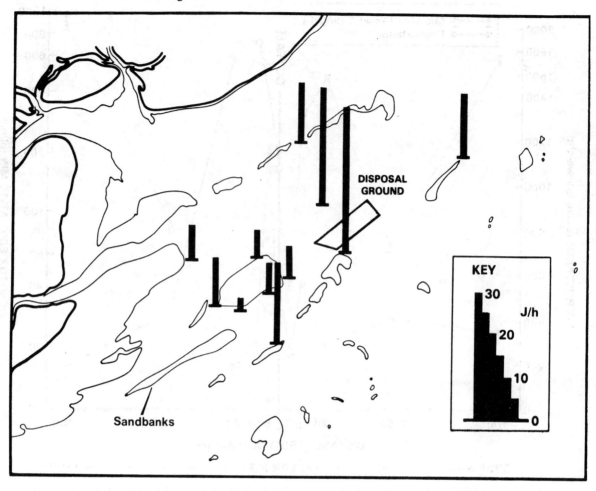

Figure 5.   The application of the mussel embryo bioassay to sewage outfalls.

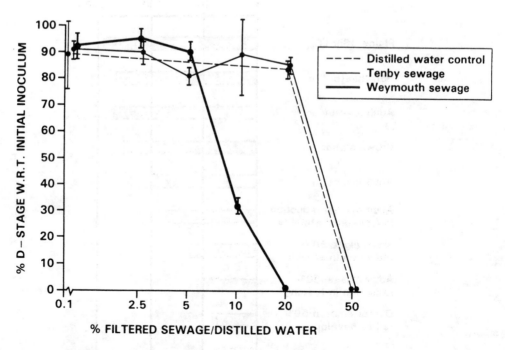

Figure 6.   Results of mussel Scope for Growth measurements (in Joules per hour) at the Thames Estuary sewage sludge disposal ground.

Chapter 12
RECENT ADVANCES IN THE FIELD MEASUREMENT OF INITIAL DILUTION AND HYDRAULIC PERFORMANCE

R Powell*
British Maritime Technology, UK
N Bennett
Southern Water Authority, UK
C Dorling
Water Research Centre, UK
*Now working for: Space Technology Systems Ltd, UK

Summary

The dilution of an effluent between the point of discharge and the point it reaches the water surface is an important factor in the design of sea outfalls. Current methods available to predict this initial dilution show good agreement when applied to still water conditions, and have been used in this restricted mode for the design of a number of long sea outfalls. In practice, true slack water rarely occurs in the coastal zone and several improved design formulae incorporating ambient speed have been proposed to allow for this. These have been based on theoretical considerations, laboratory experiments and field measurement. Predictions based on these equations do not always agree and have been shown to predict widely different values.

This paper will demonstrate the viability of field measurement in the assessment of initial dilution and hydraulic performance of outfalls. The use of field data is seen as important input to the overall understanding of the way the outfall performs and in the development of more widely appropriate equations to predict outfall performance. The paper will concentrate on the design and implementation of field techniques which may be used on existing outfalls or as part of a commissioning study for a new long sea outfall.

Specialised equipment was developed to continuously monitor the jet velocity and salinity of the effluent being discharged. These recording packages were deployed together with conventional oceanographic instruments for the duration of the site investigation. Improved methods of injecting and monitoring rhodamine dye were used to measure the initial dilution achieved on a large number of injections. The data have been collated and examples are presented in this paper, although detailed analysis and interpretation will be published in a separate paper.

A field survey has been undertaken to assess the hydraulic performance of an existing outfall and to measure the initial dilution achieved during a range of environmental conditions.

Advances in water modelling and measurement

1. Introduction

The UK has traditionally used the sea as a convenient site for the discharge of sewage effluent and, until relatively recently, outfall pipes to the sea have been constructed with little regard to the environmental consequences. This situation stemmed from attitudes prevailing during the Victorian era and the early part of this century. It is the legacy of this period which the modern generation is currently updating. Indeed, in a recent survey (1) of outfalls operated by Water Authorities an estimated 45% were constructed prior to 1940. It was also noted that 93% of UK outfalls discharge less than 100m below the low water mark.

One of the main reasons for the current trend towards scientifically designed long sea outfalls is the legislation passed by the European Community (2). This Council Directive has had a major impact on the water industry and despite some adverse comments it represents a major feature of current thinking. The document was unique in that it provided bathing water standards to which Authorities should conform.

Despite the recent criticism and comment that bacteriological standards are not the most appropriate parameter to use in the monitoring of outfall performance, a total of 389 bathing waters have been defined in the UK since the Directive was published. In order that these beaches should conform to the required standards, a considerable investment is required in cleaning up coastal and estuarine waters. This investment is expected to reach £800 million over the next 10 years.

When designing a new outfall to conform to the EC Directive, complex measurements and analyses are usually undertaken. These would generally include field studies, numerical modelling and engineering design. Aspects of these studies are covered in greater detail elsewhere (3,4), but in general, the trend in outfall design has been towards use of empirically verified models to predict behaviour of the effluent following discharge. To simplify such modelling, the process of dilution of an effluent can conveniently be divided into two distinct phases.

- Initial dilution - which occurs as the effluent rises from the outfall port to the surface.
- Secondary dilution - which occurs as the effluent slick is transported away from the discharge area.

This current paper concentrates on methods that have been developed to measure both this initial dilution directly and to assess the hydraulic performance of existing sea outfalls. The joint study of both hydraulic performance and initial dilution makes use of similar data collection equipment and support services, which results in a number of logistic advantages. Data on diffuser performance also allows a more detailed analysis of the data.

2. Initial Dilution

2.1 Introduction

For marine outfalls, initial dilution is defined by the authors as the ratio of the concentration of sewage at the diffuser port to maximum concentration at the sea surface.

A considerable number of papers have been published which report on experiments or propose methods of predicting initial dilution. The work may be divided into three groups:

- work based on field studies, (5,6)
- work based on laboratory trials, (7)
- work based on theoretical considerations. (8)

Work by other authors (9) has taken existing data and proposes alternative equations. Others (10) have reviewed the three techniques and concluded that none of the theories is ideal and that, with the current state of knowledge, it is perhaps impossible to predict moving water dilutions accurately.

This long standing diversity of opinion and the results of the own investigations which indicated that observed dilutions were at variance to predicted values was sufficient justification for Southern Water to carry out an extensive series of dye injections on the Hastings long sea outfall under moving water conditions in 1980. The original equipment and methodology is described by the co-author (6). Recent investigations have improved these techniques.

2.2 Instrumentation and Site Operations

In principle the measurement of initial dilution is a simple field technique: a tracer is injected into the diffuser port under trial, and the concentrations at the point of discharge and at the surface are measured. This apparently simple methodology is however made more complex by the environment in which the measurement must be made.

The key problems in finalising the design of the instrumentation were:

- to ensure thorough mixing of the tracer at the point of discharge to the sea;
- to measure a time history of the discharged concentration;
- to collect surface concentration data at the point where the concentration was at a maximum.
- to collect reliable support data on the ambient conditions and the nature of the flow from the diffuser ports.

The instrumentation used is illustrated as Figure 1.

In essence, the injection equipment consists of a closed loop around which effluent is circulated using a petrol powered Honda pump. The effluent is sampled from the centre of the diffuser mouth and pumped to the surface where the flow maybe sampled using an auxiliary pump. The flow round the main circuit is split, part passing through a Turner Designs Model 10 fluorimeter, with the remainder flowing around a by-pass circuit with a restricter valve. This allows a controlled rate of flow to be maintained through the fluorimeter. Rhodamine B dye (made up to a 1% solution) is bled into the effluent flow at a controlled rate before passing through the circulating pump and being injected into the diffuser. The design of the injection nozzle is such that blockages due to the 'rag' often associated with outfall discharges are minimised. Despite the care taken, blocking has proved to be a problem during the field trials, and vigilance is required of the divers who install and monitor the equipment.

A second vessel is used to measure the concentration of the dye as it reaches the surface. In order for the data to be reliable, it is important that the concentration of the rising plume is measured as soon as possible after it reaches the surface. The quantity of rhodamine injected is therefore adjusted to ensure that the surface colour is easily visible. This assists accurate tracking of the rising plume. Concentrations at the surface are measured using a Chelsea Instruments Sub-Aqua Tracka fluorimeter. This instrument, unlike the Turner Designs instrument has a totally self contained sensor which makes use of logarithmic amplifiers to allow output of the full range of dye concentrations between 10^{-6} to 10^{-10} on a single 0-10v chart recorder.

The Sub-Aqua Tracka is deployed away from the side of the vessel and set to measure dye concentration in the upper 20-30cm of the water column. The package is then towed through the visually most concentrated part of the dye patch on as many occasions as possible (Figure 2). By making continuous measurements in this way, the chance of observing the maximum concentration is greatly increased. Both fluorimeters record data on analogue chart recorders although the advantages of using digital logging techniques are currently being assessed. Information on the quantity of dye used, the pumping durations and the dye injection times are logged, together with visual notes both by the divers and scientists on the surface. Development of the instrumentation is continuing with alternative methods of measuring concentration at the diffuser port. This will allow higher concentrations to be injected, improving visibility of the dye patch at the surface.

Separate recording instrumentation is usually deployed close to the site to measure ambient current speed and direction, temperature salinity and tidal height.

3. Hydraulic Monitoring

3.1 Introduction

The design of long sea outfalls usually involves a diffuser section which generally consists of a tapering main pipe with a number of smaller diameter risers. Each riser terminates in a number of ports. The number of ports range from only one port per riser, to as many as eight. The number of risers varies between two and sixty, the main consideration being the ability to achieve suitable dilution and dispersion of the effluent.

The hydraulic design of long sea outfalls is concerned primarily with achieving a balanced flow from each of the ports over a range of input flows and tidal heights. Typically outfalls have to pass flows from a winter population which may be tripled in the summer months due to visitors. This makes the sizing of outfall pipelines and diffusers difficult. This problem is further exacerbated by a lack of information on headloss coefficients for the typical bends and junctions found in outfall diffuser pipelines. The purpose of the surveys was to check the operation of the outfall and to calibrate computer models developed by WRC to predict port velocities and headlosses for a range of flows and tidal conditions.

3.2 Instrumentation

Field trials usually involve the collection of effluent flow data at the landward section of the outfall. This is generally achieved in the outfall headworks by monitoring discharge flow using, either in-line doppler flow meters or clamp-on ultrasonic flow meters. The effluent 'head' is also recorded. This is the level of effluent required, above the sea level, to 'drive' effluent along the outfall pipeline and through the diffuser.

At the diffuser section of the outfall the effluent velocity from the ports is recorded. During initial field trials this was achieved using an electro-magnetic current meter fitted to the centre of the port by divers, with the data being recorded on-board a survey vessel moored over the diffuser via umbilicals. This system provided good data but suffered from several operational problems. The intrusive nature of the probe into the flow resulted in sewage debris collecting on the sensor head. Secondly, data could only be collected whilst the survey vessel was moored over the diffuser. Data could therefore only be collected for one or two hours per day, and more importantly the operation was very sensitive to weather conditions.

To improve general applicability, WRC in conjunction with Detectronic Ltd, developed a submersible package capable of recording both velocity and salinity of the flow. The system consists of an ultrasonic Doppler probe and a salinometer. The Doppler probe has a range of 0-3m/s and the salinometer a range of 0 - 35 ppt (parts per thousand). The senors are powered from an adapted Detectronic flow survey unit which contains a solid state logger. The sensor head is attached to the port by divers and relays information to the underwater logging package via a short cable.

The logging package is weighted to prevent movement on the sea bed by wave or current action. Each package has a 32K solid state memory and is capable of recording data at a slow logging rate of between thirty seconds and fifteen minutes with an added facility of fast logging when a pre-set velocity is exceeded. For most surveys the logging interval is set at two minutes with a fast logging rate of ten seconds when the pre-set trip-velocity of 1 metre/second was exceeded. When the velocity falls below the trip velocity the instrument resumes logging at the user specified slow logging rate. When the memory is full the unit 'auto-powers' down (sleep mode) retaining the data.

The logger is battery powered, incorporating a real time clock and is housed in a UPVC pressure casing. The entire package has been pressure tested to a pressure equivalent to 30m of water. Field trials of the package has resulted in a number of improvements in the design. In particular the use of better underwater connectors has eliminated water ingress, which had previously resulted in some memory loss. Development of the package is continuing with the applicability of alternative flow measuring systems, such as time-of-flight sensors, being evaluated. The use of a such devices should provide data on the direction of flow which, with the present system, has to be evaluated from the salinometer recordings.

4.0   Field Trials on the Hastings Long Sea Outfall

Previous work on the older Hastings Outfall had indicated that the outfall had five blocked diffuser ports. Modelling studies indicated that this blockage resulted from poor flow characteristics. To overcome these problems, and achieve a self cleansing system it was determined that minimum pipeline flushing velocities should be raised to 1.5m/s. A continuously tapering diffuser and an area ratio (i.e. the total area of ports to the area of the main outfall pipe) of 0.75 would ensure a velocity of 2m/s through each port. A head tower was prescribed to maintain these design velocities.

The design of the second Hastings outfall thus varied significantly from previous design practice as a result of the findings from the 1980 dilution study. However, design is one facet, practice another. Indeed, a lack of corroborative design data and too few field studies on operational outfalls has been apparent for many years.

There must be a responsibility within the engineering sector to check, report on and learn from past designs. The co-author has always been anxious to monitor and test the performance of his outfall design under full operational conditions and it was to this end that a field study was carried out in August 1987.

The objectives of the field trial were to:

- obtain information on the hydraulic performance and initial dilution of the Bulverhythe outfall

- provide guidance on the future operation of the outfall

- collect initial dilution data to enhance existing information and refine existing empirical formulae

- provide feedback for improving the hydraulic design of long sea outfalls

- improve procedures for monitoring long sea outfalls.

The study was carried out in two phases: installation and calibration of the hydraulic packages, and the initial dilution study.

For the first phase, five packages were deployed on ports 1,4,9,13 and 18 respectively (Figure 5). 'Top hats', secured to the ports with locking screws were used to mount the sensors at ports 1,4,9 and 13. However an oversized pipe prevented the use of this method on port 18. Here the sensors were mounted on a sprung steel hoop which was inserted into the port and allowed to expand against the internal wall. A typical diffuser port is shown on Figure .

All the packages, with the exception of port 18, were deployed on 22 August. The package for port 18 was fitted on 23 August. The packages on ports 9,13 and 18 were left in place until 26 August when they were recovered. The packages on ports 1 and 4 were recovered on 25 August.

The logging intervals chosen for the survey were two minutes at the slow logging rate, and 15 seconds at the fast logging rate. The fast logging rate was initiated when the recorded port velocity exceeded 1m/s.

Instrumentation in the headworks consisted of recording the effluent level in the header tank and the flow discharged through the outfall pipeline. The level was recorded using the permanently installed ultrasonic depth recorder and the flow using an ultrasonic Doppler type flow meter temporarily bonded to the outside of the outfall pipe for the duration of the trial. Figure 4 shows the layout at the headworks tower. A pressure transducer was also installed in the surge pipe to record the effluent level when valve 'A' was opened and the effluent released.

The procedure during the field trial was to fill the header tank to a pre-determined level, normally five metres, and then to release the flow through the outfall pipeline. Calibration checks were undertaken insitu at the diffuser section on two occasions by divers deploying electro-magnetic current meters in each of the diffuser ports in turn. A special clamping arrangement was devised in the 'top hat' so that the meter could be held in the centre of the flow. The data were recorded on the survey vessel, and subsequently compared with the data recorded by the flow packages. Calibration checks were also carried out by comparing flows from the ports with the total flow recorded at the headworks tower.

During the second phase, the objectives were primarily the measurement of initial dilution. A total of 57 dye injections were carried out on four separate diffusers between 27/8/87 and 7/9/87. For the purpose of this study only 2 packages were installed, one on port 1, with the second on the diffuser undergoing tests.

For each sample run the following procedure was adopted. When the effluent in the onshore tower was approaching the desired level (after the initial tests this was standardised to 5m to ensure an adequate time for the dye injection), the circulating pump was started and the divers checked the sampling and inlet nozzle for blockages. The time that the discharge was commenced was noted and the dye injection equipment prepared. Dye injection was started approximately 1 minute after the onset of flow from the diffusers. This was to assist in the flushing of any higher density saline water that may have intruded into the outfall. Dye injection was usually continued for 3-4 minutes with, whenever possible, a steady rate of input being maintained. After each trial the quantity of dye used was estimated. Valve 'A' was not closed when the tower was empty, and the flow down the outfall was, subject to ambient effluent flow rates, continued for 5-10 minutes to reduce the effect of negative flows that had been noted during the hydraulic study (Figure 6).

During the injection the chart recorder trace was carefully monitored to ensure that the timing was correct and that periods when the high dye concentration resulted in internal absorption of the emitted fluorescence were noted. This was found to be a problem during some injections and is caused by the strong colour of the dye masking the emitted fluorescence from the photo-multiplier. The occurrence of this was noted by the operator and appropriate corrections made to the observed concentrations by reference to an appropriate calibration curve.

When the dye appeared on the surface the auxiliary vessel tracked the patch. It was intended that the fluorimeter should pass through the centre of the rising plume as it reached the surface. However, to ensure that dye was not rising undetected into existing patches, the vessel also tracked the main surface slick. When tracking the main patch the fluorimeter was towed through the visually most concentrated area of dye (Figure 2).

Data from the study were presented graphically (Figure 3). The dye concentration (expressed as parts per 10^{-7} Rhodamine B) from both the diffuser and the surface tracking have been included. The diffuser concentrations have been read off at 10 second intervals and plotted directly, whereas for data collected by the auxiliary vessel only the peak concentration from each run through the patch have been included. Subsidiary data such as tidal state, diffuser number and jet velocity have also been listed to assist interpretation.

During the analysis of the data, a number of conventions were adopted:

- Mean Discharge Concentration at the diffuser port.

  The mean discharge concentration has been calculated as the arithmetic mean of the data points covering the majority of the injection. Low concentration data at the start of each injection and as the run is completed are not included. Similarly, data points which are uncertain due to blockages or other problems have not been included.

- Estimated Initial Dilution

  This has been calculated as the difference between the MEAN discharge concentration and the MAXIMUM concentration measured at the surface. This is considered to give the most appropriate estimate of initial dilution, but it should be noted that different methods may be appropriate for particular injections. For example, if a burst having a high dye concentration is discharged, it is possible that this will temporarily increase the concentration measured by the sampling vessel at the surface, resulting in an underestimate of the initial dilution achieved. The use of this definition has been termed the lower limit concept of initial dilution (11).

During survey operations additional data were collected. Ambient current speeds and directions were collected using two Aanderaa RCM4S recording current meters, deployed close to the seabed and at mid-depth. Tidal information was collected using an Aanderaa WLR5. Ambient wind speed/direction and sea state observations were made hourly by the field staff.

5. Applications of the Data and Conclusions

In the case of the Hastings study the data confirmed a number of points.

- Initial dilution, the site data confirmed that the initial dilution was within the design predictions.

- Dilution of the sewage discharged under a variety of flow conditions

Knowing the port velocities and values of the ambient currents, Bennett's formula (11) was used to predict the initial dilutions. Example results from this comparison are shown in Table 1. A more detailed analysis of data will be presented in a later publication.

The data recorded from the outfall ports showed that effluent discharge preferentially through the landward ports. The ratio of discharge through port 1 to that through port 18 was approximately 2:1. The maximum port velocity recorded was 4m/s through port 1 which was equivalent to a flow of 71 litres/second. It was predicted that this distribution of flow in the diffuser section would produce effluent velocities in the main pipe of 1.4m/s at the landward end, and 0.5m/s at port 18. These are expected to be sufficiently high to prevent progressive sedimentation in the diffuser section of the outfall. The distribution of flow through the diffuser section for a main pipeline flow of 800 l/s is given in Figure 7.

Further data were also collected on flow reversals in the outfall. This showed that the sudden release of effluent from the header tower creates a flow oscillation in the outfall. These oscillations cause salt water to be drawn into the diffuser ports when the outflow ceases. Typically maximum reverse velocities at port 1 of 1.2m/s were recorded for a period of 90 seconds. The maximum volume of salt water that would flow into port 1 during each cycle would be $5.3m^3$. The salinometer data suggests that the main pipeline velocities are sufficiently high to eject this salt water on the next cycle. The reverse flow and probable intrusion of salt water is shown graphically on Figure 7.

The data recorded during the hydraulic field trial have shown that the distribution of flow through the outfall diffuser is as predicted by Southern Water at the design stage. It has also been proved that main pipeline velocities of between 1.4m/s and 0.5m/s are achieved and these velocities are considered to be sufficiently high to prevent excessive sedimentation in the pipeline. This has proved valuable information at the commissioning stage of the outfall to ensure that its future operation is as predicted at the design stage.

The importance of data of this type at the commissioning stage of the outfall cannot be overstressed. From this present study it is clear that the outfall was performing in accordance with the design specification although it was noted that sedimentation occurred when the header tank was not used in the design mode i.e. when the flow was diverted to by-pass the tower.

In other cases the studies may be used to determine an appropriate operational mode for an outfall. Many outfalls are designed to cater for expected flows during its design life. The expected design flow may be twice the flow at the date of commissioning. To ensure the outfall operates adequately during its early life a number of onshore risers are blanked-off until flows increase. The hydraulic monitoring of outfalls at commissioning ensures that the correct number of ports are left open. In more general terms outfalls are often designed on the basis of hydraulic models and initial dilution predictions that have not been validated. At the commissioning stage of an outfall construction it is important to verify that any assumptions made during the design process can be verified. In particular, the balance of port velocities to prevent sedimentation or saline intrusion should be substantiated and the predicted initial dilutions anticipated should be verified.

The development of the techniques above to provide a viable field procedure allowed the successful execution of a field study at Hastings. The study was conceived with the following aims:

-    to confirm outfall design predictions,

-    to aid future designs,

- to ensure the correct operation of the outfall during its design life,

- to improve public awareness.

The former three have been covered earlier in the paper the last may need some explanation. There is a growing public awareness of the discharge of sewage to coastal waters. The monitoring of a new outfalls programme at the commissioning stage shows the operating authorities' willingness to adopt a professional attitude to outfall design.

6.   References

1.   Water Research Centre: "Site Investigation for Outfall Design, Construction and Monitoring", in preparation.
2.   Official Journal of the European Communities Council Directive and Annex on the Quality of Bathing Water. OJ L31/1 (5.2.76).
3.   Neville-Jones, P.J.D., Dorling, C.: "Outfall design guide for environmental protection - a discussion document". Water Research Centre External Report, ER 209E.
4.   Dorling, C.: "Report on the Survey of Outfalls to Coastal and Estuarine Waters in the United Kingdom". Water Research Centre External Report ER 281E.
5.   Agg, A.R. and Wakeford, A.C.: "Field studies of jet dilution of sewage at sea outfalls". Instrn Publ Hlth Engrs J, 1972, 71, 126-149.
6.   Bennett, N.J.: "Initial dilution : a practical study on the Hastings long sea outfall". Proc Instn Civ Engrs, Part I, 1981 70, 113-122.
7.   Hydraulics Research Station: "Horizontal outfalls in flowing water". Report EX763, Hydraulics Research Station, Wallingford, 1977.
8.   Fan, L.N.: "Turbulent buoyant jets into stratified or flowing ambient fluids". Report KA-R-15, W.M. Keck Lab or Hydraulics and Water Resources, California Institute of Technology, 1967.
9.   Lee, J.H.W. and Neville-Jones, P.: "Sea outfall design - prediction of initial dilution". Proc Instn Civ Engrs, Part 1, 1987, 82 Oct., 981-994.
10.   Sharp, J.J. and Moore, E.: "Estimation of dilution in buoyant effluents discharged into a current". Proc Instn Civ Engrs, Part 2, 1987, 83, Mar., 181-196.
11.   Bennett, N.J.: "Design of sea outfalls - the lower limit concept of initial dilution". Proc Instn Civ Engrs, Part 2, 1983, 75, 113-121.

TABLE 1

Sample Initial Dilution Comparisons

Diffuser No.	Water Depth (m)	Ambient Current (m/sec)	Discharge (m³/sec)	Initial Dilution (S) Measured	Bennett	Lee
3	17.6	0.133	0.0507	144	168	171
3	16.9	0.273	0.0491	*	269	508
3	15.5	0.286	0.0535	253	231	411
3	16.6	0.145	0.0548	209	154	147
3	17.2	0.061	0.059	284*	83	149
3	17.1	0.226	0.061	972*	195	347
3	16.9	0.205	0.0465	459	234	403
3	16.5	0.200	0.0375	84*	274	465
3	11.0	0.211	0.0403	168	164	203
3	11.2	0.040	0.0378	125	57	98
3	11.8	0.227	0.0292	230	256	346
3	12.4	0.320	0.026	798*	382	630
3	13.1	0.385	0.0179	425	663	1181
3	13.7	0.400	0.0361	823	366	665
3	15.2	0.320	0.0359	695	357	659
3	16.2	0.190	0.0361	412	269	442
9	11.3	0.015	0.0428	110	26	92
9	11.6	0.140	0.0447	301	120	93
9	12.0	0.265	0.0415	328	207	294

NOTES:

1) The range of recorded port discharges was 0.0179 - 0.0622 m³/sec. This represented a range of port velocities of 1.01 - 3.52 m/sec.

2) The specified design port velocity was 2 m/sec. at H.W., with a full head in the tower (5m).

3) The specified design main pipe velocity was 1.5 m/sec.

4) * This denotes that difficulties arose during the injection of dye which may have affected the measurements.

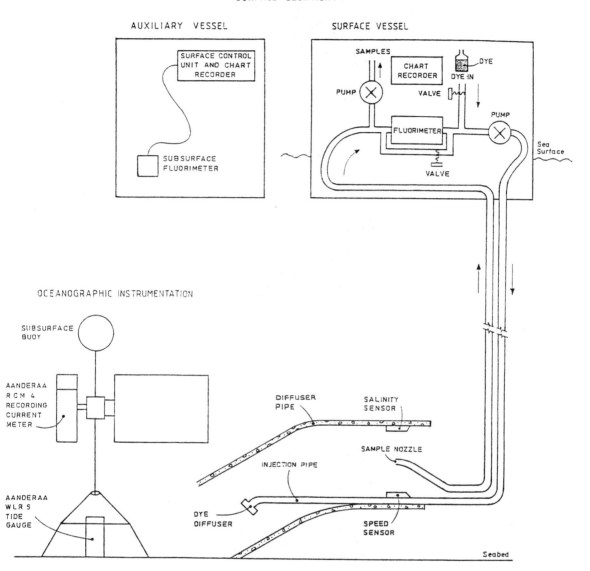

SURFACE EQUIPMENT

AUXILIARY VESSEL

SURFACE VESSEL

SURFACE CONTROL UNIT AND CHART RECORDER

SUBSURFACE FLUORIMETER

SAMPLES

CHART RECORDER

DYE

DYE IN

PUMP

VALVE

PUMP

FLUORIMETER

Sea Surface

VALVE

OCEANOGRAPHIC INSTRUMENTATION

SUBSURFACE BUOY

AANDERAA RCM 4 RECORDING CURRENT METER

DIFFUSER PIPE

SALINITY SENSOR

SAMPLE NOZZLE

INJECTION PIPE

AANDERAA WLR 5 TIDE GAUGE

DYE DIFFUSER

SPEED SENSOR

Seabed

DYE INJECTION AND SAMPLING EQUIPMENT   Figure 1

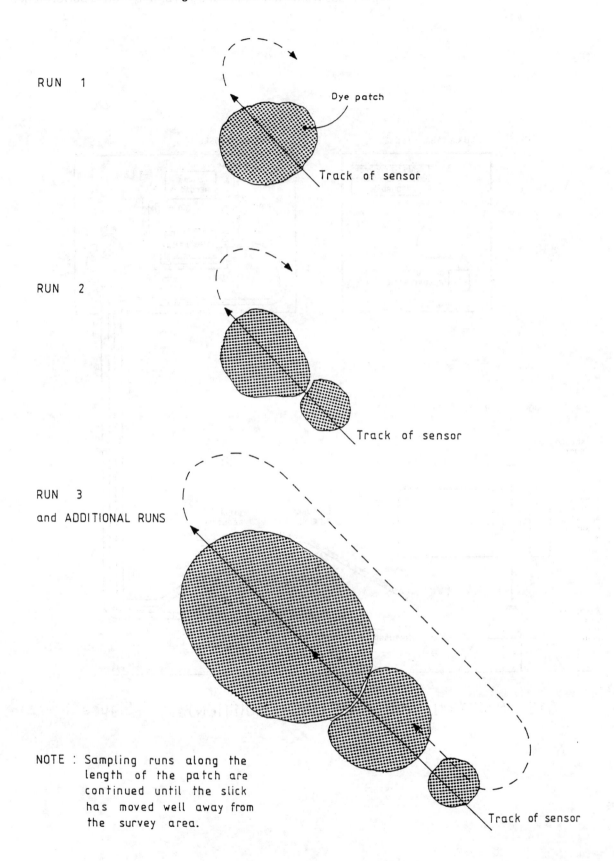

RUN 1

Dye patch

Track of sensor

RUN 2

Track of sensor

RUN 3
and ADDITIONAL RUNS

NOTE : Sampling runs along the
length of the patch are
continued until the slick
has moved well away from
the survey area.

Track of sensor

AUXILIARY VESSEL SAMPLING PROCEDURE   Figure 2

INJECTION No: 9   DIFFUSER No: 3   DATE: 29/8/87

TIDAL STATE HW-5:30 WATER DEPTH: 11m WIND: 1-2 NE SEA: CALM

CURRENT SPEED   21.1 cm/sec   CURRENT DIRECTION: 257°T

EFFLUENT DISCHARGED FROM HEADWORKS: 0911:45-09/9:00 BST

DYE INJECTION INTO DIFFUSER: 0913:30 - 0918:20 BST

DYE USED: 1 LITRE   DYE ON SURFACE: 0914:20 BST

MEAN JET VELOCITY: 2.28 m/sec

DIFFUSER CONCENTRATIONS (×10^{-7})   SURFACE CONCENTRATIONS (×10^{-7})

MAXIMUM: 13.2   MAXIMUM: .065

MEAN: 10.094 (0913:50-0918:20)   MEAN: .025

SD: 1.94   SD: .013

ESTIMATED INITIAL DILUTION: 168

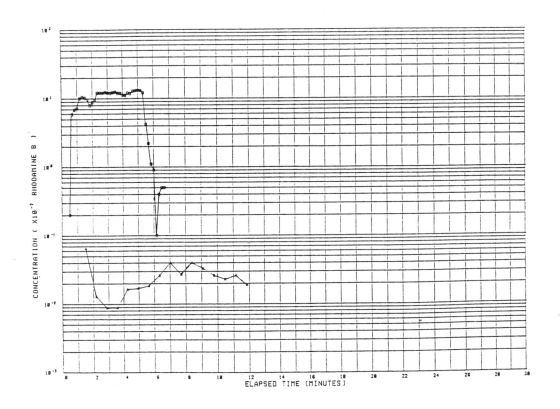

INITIAL DILUTION STUDY - SAMPLE RESULTS   Figure 3

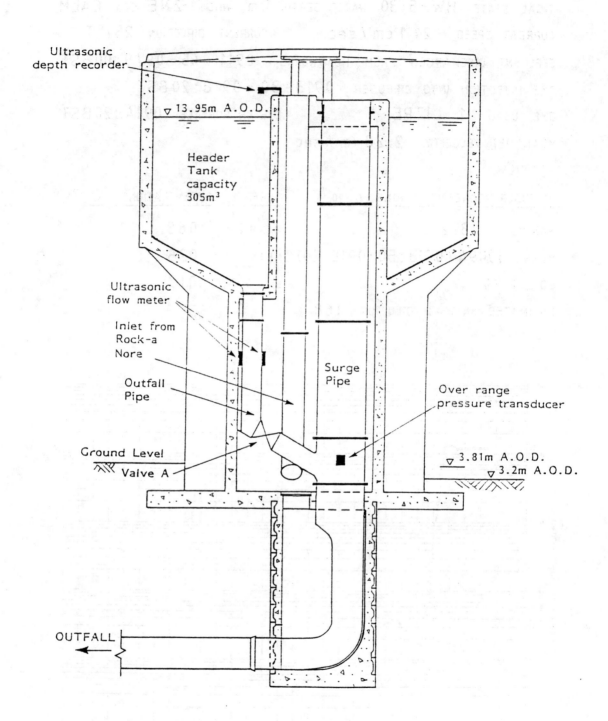

Ultrasonic
depth recorder

▽ 13.95m A.O.D.

Header
Tank
capacity
305m³

Ultrasonic
flow meter

Inlet from
Rock-a
Nore

Outfall
Pipe

Surge
Pipe

Over range
pressure transducer

Ground Level

Valve A

▽ 3.81m A.O.D.
▽ 3.2m A.O.D.

OUTFALL

LAYOUT OF HEADWORKS TOWER   Figure 4

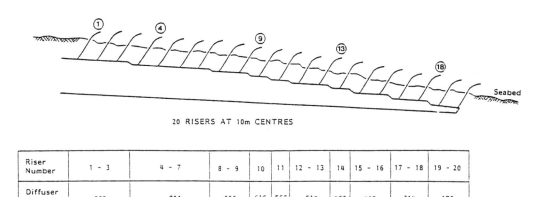

20 RISERS AT 10m CENTRES

Riser Number	1 - 3	4 - 7	8 - 9	10	11	12 - 13	14	15 - 16	17 - 18	19 - 20
Diffuser Pipe $\varnothing_i$	863	711	666	615	565	514	463	412	314	176

LONG SECTION OF DIFFUSER SHOWING LOCATION OF
RECORDING PACKAGES DURING HYDRAULIC FLOW SURVEY   Figure 5

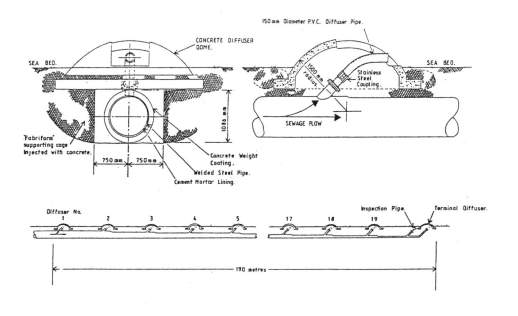

A TYPICAL DIFFUSER PORT   Figure 6

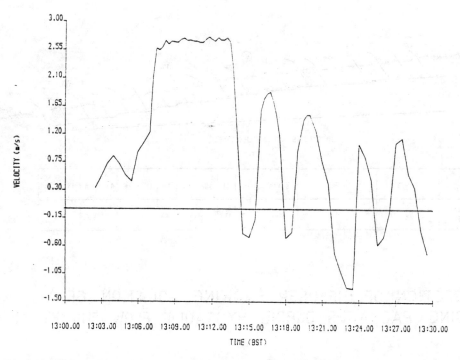

FLOW REVERSAL IN DIFFUSER FOLLOWING SHUTDOWN
OF OUTFALL VALVE

Figure 7

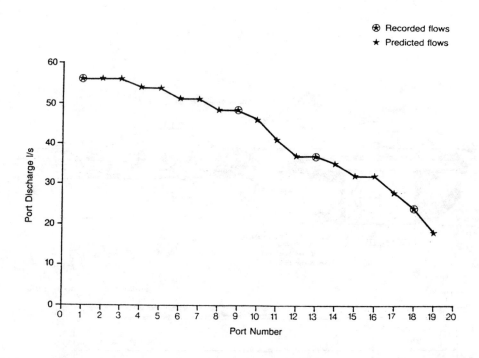

TYPICAL DISCHARGE THROUGH DIFFUSER PORTS AT
A FLOW OF 800 l/sec

Figure 8

Chapter 13
USES OF REMOTE SENSING IN COASTAL WATER MONITORING

I S Robinson
Southampton University, UK

Summary

Remote sensing from airborne and space platforms is providing a new
perspective for monitoring water quality conditions in estuaries and
coastal waters. This paper reviews the capabilities and limitations
of remote sensing as a marine measurement technique and points out the
unique contribution it can make in studies of coastal waters,
complementing conventional monitoring methods. Its ability to provide
a synoptic, spatially detailed view can be exploited in new ways,
despite the restricted capacity for high frequency time series of
measurements. The size, shape and orientation of hydraulic phenomena
such as discharge plumes, fronts, eddies, highly turbid suspended
sediment patches and even surface wave refraction can readily be
measured.

A brief overview is provided of the range of methodologies available,
using sensors operating in the visible, infra-red and microwave parts
of the spectrum. The space-time sampling characteristics of the
methods are defined, identifying the trade-off between spatial and
temporal resolution. An outline is given of the processing steps
necessary to convert the raw data into a product of use to the coastal
studies engineer or estuarine scientist.

Examples are presented of different types of data product, including
sea surface temperature maps based on satellite infra-red sensors,
sediment and discharge plume images from satellite ocean colour
scanners, and detailed surveillance of an estuarine plankton bloom
using airborne colour scanners. The potential for the engineering
application of these data is discussed in relation to the monitoring
of coastal civil engineering works, the siting of power station
outfalls, the dispersion of effluent discharges and the provision of a
baseline against which to compare the environmental impact of coastal
modifications.

With the anticipated launch of several new earth monitoring satellites
in the 1990s the future for coastal remote sensing looks promising, if
commercial applications can be successfully developed.

1.   Introduction

Whilst for many years aerial photography has been available to the civil engineer for surveying, cartographic and hydrographic purposes, remote sensing has only recently become a measurement tool of useful accuracy and precision, following the technological impetus provided by the development of space remote sensing techniques. For a decade, ocean scientists have had access to data from satellite sensors designed especially for marine use, and have learned also to exploit data from land-use sensors. Now that new scientific methodologies have been produced to interpret remotely sensed marine data, it is time to develop ways of using them to meet the needs of offshore engineering activities.

This paper offers an overview of those techniques of remote sensing from satellites and aircraft which, actually or potentially, have application in the modelling and measurement of estuaries and coastal waters. Because the range of methods is quite wide, the subject is discussed in broad terms rather than analysing a particular technique in detail, although specific examples are given of certain applications. The paper aims to demonstrate that remote sensing provides instruments for measuring the sea which are complementary to those used in situ by marine scientists and engineers. Its objective is to encourage coastal engineers to consider how remote sensing methods may be used in their particular field of application, and to stimulate further collaboration between remote sensing specialists and the wider "user community" of coastal water quality engineers and scientists.

2.   Remote sensing methods

2.1   Fundamentals

Marine remote sensing methods utilise the visible, infrared and microwave parts of the electromagnetic spectrum, wherever "atmospheric windows" allow the radiation to penetrate without appreciable absorption or scattering in the atmosphere (Figure 1). Consequently a wide range of different types of active and passive sensors have been developed to make measurements of such diverse ocean properties as colour, temperature, surface roughness and surface slope. These are described in comprehensive texts on satellite oceanography (1,2,3,4) but here we consider only the methods applicable to the coastal zone and rule out techniques such as satellite radar altimetry, scatterometry and microwave radiometry which cover too wide a field of view to adequately resolve nearshore variability.

2.1.1   Visible wavelength methods

Visible wavelength techniques (Figure 2) measure solar radiation reflected from the upper layers of the sea. Some light is reflected at the surface itself, and this can be used to detect wave patterns etc. but is usually treated along with the light backscattered from the atmosphere as needing to be removed by "atmospheric correction" in order to obtain the signal which is the water-leaving radiance. Sunlight penetrating the sea is both absorbed and scattered by the seawater itself and its optically active constituents, such as suspended inorganic particulate matter, phytoplankton and yellow substance (dissolved organic detrital material). Clear seawater absorbs most strongly at the red end of the spectrum, and preferentially but weakly scatters blue light, so that from above it appears a dark blue colour. The presence of backscattering particulates greatly increases the light reflected back to the sensor, but if there is chlorophyll present its characteristic absorption spectrum removes the blue light to give a green appearance. Similarly yellow substance has a characteristic absorption spectrum (Figure 3).

Thus by careful analysis of the spectral properties of the subsurface reflectance, R_λ, it ought to be possible to make measurements of the concentration of water constituents such as yellow substance, suspended particulates and chlorophyll concentration. The inherent optical properties, absorption a_λ and backscattering coefficient b_λ, are the sum of the contributions from the individual water constituents. However, although approximately $R_\lambda = b_\lambda/a_\lambda$, the exact value of R_λ and its variation with colour depends on the illumination and viewing geometry. Thus a simple inversion of the theory to yield a calibration of sea colour in terms of water quality parameters is not possible, but empirical calibration can be performed. For example (5) in the open ocean $\log_{10} C$ can be estimated to an accuracy of 0.5.

Until recently sensors sampled in no more than four or five wavebands across the visible part of the spectrum, and useful calibration was feasible only in waters containing a single type of constituent (e.g a phytoplankton bloom but no suspended sediment). Advanced sensors are now being developed which sample in more than a hundred narrow bandwidths less than 5nm wide. With such spectral precision, Grassl (6) considers that it should be possible to resolve the contributions of different water constituents even though the subtle changes of colour elude the human eye. With such spectral sensitivity it is also possible to detect the strength of the chlorophyll fluorescence peak above the background at 685nm, and hence obtain an independent measure of chlorophyll concentration.

The depth to which visible light penetrates depends on the optical attenuation coefficient. The stronger the backscatter and absorption, the weaker the penetration. In the turbid waters of the Severn Estuary, for example, the reflected light observed by a sensor comes from within the top metre of the water column, whereas in the clearest waters the signal is representative of conditions in the upper 20 m. Care must therefore be taken in interpreting any measurements of water quality parameters made from remote colour sensors, since they cannot reveal the effect of vertical structure. Another factor which must be accounted for is the bottom reflection which alters the apparent water colour if the sea is shallower than the optical penetration depth. This can, of course, be put to good use in bathymetric surveying applications.

2.1.2   Infra-red sensing.

Seawater strongly absorbs electromagnetic radiation at increasing wavelengths beyond the red end of the spectrum, and sensors operating in the 700-800 nm waveband record zero water-leaving radiance except in the most turbid regions where there is sufficient material in suspension to backscatter light from the upper few centimetres. Thus near-infra-red measurements are of no marine value, but in the thermal infra-red ($>7\mu$m) region, sensors are used to measure sea-surface temperature. Because the solar radiation at these wavelengths is almost completely absorbed by the sea, the measured upwelling radiation is that emitted by the sea itself, in proportion to the fourth power of the temperature. Thermal i-r radiance can readily be calibrated in terms of the brightness temperature of the sea surface. Using two or more thermal wavebands, corrections for the atmospheric absorption can be made. Hence surface temperatures can be measured with an absolute accuracy better than 1 degC and a sensitivity within an image of 0.2 degC.

It is the skin layer, the upper few tens of microns, which controls the radiation. Thus in stratified water the temperature measurement is representative of the upper layer only. Moreover, as figure 4 illustrates, there are two possible effects which can prevent the remote sensor from seeing what would be measured from a boat as the "bucket" sea surface temperature. These are the diurnal thermocline which can develop on calm sunny days but breaks down at night, and the temperature deviation caused by heat flow through the surface skin where turbulent heat transfer processes are inhibited. The skin may be a few tenths of a degree cooler than the temperature measured by a probe in the water. Such factors do not seriously hinder the use of remotely sensed sea surface temperature (7), but need to be considered when image data are being interpreted.

2.1.3 Radar backscatter.

Another remote sensing methodology of particular value in the coastal zone is the use of active radars. Radar pulses are emitted obliquely and an image is derived of the "radar backscatter cross-section" which is a measure of how much energy is reflected from different parts of the sea surface. This measurement is harder to interpret than colour or brightness temperature, but such images are capable of revealing a variety of phenomena of interest to coastal engineers. There is still uncertainty about the physical mechanism best used to describe radar interaction with the sea surface. It has often been assumed that Bragg scattering occurs, but it is also likely that a small amount of specular reflection occurs from normally oriented surface facets, and the true mechanism is undoubtedly quite complex. Nonetheless, it is clear that the rougher the surface, the stronger the backscatter, and probably the interaction is most effective when the wavenumber of surface roughness is of the same order as the radar wavenumber horizontal component. Thus. for example, a C-band radar at 23° incidence will interact strongly with 8cm waves on the surface. It is how the surface roughness at this and longer length scales is modulated by dynamical processes, or the presence of surface films etc, which enables the radar to image those modulating features.

In the open ocean, the magnitude of radar backscatter can be used to measure the wind speed, but such an application is not readily adaptable to nearshore regions. There are several other ways of using radar which also lie outside the scope of this paper, including the measurement of the doppler shift of return pulses which enables surface currents to be measured from shore-based HF radars.

2.2 Techniques of marine remote sensing

2.2.1 Platforms

Aerial remote sensing can be performed from aircraft or satellite platforms orbiting in space. Because most of the processes in estuaries and near the shore vary over short length scales (eg. <100m) it could easily be assumed that aircraft offer the most suitable platform for marine coastal surveillance. They can fly quite low to avoid problems of high cloud and to achieve very high spatial resolution. They can also repeat overflights at frequencies of several per hour if required. However, because of the low vantage point the viewing angle varies considerably across an image, and there is geometric distortion which is sensitive to fluctuations in the aircraft's attitude. Away from the coastline it becomes increasingly difficult to geographically coordinate the image.

Satellite platforms overflying the UK are in near polar orbit, inclined at about 100° to the equatorial plane, at a height of between 700 and 900 km. Thus the whole of a particular coastal area is viewed with the same geometry, and geometric correction is much more accurate from a stable platform. With a period of around 100 min the satellite orbits about 15 times per day. Orbits are usually arranged to be sun synchronous, so a particular latitude is always crossed at the same local (solar) time. The frequency of overpasses depends on how wide a swath the sensor views on either side of the sub-satellite point. A wide enough swath will overlap with the previous orbit, ensuring each point has at least one daytime overpass per day (and one at night), as shown in figure 5a. A narrow swath sensor (figure 5b) must wait several days to be able to fill in the gap between swaths of adjacent orbits, leading to a revisit time of typically 15 to 30 days.

2.2.2 Sensors

Visible and infra-red radiance measurements are made by scanning radiometers. An optically focussed field of view scans normally to the flight direction and builds up an image of pixels (Figure 6). Radiances are stored for several spectral bands for each pixel. It is worth emphasising that the resulting multi-dimensional image dataset is far more than a map or chart, since it consists of up to several million individual precise measurements of radiance, which can each be interpreted in terms of the desired marine variable. Normally some facility for calibrating the sensor is incorporated into the scan-cycle. The same basic principles are employed for airborne and spaceborne sensors, although the latter have to be engineered to very high standards of reliability, operate in a more uniform environment and tend to yield more precise measurements than the former.

The newest generation of sensors comprise CCD arrays. A two-dimensional array enables instantaneous cross-track scanning with one axis of the array and, after suitable angular dispersion of different wavelengths, spectral scanning in the other axis. Active, laser devices are also being deployed from aircraft. The laser beam is used either to penetrate through the sea to the sea bed and hence provide a bathymetric tool, or alternatively it can be used to stimulate light emission which is measured with a sensitive spectrometer in order to identify the water constituents.

Since radar has too long a wavelength to be focused optically high spatial resolution on radar images must be achieved by signal processing. Range-gating achieves high resolution in the radar pointing direction, normal to the flight direction. Resolution along the flight direction (azimuth) can be obtained from aircraft by constructing a narrow beam and emitting frequent pulses. The distance to the ground is too great for this to be feasible from a satellite, but high azimuth resolution can also be achieved by synthetic aperture radars (SARs) which construct the final image by complex processing of many returns, using phase and amplitude information. Because the platform speed affects the image reconstruction, airborne and satellite SARs have very different characteristics.

Table 1 lists some typical remote sensing instruments of relevance to coastal studies, and describes their spectral and spatial sampling capabilities.

2.3   Space-time sampling characteristics

Figure 7 summarizes the space-time sampling frequencies of various kinds of sensor, located as different areas on the figure. The left hand boundary represents the temporal sampling frequency. For sensors affected by cloud cover this is the most optimistic possibility for cloud-free conditions. The right hand boundary indicates the maximum span of time over which sampling can be continued, which for an operational satellite sensor may be indefinite provided a series of space vehicles is provided.

The lower boundary corresponds to the spatial resolution cell size, given as a linear (left ordinate) or as an area (right ordinate) scale. The upper boundary denotes the maximum length or area that can be observed on a single (nearly-) synoptic image. The dashed lines denote the non-synoptic spatial coverage which is possible from the total orbit coverage of the satellites carrying the sensors.

Comparison between the high and medium resolution satellite sensors shows the trade-off between space and time resolution. This is incompatible with the normal requirement in the marine environment to observe smaller scale processes at higher temporal frequency. Clearly satellites cannot adequately resolve the time variability of most estuarine and coastal processes. Aircraft can fill the sampling gap to some extent, but in situ measurements from vessesls, buoys or fixed platforms are necessary to define the detailed time-evolution of a phenomenon. However, figure 7 also clearly demonstrates that in situ methods are unable to provide a synoptic spatially detailed view of a region. Thus it is evident that remote sensing and conventional measurements should be regarded as complementary techniques, and their application should be approached in this light.

2.4   Data processing tasks

Before remotely sensed data can be applied, various stages of processing must be performed depending on the required image product. Generally visible wavelength and infra-red data are supplied by satellite data agencies, or from aircraft scanners, in digital form based on the radiances observed at the sensor. Figure 8 summarises in a general way the processes which must be applied to convert the raw data into a resampled image (in regular geographic coordinates) of an oceanic variable such as surface temperature, chlorophyll concentration or suspended sediment load. Alternatively an intermediate optical measurement (e.g. blue/green colour ratio) may be the final product for situations where chlorophyll or sediment algorithms cannot be relied upon. Often the quickest and easiest product to acquire is an analogue image of the uncorrected raw data. This can provide a rapid overview of the utility of the data, and is often used for qualitative analysis, but to use only this is to ignore the wealth of information contained in the digital signal.

3.   Application of remotely sensed data to coastal studies

3.1   Phenomena which can be observed by remote sensing

To become a tool for applied marine science, remote sensing must be capable of directly observing processes of interest in coastal regions. Given the fundamental measurements which can be made from airborne or satellite remote sensors, i.e. colour, brightness temperature and radar roughness, table 2 lists the various marine phenomena which can be detected. There are many features which have a colour or thermal signature, particularly those relating to the inflow, entrainment and dispersion of one water type in another, e.g. river outfall plumes into the sea (Figure 9), polluting discharges into estuaries, or tidal intrusion fronts in estuaries and rivers. Provided the two bodies of water have different colours or temperatures, the near-surface expression of their mixing characteristics can be studied remotely, although vertical structure can only occasionally be inferred, e.g. from comparisons between thermal and colour signatures.

The fate of material in the water can also be followed by its colour signature, e.g. the development of phytoplankton blooms (Figure 10), or the resuspension of bottom sediments into the upper layers and their subsequent transport and dispersion (Figure 11). The more buoyant a tracer, the more likely it is to have a surface signature. Material actually at the surface in the form of slicks and floating rubbish has a different effect. It may change the surface reflection properties rather than the underlying sea colour. It may reduce the thermal emissivity of the surface and thus reduce the apparent brightness temperature, although in such a case it would tend to absorb more solar radiation and so its actual temperature could be raised above that of the surrounding sea. Surface slicks also

significantly change the surface roughness, and thus can appear on
radar images. Oil spills can be readily detected from aircraft, but
not so easily from satellites. A range of complementary techniques
including ultra-violet, infra-red and microwave sensors have been
developed for operational monitoring of oil spills (8). This is a
specialist subject in its own right and falls beyond the scope of this
paper.

Applications of radar images are the least easy to predict, since the
subtleties of surface roughness variability are not so readily
understood. It appears that under the appropriate surface wind
conditions, small changes in the near-surface currents can modulate
the surface roughness to an extent which is detectable on radar
images (9,10). Such current structure may be due to the underlying
bathymetry, in which case the radar image reflects aspects of the
sub-sea topography (Figure 12), or may be due to internal wave motions
or other dynamical processes. Long surface waves themselves modulate
the radar detectable roughness ripples, so that radar images can
provide instantaneous views of wave refraction patterns impinging on
coastlines. There remain other features on radar images yet to be
fully explained, which may lead to further applications.

3.2   How to exploit remotely sensed data

To achieve the maximum potential of remotely sensed data it is
important to identify what unique contributions they provide in
comparison with other measurement techniques, and to exploit that
uniqueness to the full. Evidently remote sensing can only measure a
restricted set of variables, limiting the phenomena which can be
observed. It is impossible with satellites, and difficult with
aircraft to obtain high frequency time series, and except with radar
sensors the problem of cloud cover precludes the possibility of
guaranteed operational sampling. What remote sensing methods can
provide is a synoptic spatial overview of an area, with high spatial
resolution over two or three decades of length scales. In the case of
satellites this can be repeated consistently by the same sensor over
spans of several years. The synoptic view cannot be achieved by any
other means, and it is difficult to maintain a consistent dataset by
conventional measurements over several years.

Consequently the most significant applications of remote sensing
should be those which use the detailed spatial information, and/or
which monitor changes over several years. Because these are new
possibilities, it may require a process of education before applied
marine scientists and coastal engineers learn to derive the full
benefit in commercial applications. Many processes are conventionally
studied in terms merely of their time evolution, corresponding to the
time series at a few discrete points which has been the conventional
sampling method, with little attempt to consider the spatial
structures of the process. The addition of an occasional spatially
detailed view to a programme of in situ observations provides the
opportunity, inter alia:
-to place the discrete sampling points in their spatial context;
-to measure length scales of processes directly instead of by
 inference from time scales;
-to understand and if necessary correct for the aliasing of temporal
 frequency measurements by the advection of spatial structures past a
 sampling station;
-to detect changes in spatial patterns of a tracer over time spans
 from hours to years.

3.2.1   Complementing field measurements

Remote measurements on their own often lead only to qualitative
interpretations, but when used to complement a field sampling
programme (which can also be used to calibrate the image data) the

total benefit can be far more than the sum of either technique applied by itself. For example:
-the study of a water quality or sedimentation problem in a previously unsurveyed location can avoid a costly preliminary large-area coarse survey by analysis of a few historic satellite images of the region to identify the problem areas and to direct the detailed survey work;
-remote sensing can reveal quite subtle variations of water quality (in ocean colour images) or temperature (in thermal i-r images) to reveal the extent of a dynamical feature, which might have too small a signature to detect above the noise of in situ measurements. It is the spatial coherence of such subtle changes which enables the eye to detect patterns even when there is noise of comparable magnitude.

3.2.2   In relation to estuarine and shelf sea models

Another primary way to exploit remotely sensed measurements is through models (11,12). There is a particularly close match between the character of image datasets and two-dimensional finite difference model fields:
-Both represent variables as an average over a cell.
-The cells are contiguous and arranged in a rectangular grid, providing spatial structure over a large area.
-The whole field is viewed (nearly) instantaneously.
-A representative variable is used to incorporate any depth dependance (in the model it may be the depth average, in the image it is the near-surface value)

Models and remote sensing can be used to complement each other in the study of a problem. The model normally has a time-step much more rapid than the image sampling frequency, and can therefore be used to interpolate in time between infrequent images. The image can be used to provide realistic initial conditions for a model. It can also test the validity of model predictions of the evolution of spatial patterns, far more thoroughly than could a few point measurements which may not even be truly representative of the cell to which they are allocated.

The matching of models and image data has not yet been extensively attempted, but Figure 13 shows some comparisons from an attempt (13) to match a model of the (vertically well mixed) thermal structure of the English Channel to satellite-derived temperature fields. Figure 13a (23 Oct 1983) was used as initial condition for the model which simulated several months of heat transfer processes. Comparison between satellite observations (b) and best-fit model predictions (c) of temperature are shown for 4 Dec 1983 after the model has run for some six weeks. Some problems were encountered in parameterising the surface heat flux but the comparison of the modelled temperature field with satellite data was able to determine the most suitable model representations of advection and anisotropic diffusion which would not have been possible using sparse in situ measurements. Such information can now be used in water quality and pollutant transport models of the S. North Sea / English Channel, an example of an indirect application of remote sensing to an operational problem.

4.   Further examples of marine remote sensing applications.

We now briefly present some further examples of studies where remote sensing has been applied to coastal and shallow sea problems.

4.1   Visible wavelength (sea colour) data

Figure 14 is a CZCS image of the English Channel which has been analysed (14) as part of a series, along with in situ drifter experiments, to reveal patterns of residual flow and their seasonal changes. This information is relevant to the prediction of tracer

dispersal in the English Channel, and the possible tendency for
coastal discharges to be inhibited from dispersing out to the centre
of the Channel.

At a higher resolution, Landsat TM images of the Dover Straits show
patterns of water turbidity aligned with the tidal streams, probably
due to suspended particulate material. Several images from the
archive of over ten years Landsat data have been used to provide a
baseline of water turbidity conditions off Dover before engineering
work started on the Channel Tunnel. It is considered feasible (15) to
use remote sensing to monitor any changes to the overall turbidity in
the area which might occur when excavated material is deposited on a
coastal site. An example of how high concentrations of suspended
material can act as dynamical tracers is shown in figure 15, where the
eddy off Dungeness is plainly revealed in a TM image.

Figure 16 is a series of chlorophyll distribution maps for a small
area off swansea Docks within Swansea Bay, based on image data from an
airborne scanning radiometer (16). Calibration of the images was
achieved using in situ samples collected by helicopter in an exercise
coordinated by the Welsh Water Authority in 1984. Four overflights
were made during a single tidal cycle, and help to detect the
occurence of blooms associated with sewage outfalls.

Figure 17 illustrates a new and potentially useful application of
large area CZCS data. Two images recorded 24.5 hours apart have been
compared (17). Using a maximum correlation matching routine, the
movement of patterns in the water turbidity field from one day to the
next have been defined and plotted as flow vectors. Care is needed in
interpreting these vectors, but they promise a much more spatially
dense measurement of residual flow fields than can be achieved by in
situ methods.

4.2 Thermal infra-red data

An example has already been presented of how thermal data can be used
to assist in the construction of tracer and water quality models.
Figure 18 illustrates how AVHRR data can be used to locate thermal
fronts, in this case the development in May, 1980, of the boundary
between stratified and unstratified water in the Irish Sea. As well
as affecting water quality considerations, the presence of
stratification in a tidal sea may influence the vertical structure of
tidal currents and hence the loading on offshore structures. Thermal
images of the North Sea (Figure 19) show a number of features,
including a warm coastal fringe off the English coast due to the
Humber and Wash discharges, and the more extensive discharge plume of
the Rhine. A sequence of images is being studied to gain
understanding of how this water eventually disperses into the main
body of the North Sea.

At a more local scale, airborne thermal images can be used to locate
the surface expression of a power station cooling water discharge
plume, and its movement over a tidal cycle.

4.3   Radar images

Application of radar images to coastal marine problems is still in its
infancy, with so far only the experience of Seasat and a few airborne
experiments with which to work. In stratified waters SAR images can
reveal the horizontal structure of internal waves, their associated
currents modulating the roughness so that the radar can see the waves
which occur on the interface up to 50m below the surface. Their
detection in this way has operational applications for offshore
operations which could, for example, be affected by the loading of
internal waves on underwater structures.

Surface wave modelling will be able to make use of SAR images of refraction patterns. Updating of maps of shifting sandbanks is also a possibility in those locations where flow conditions enable bathymetry to be imaged. Other possible applications of radar images yet to be developed include the tracing of slicks associated with fronts, plumes and polluting discharges.

5. Conclusion

This paper has sought to outline the broad scope of possible ways in which remote sensing can contribute to offshore engineering and coastal pollution studies. As yet there are few examples of wholehearted adoption of the techniques for other than purely scientific experiments. However, if the synoptic spatial sampling capabilities of remote sensing are exploited in conjunction with conventional methods, there is much to be gained, in terms both of new monitoring capabilities and more cost effective achievement of existing goals. Whilst the overall cost of placing satellites in orbit is very high, it is shared over a multitude of land, ocean and atmospheric appplications, and the cost of image data to the user is very modest if it can be used in a quantitative manner.

Following a decade in which experience has been gained with prototype sensors, the 1990s will see several space agencies launching semi-operational marine monitoring satellites. For a modest investment in data purchase, development of new interpretation techniques and training, the marine technology and environmental sector can have access to this promising technology. If a clear forecast can be made now of commercial user requirements, there is still the opportunity to influence the design of the next generation of sensors including, for example, much more spectrally sensitive colour scanners which may be able to monitor a wider range of polluting substances in the sea.

6. Acknowledgements

The author is grateful to his colleagues in the satellite oceanography group at Southampton University for their contributions to the processing of some of the data presented in this paper.

7. References

1. Robinson, I.S.: "Satellite Oceanography". Chichester, Ellis Horwood, 1985, 455pp.

2. Stewart, R.H.: "Methods of Satellite Oceanography". University of California Press, 1985.

3. Maull, G.: "Introduction to Satellite Oceanography". Dordrecht, Martinus Nijhoff, 1985.

4. Saltzman, B.: "Satellite Oceanic Remote Sensing", Advances in Geophysics, Vol. 27. Orlando, Academic Press, 1985.

5. Gordon, H.R. and Morel, A.Y.: "Remote Assessment of Ocean Color for Interpretation of Satellite Visible Imagery. A Review". Lecture Notes on coastal and estuarine studies, 4, 1983, New York, Springer Verlag.

6. Grassl, H.: "Use of chlorophyll fluorescence measurements from space for separating constituents of sea water". In: "Ocean Colour - Report by ESA Ocean Colour Working Group", Paris, European Space Agency, SP-1083, 1987, p103.

7. Robinson, I.S., Wells, N.C. and Charnock, H.: "The sea surface thermal boundary layer and its relevance to the measurement of sea surface temperature by airborne and spaceborne radiometers". Int. J. Remote Sensing, 5, 1984, pp19-45.

8. see Hurford, N. and Tookey, D.: "A detailed evaluation of the maritime surveillance system for oil slick determination". Oil and Chemical Pollution, 3, no3, 1987, 231-244,
   and Schriel, R.C.: "Airborne surveillance: the role of remote sensing and visual observation". Oil and Chemical Pollution, 3 no3, 1987, 181-190.

9. Fu, L-L. and Holt, B.: "Seasat views oceans and sea ice with synthetic-aperture radar". NASA JPL Publication 81-120, 1982.

10. Kasischke, E.S., Meadows, G.A. and Jackson, P.L.: "The use of synthetic aperture radar imagery to detect hazards in navigation". ERIM Report No.169200-2-F, Washington, Defence Mapping Agency, 1984.

11. Robinson, I.S.: "Applications of remotely sensed image data to marine modelling". In "Focus on Modelling Marine Systems", (ed. A.M. Davies), C.R.C. Publishers, Vol 1, 1988.

12. Balopoulos, E.Th., Collins, M.B. and James, A.E.: "Satellite images and their use in the numerical modelling of coastal processes". Int. J. Remote Sensing, 7, 1986, p905.

13. Daniels, J.W.: "Dispersal in shallow seas inferred from remotely sensed infra-red imagery". PhD Thesis, Umiversity of Southampton, 1986.

14. Boxall, S.R. and Robinson, I.S.: "Shallow sea dynamics from CZCS imagery". Advances in Space Research, 7, 2, 1987, pp37-46.

15. Robinson, I.S. and Boxall, S.R.: "Remote sensing of suspended sediment - A study of its feasibility for monitoring the impact of channel tunnel spoil disposal". Unpublished contract report for Eurotunnel, Southampton University Department of Oceanography, Jan 1987, 53pp.

16. Garcia, C. and Robinson, I.S.: "Chlorophyll a mapping using the Airborne Thematic Mapper off the South Gower Coastline". Submitted to Int. J. Remote Sensing.

17. Garcia, C. and Robinson, I.S.: "Measurement of flow velocities from successive CZCS images using a maximum correlation technique". Submitted to J. Geophysical Res.

Table 1: Sensors relevant to coastal marine applications.

Sensor	platform	wavebands μm	spatial resolution	swath width
Thematic Mapper (TM)	Landsats 4 and 5	0.45-0.52 0.52-0.60 0.63-0.69 0.76-0.90 1.55-1.75 2.08-2.35 10.4-12.50	30m 120m	185km
Multi-spectral scanner (MSS)	Landsats 1 to 4	0.5 - 0.6 0.6 - 0.7 0.7 - 0.8 0.8 - 1.1	80m	185km
Coastal Zone Color Scanner (CZCS)	Nimbus-7	0.433-0.453 0.510-0.530 0.540-0.560 0.660-0.680 0.70 -0.80 10.5 -12.5	825m	1600km
Advanced Very-High Resolution Radiometer (AVHRR)	TIROS N, NOAA 6,8, 10	0.58 - 0.68 0.725- 1.10 3.55 - 3.93 10.5 -11.5	1.1km	2580km
AVHRR/2	NOAA 7,9	0.58 - 0.68 0.725- 1.10 3.55 - 3.93 10.3 -11.3 11.5 -12.5	1.1km	2580km
High Resolution Visible Sensor (HRV)	SPOT	multi-spectral mode 0.50 - 0.59 0.61 - 0.68 0.79 - 0.89 panchromatic mode 0.51 - 0.73	20m 10m	117km 117km
Synthetic Aperture Radar (SAR)	Seasat	Active radar	25m	100km
Daedalus Airborne Thematic Mapper	Aircraft	0.42 - 0.45 0.45 - 0.52 0.52 - 0.60 0.605- 0.625 0.63 - 0.69 0.695- 0.75 0.76 - 0.90 0.91 - 1.05 1.55 - 1.75 2.08 - 2.35 8.5 -13.0	2m-10m	1.5km-8km depending on aicraft height

Table 2: Coastal phenomena observable by remote sensing.

Primary measurement	Measurable quantities	Phenomena detected
Visible wavelength radiance (ocean colour)	Chlorophyll a conc. Suspended sediment Yellow substance Optical attenuation coefficient.	Phytoplankton blooms Sediment plumes Dispersion of land runoff Sewage discharge plumes Fronts Eddy structures Shallow bathymetry
Infra-red radiance (brightness temperature)	Sea surface temperature	Thermal effluent plumes Dispersal of river discharge into the sea Fronts Eddies Oil spills and slicks
Radar backscatter image	Radar cross-section	Swell waves Internal waves Surface slicks Ship wakes Bathymetric features

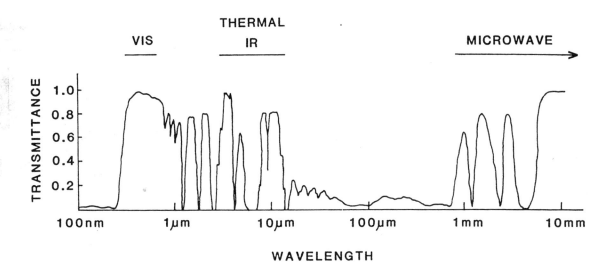

1. Approximate transmittance of electromagnetic waves through the atmosphere, showing the visible & near i-r, the thermal infra-red and the microwave atmospheric windows.

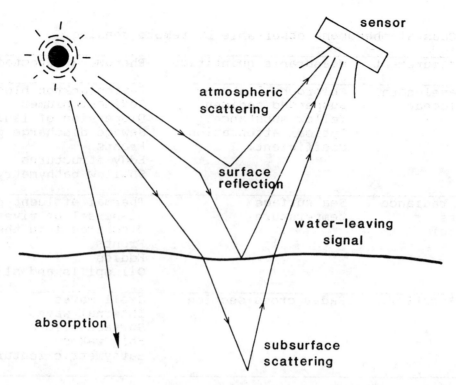

2. Sketch of optical pathways for sunlight reaching a visible wavelength sensor.

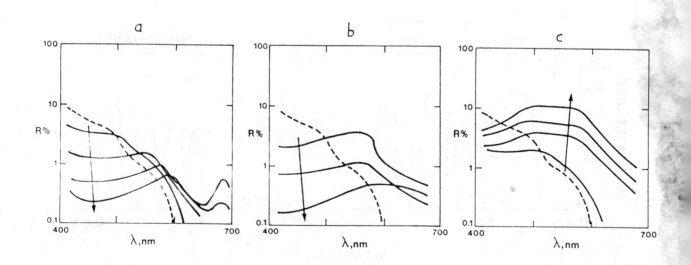

3. Typical reflectance spectra of sea water whose optical behaviour is dominated by (a) phytoplankton, (b) yellow substance, (c) suspended sediment. The different lines correspond to different concentrations of water content, increasing in the direction of the arrow. The dashed line indicates the clear sea water reflectance spectrum.

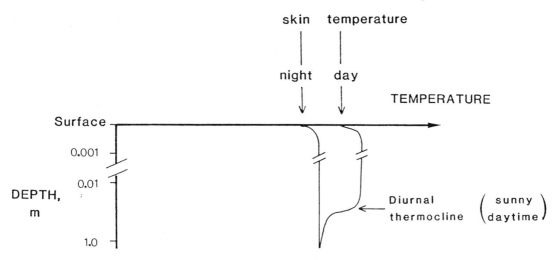

4. Typical thermal structure in the upper metre of the sea.

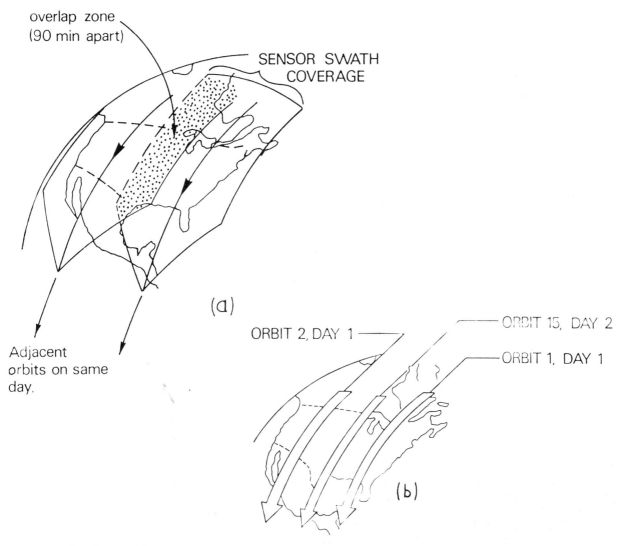

5. Typical swath coverage of adjacent orbits for (a) medium resolution wide field of view sensor, (b) high resolution narrow swath sensor.

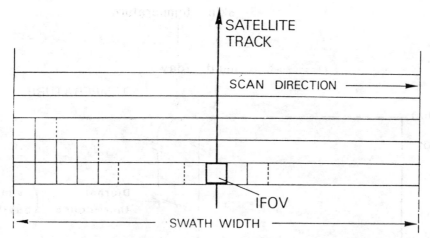

6. Image constructed from successive instantaneous fields of view by sensor scanning across the travel direction of the platform.

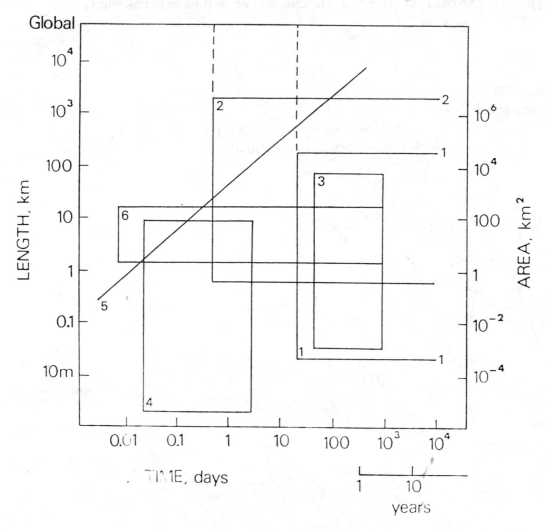

7. Space-time sampling characteristics of different sensors and instrument platforms (see text for further explanation)
(1) High resolution scanning sensor in polar orbit
(2) Medium resolution scanning sensor in polar orbit
(3) Satellite synthetic aperture radar.
(4) Scanning sensor on light aircraft
(5) Measurements from research vessel (no synoptic view, linear transect only)
(6) Measurements from an array of buoys.

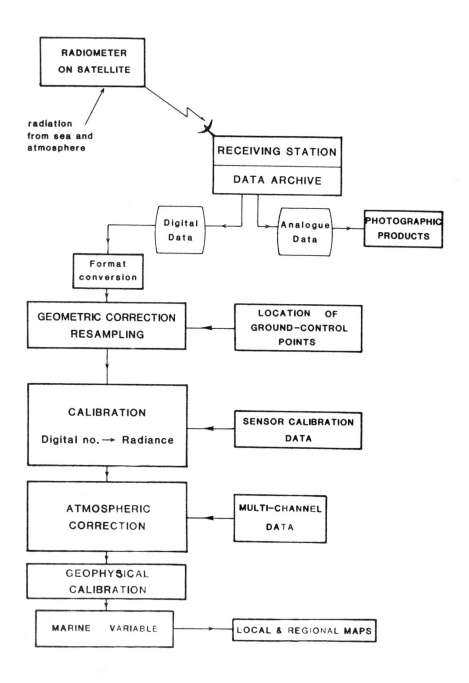

8.   Schematic of the processing tasks required to convert raw data from a scanning sensor into a geolocated, calibrated image representing a marine variable.

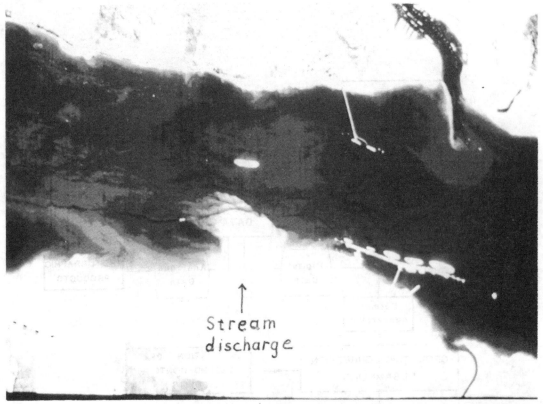

9. Image from thermal infra-red channel of airborne thematic mapper over Southampton Water, revealing amongst other features the dispersal plume from a small stream discharging from the south-western coast. In this image the darker shades correspond to lower emitted radiance i.e. cooler water.

10. Image from a visible channel of the airborne thematic mapper, the patchiness of which reveals the spatial distribution of the "red tide" organism, Mesodinium Rubrum, in Southampton Water, June 1987. This organism can be most clearly identified using a colour composite combination of several spectral bands.

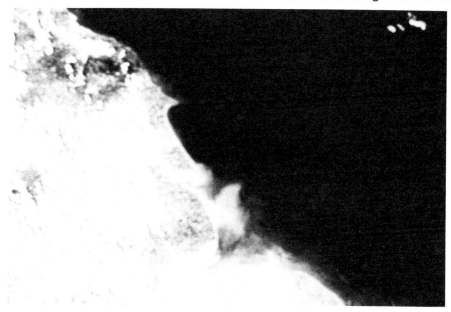

11.   Channel 1 AVHRR image (visible wavelengths) of the North Sea off
the Humber, 6th May 1988.   The patterns of lighter shading are mostly
due to the enhanced reflectance from suspended sediment.

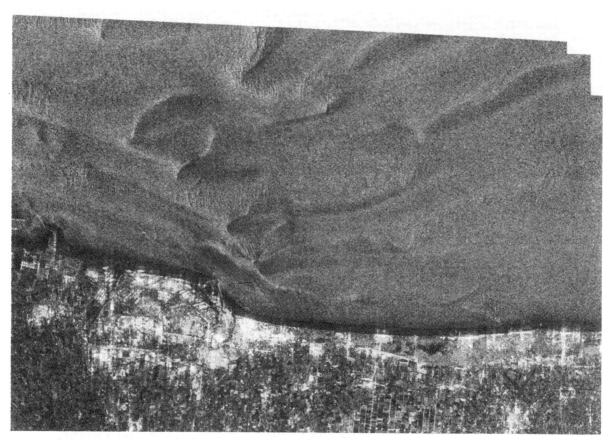

12.   Seasat synthetic aperture radar image revealing the signature of
bathymetric features in the Straits of Dover off Dunkerque.   Image
processed by R.A.E. Farnborough.

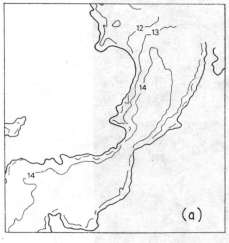

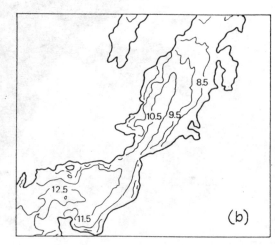

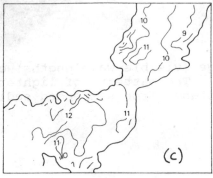

13.   Sea surface temperature maps based on satellite data and numerical models of thermal structure.
(a) SST map based on NOAA-7 AVHRR data, 0400, 23 Oct 1983. Used as initial condition for thermal model of E Channel / Southern Bight.
(b) SST map based on NOAA-7 AVHRR data, 1500, 4 Dec 1983.
(c) Model-derived SST distribution for 4 Dec 1983.

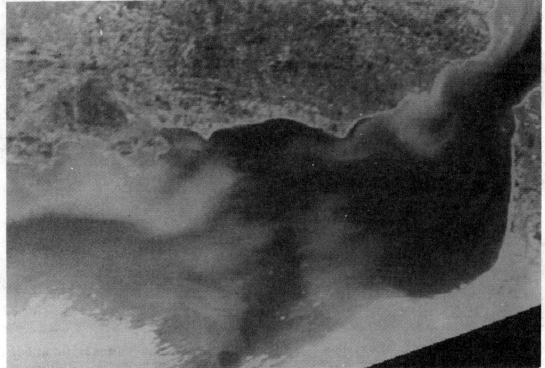

14.   CZCS channel 3 image of English Channel, the colour signature primarily due to suspended sediment, indicating significant variability normal to the coast. (May 1980)

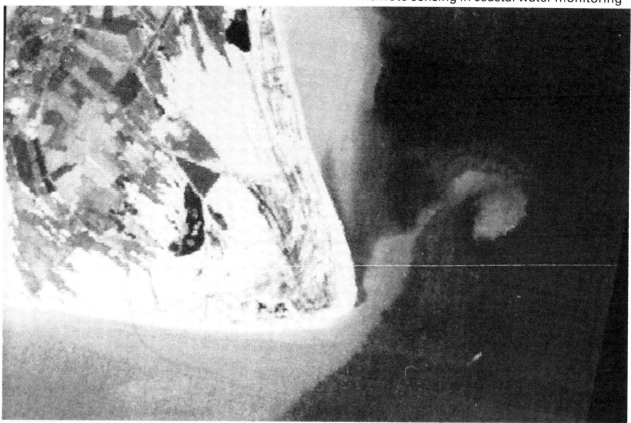

15.   Landsat Thematic Mapper image (Channel 3) showing a strong eddy feature with a turbidity signature off Dungeness.

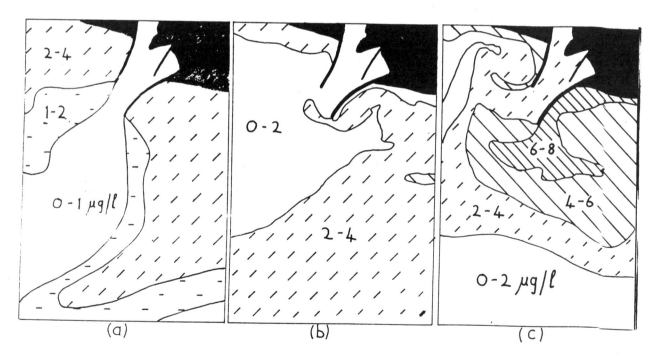

16.   Maps of chlorophyll concentration at the mouth of the River Tawe in Swansea Bay, based on airborne thematic mapper data calibrated in relation to synoptic in situ measurements, 8th June, 1984. (a) One hour before low water, (b) 1.5 hours before high water, (c) High water

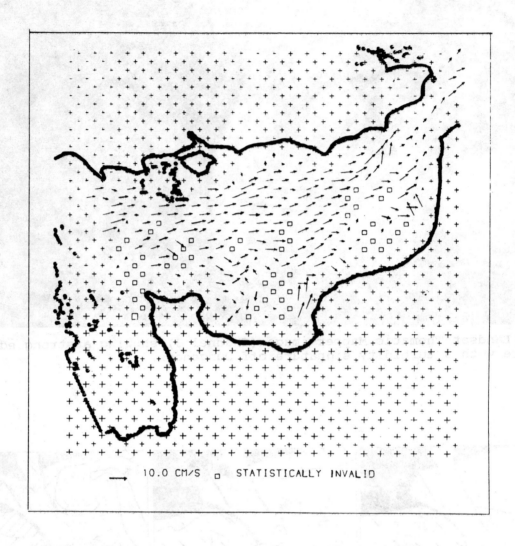

17.   "Residual Current" vectors in the English Channel, as deduced
from the movement of tracer patterns between the CZCS channel 3 images
of 10th and 11th May, 1980.

18.   AVHRR-derived SST map of the Irish Sea, 16 June 1980,
identifying the location of thermal gradients including the seasonal
stratification front south west of the I. of Man. In this image
warmer temperatures appear darker.

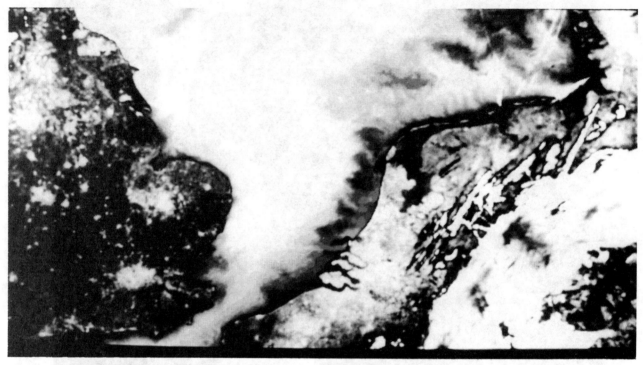

19.   AVHRR-derived SST map of the North Sea, 30 May 1985, showing the plumes and fringes of warmer water (darker shades) resulting largely from river discharges.

Chapter 14
THE APPLICTION OF NUMERICAL MODELS FOR SIMULATING THE WATER FLOW AT THE UPSTREAM SIDE OF STRUCTURES

F A van Beek and H W R Perdijk
Delft Hydraulics, The Netherlands

Summary

In hydraulic engineering and consultation practice mathematical models for the simulation of complex water flow patterns are widely applied, especially for so-called "far-field" problems, such as flow in rivers and estuaries. For this purpose one- and two-dimensional, and sometimes (quasi-)three-dimensional models are used, often in combination with models for the simulation of other phenomena, like sediment transport.

Although several two- and three-dimensional modelling systems are available for "near-field" problems, such as flow around piers, barriers and sluices, it is felt that more insight is needed into the practical applicability of mathematical models. An attractive approach to this problem is the comparison between mathematical and physical model results.

This paper reports on measurements and simulations of the flow in a straight canal with a pier in the middle of the downstream side. The experience thus obtained has lead to more knowledge on the accuracy of the simulation and, in coherence, on practical applications in mathematical modelling of near-field flow patterns.

1. Introduction

Mathematical modelling of the water motion has become widely applied for engineering purposes for several years now and the perspectives for practical application are promising. With the increased use of these models the results become more and more reliable and also the understanding of the physics of water motion is improving, because the mathematical models can be seen as complementary to the physical models. Nevertheless, substantial research efforts remain necessary to solve questions as to the physical, mathematical and numerical basis of these models.

Mathematical models for the simulation of the water motion can be classified according to their space dimensions and their spatial orientation. Consequently, the following models are distinguished: one-dimensional, two-dimensional vertical, two-dimensional horizontal and (quasi) three-dimensional. For application in engineering practice, the simulation of water layers of different density and the combination with specific models for the simulation of the temperature, sediment transport or sediment-concentration distribution are useful options.

The application of the models can be divided into far-field problems and near-field problems. Extended experience with various practical situations in the far field is available, such as the water and sediment transport in estuaries, rivers and canals and results have been verified with situations in nature. Here one-dimensional and two-dimensional models are mostly used. In such cases only a rough description of the structures is sufficient.

Although two- and three-dimensional models are already applied for near-field situations, it is felt that in particular applications for a detailed flow pattern near structures and comparisons with physical models and nature is still insufficient. One of the restrictions to applying mathematical models in engineering practice is that calibration constants still have to be established, so that forecasting without calibration by data from the past is as yet difficult. Therefore tests should be carried out to explore the applicability for the detailed flow pattern near structures. Results will show the relevance of these applications and should justify investments in the further development of these models.

The present paper consists of three parts:
1. consideration of the possibilities of mathematical models for applying to near-field problems;
2. presentation of the mathematical modelling system ODYSSEE used in the study;
3. discussion of the results of simulations at an elementary flow geometry (intake to sluice or power station).

2.   Application of mathematical models for flow around structures

The flow around structures can be distinguished in the far-field and in the near-field flow. Far-field flow can be seen as not influenced by the structure itself, both upstream as well as downstream of the structure. Logically, the near-field flow will be influenced by the structure.

Knowledge of the far-field flow will be necessary for overall flow patterns, dispersion or boundary conditions for the structure such as water levels, discharges, sediment transport etc. Usually one-dimensional or two-dimensional models can provide this information.

The near-field flow can be divided into the flow upstream of the structure, the flow through the structure and the flow downstream of the structure. Most of the practical applications concern complex situations. In these situations one-dimensional models do not have a strong predictive character, which is caused by the number of calibration constants.

For flow patterns upstream of the structure two-dimensional models can be applied successfully as far as the flow pattern is mainly two-dimensional. Problems have to be expected for flow patterns with horizontal as well as vertical eddies, such as the flow around a pier. In those cases complementary studies in physical models will be necessary to assess the simulated flow patterns of the two-dimensional models or models describing the phenomena in three dimensions are needed. In principle this will be achieved by solving the constituting differential equations on a three-dimensional grid, but in some cases it is sufficient to introduce only a quasi three-dimensional model. Although the work on three-dimensional models is in full progress, the fully three-dimensional models are not yet widely applied in engineering practice.

For the internal flow pattern, the wall influence, the small-scale turbulence and the occurrence of flow separation seem to be of great importance. The internal flow geometries are often quite complex. Although the application of two-dimensional models can be seen as a step forward, in particular for this type of problems the three-dimensional models will appear to be very suitable.

Application of two-dimensional models is also possible for the flow downstream of the structure, particularly when the occurring eddies mainly have a horizontal axis (2DV) or mainly a vertical axis (2DH). In these cases the energy dissipation is important, thus requiring a suitable turbulence model. For a combination of eddies with vertical and horizontal axis also models are needed describing the phenomena in three dimensions.

The application of these models in practical engineering problems demands sufficient knowledge of the physical phenomena, the mathematical description with their simplifications and assumptions, and the used numerical solution methods. Moreover, the model must be sufficiently tested, numerically as well as with results obtained from physical scale models or from nature. Finally, these tests will also give insight into the limitations of the mathematical modelling system.

In this paper attention will be given to near-field problems at the upstream side of the structure. An elementary flow geometry is modelled both in a physical model as in a mathematical model for the simulation of the flow in two dimensions. The present study is a part of a long-term research programme, in which the reliability of two- and three-dimensional mathematical models will be tested and improved.

3.   The mathematical modelling system ODYSSEE

3.1 Description

The applied mathematical model was developed in a close co-operation between the Laboratoire National d'Hydraulique in Chatou, France and DELFT HYDRAULICS. The program ODYSSEE is discussed extensively in (1). Here only a brief summary will be given.

The program solves the Navier-Stokes equations for unsteady flow in two dimensions. As a variant, the equations for depth-averaged flow in shallow water can be treated. For the latter the surface is treated as a rigid lid (the free surface is replaced by a free-slip plate). Turbulence closure can be obtained either by means of a constant eddy viscosity or a k-ε turbulence model. It is known from literature (2) that the k-ε turbulence model is able to achieve reasonable approximations for the turbulence distribution for a fairly wide range of flow situations.

The numerical method is based on finite differences, taking care of source terms, convection, diffusion and continuity. A boundary-fitted curvilinear grid is used including local grid refinement, allowing an accurate representation of the boundaries. A pre-processing program takes care of the grid generation. Graphical output in various forms can be obtained, which is essential for using the mathematical model in an effective way.

The usual types of boundary conditions are used, with velocity components specified at inflow, no-slip, free slip or law-of-the-wall conditions at fixed boundaries, and "free" conditions at the outflow.

3.2 Experience with the program

The ODYSSEE program is regularly applied for both research as well as engineering problems. The applications concern horizontal depth-averaged flow (also called 2DH) as well as vertical two-dimensional flow (also called 2DV). For both applications far-field flow patterns and near-field flow patterns were computed.

Examples of depth-averaged flow for far-field studies are, for instance: flow pattern in a river junction in the River Meuse, flow pattern in river bends, also coupled with a sediment model to compute the development of the bed profile (3), and a flow pattern near the harbour of IJmuiden.

Examples for two-dimensional flow patterns are for instance: the flow pattern around a cylinder, in a latter stage also combined with a sediment model to compute the scour development of the sea bottom (4); the development of a jet flow downstream of a gated opening.

The need for further experience and verification of mathematical models remains obvious. They are a necessary step to develop them to operational models. Particularly, if these models are used for near-field problems. Here an assessment of the occurring three-dimensional effects is indispensable, which is a common flow pattern upstream of a structure. The relevance to study such a flow pattern will be established by the following example.

4.   Tested flow geometry (2DH)

With the mathematical model the water motion has been simulated as a depth-averaged flow and with the same dimensions as of the used physical scale model. The initially chosen flow geometry is a straight canal, provided with a pier in the middle on the downstream side. The width of the pier is a third of the width of the canal (see Fig. 1). One of the downstream openings can be closed with a gate. With this simple geometry it is likely to compute the flow pattern with a horizontal two-dimensional mathematical model. This because flow velocities in the vertical direction were small compared to the horizontal velocities. A comparison between the results of the physical scale model and the mathematical model is the starting point for computing flow patterns upstream of locks, combined with a discharge sluice or in the intake canal of a low-head hydro-power station along a river.

The flow geometry was schematized in a mathematical model. The applied grids are shown in Fig. 1. Only locally the grid was refined to minimise the number of grid points. The physical boundaries of grid B1 lie within the boundaries of grid B2, around the corner marked with symbol "P". Around that corner, grid B1 has been refined substantially.

The flow geometry was also built in a physical scale model. The overall flow patterns were established by means of photos; flow-velocity profiles were measured upstream of the model and in cross-sections near the left wall and in the right canal by means of propeller-type flow meters.

5.   Description of tests in the physical and mathematical model

The tests here described were performed with a rounded pier (Fig. 1, grid A) and with a rectangular pier (Fig. 1, grid B1 and B2). The flow pattern and water velocities were measured and computed for two discharges ($Q = 0.1$ m^3 and 0.185 m^3) with different water levels (respectively 0.4 m and 0.3 m). The flow patterns in the physical model appeared to be almost the same for both discharges. Therefore, only the tests with the low discharge will be considered.

Table 1. Test variants

	rounded pier	rectang. pier	discharge	water depth	grid	used turb. model
Test A	x		0.1 m^3/s	0.40 m	A	k-ε
Test B1		x	0.1 m^3/s	0.40 m	B1	(1)
Test B2		x	0.1 m^3/s	0.40 m	B2	k-ε

(1): constant eddy viscosity

In both physical and mathematical models the upstream boundary was chosen at such a distance from the pier, that a small difference of the inflow hardly has any influence on the flow pattern. This was checked by changing the inflow conditions in the mathematical model as well as the physical model.

The time-step of the mathematical model has been set up in such a way, that numerical stability criteria were satisfied: the Courant-number did not exceed the value of 5, and the diffusion parameter was smaller than 1 at any place in the mathematical models. With the exception of test B2 a two-dimensional k-ε turbulence closure for the computation of the eddy viscosity has been applied. Here the standard form of the k-ε equations was used (2).

In the tests with the rounded pier (see Fig. 2) attention was paid to:
- separation and pressure points, which are important for the overall flow pattern;
- flow velocity distribution in the right canal.

In the rectangular-pier tests (Fig. 6) attention was paid to:
- separation phenomenon at corner "P";
- dimensions and velocities in the eddy downstream of "P".

6.   Tests with rounded pier

6.1   Introduction

In general the mathematical simulation of the overall flow pattern shows a good resemblance with the pattern in the physical scale model (Fig. 2 and 3). The locally appearing differences are caused by numerical schematization and the simulation of the detailed flow pattern, such as flow separation, pressure points and flow distribution. The influence of these differences onto the accuracy of the simulations of the measured flow pattern will be considered below.

6.2   Discussion on details of the flow simulation

a. Separation from the left wall

The separation point of the main flow from the left wall is caused by pressure differences along the wall. In fact, the momentum of the main flow must be in equilibrium with the water pressure in longitudinal direction. Since the left canal was closed, the pressure will increase in flow direction. The augmentation of the pressure causes a decrease of the water velocity in longitudinal direction and on the moment that the velocity equals zero separation occurs. This local phenomenon has been studied by comparing measured and computed velocity distribution profiles perpendicular to the left wall.

In the mathematical model the flow pattern has been simulated with no-slip and law-of-the-wall boundary conditions. Fig. 4 shows the results of these computations. In the no-slip simulation the separation point was located at a distance of about 0.50 m downstream of the measured location. In the law-of-the-wall simulation this distance was even 1.75 m. In both computations large differences between measured and computed velocity distribution profiles are visible.

Afterwards grid A was refined towards the left wall such, that the minimum mesh width perpendicular to the wall was 0.02 m. Fig. 4 shows the results of the simulation with the adapted grid and the same law-of-the-wall boundary condition as in the simulation with the original grid A. The velocity profiles are more in agreement with the measured ones. The simulated separation point still lies 1.10 m downstream of the measured separation point, but the simulation of the separation phenomenon has been improved. It is expected, that a further improvement can be obtained by a further refinement of the grid.

b. Flow near the rounded pier

The pressure point is located on the pier where the main flow and the eddy flow split up. This point on the pier is approximately the same in the physical and in the mathematical model. The location is mainly caused by a force equilibrium, which appears to be well simulated by the mathematical model.

Both the mathematical model and the physical model did not show a flow separation from the rounded pier into the right canal, which is in accordance with the loss coefficient for this kind of inflow given by (5). In the physical model a velocity reduction close to the pier was observed, which caused somewhat higher flow velocities towards the right wall. Downstream of the rounded part of the pier the flow had a tendency towards the left wall, caused by a redistribution of the flow over the width. However, the mathematical model shows higher flow velocities near the rounded pier, like a potential flow. It is expected, that this can be improved by using smaller grid meshes close to the pier.

c. Velocity distribution in right canal

The measured velocity distribution in the right canal agrees well with the simulation using the law-of-the-wall boundary condition (Fig. 5). Larger deviations are visible between measurements and the results with a no-slip boundary condition. This is mainly caused by the choice of the refinement of the grid. Physically, the influence of the boundary layer is small with the relatively high flow velocity. Using the chosen grid, however, the model is not capable to compute such large velocity gradients. The result with the law-of-the-wall condition is better because this condition allows a velocity on the wall. It may be expected that increment of the number of grid points will also improve the results with the no-slip condition, although this has not yet been tested.

7. Tests with the rectangular pier

7.1 Introduction

The simulation of the overall flow pattern shows a good resemblance with the measured flow pattern, provided the separation phenomenon at the right corner of the pier (point "P" in grid B1 an B2, Fig. 1) is correctly simulated. Therefore, much attention was paid to the physics and numerical simulation of this particular phenomenon.

7.2 Separation and recirculating flow

In the physical model the flow separated at the right corner "P" of the rectangular pier. Downstream of this separation point an area with recirculating flow is visible (Fig. 6). The maximum width of this area is about 0.35 m and the reattachment point at the downstream side of the recirculation area varies between 2.50 to 3.00 m downstream of point "P".

In (6) some fundamental experiments with mathematical models on the flow in forward-step geometries are described. It is shown that for a correct simulation of the viscous-dominated flow around the corner extremely refined grids will give good results. Following the recommendations in (6), the performance of the applied mathematical modelling system was checked in this respect. Afterwards, tests were carried out with larger mesh sizes in order to reduce the number of grid points by maintaining a correct simulation of the separation and recirculation flow. The results of both simulations have been compared with measurements in the physical model (Fig. 6).

Test B1 (Fig. 7) shows a good resemblance with the experiments in (6). Only at short distances from the right corner "P" high gradients in the vorticity distribution are visible. The dimensions of the recirculation flow (Fig. 9) are somewhat smaller than in the measurements. These dimensions, however, are influenced by the value of the eddy viscosity. By tuning this coefficient, the eddy dimensions can be fitted to the measurements.

In order to reduce the number of computational grid points, a grid with a larger mesh size (grid B2) was tested. Because the separation phenomenon is caused by the "infinite" vorticity in the right corner, the chosen grid size appeared to be too coarse for the simulation of the diffusion of this vorticity to the other grid points. Therefore, a "vorticity source" has been introduced in a computational point quite close to the right corner "P". By this the influence of the high vorticity quite close to the right corner is taken into account.

Test B2 (Fig. 8) gives the results of the simulation with the vorticity source and with the k-ε turbulence model, a model which computes the eddy viscosity distribution in the computational grid. Using this turbulence model, the eddy dimensions were computed by the mathematical model itself. The eddy dimensions in this simulation and in the physical model are in good agreement. The velocities in the recirculation flow, however, are larger than the measured ones. In Fig. 9 this is illustrated by comparing the computed and measured flow velocities.

8.   Conclusions

Rather simple flow geometries were used to study elementary flow phenomena, such as separation, reattachment, eddies and velocity distributions. In this way, the combination of physical and mathematical models appeared to provide more insight into the accuracy and applicability of mathematical models in engineering practice. In this study two types of flow separation and flow velocity distributions are investigated.

The mathematical model is able to simulate separation of the flow as an effect of the equilibrium between momentum and pressure forces. Therefore, a sufficiently refined grid near the wall is necessary. Also for the viscous-dominated separation at right corners an extremely refined grid should be applied. Moreover, it is found that simulation in a rather coarse grid will give acceptable results, provided a vorticity source is introduced close to the separation point. This implies, that both for a refined grid as well as for a coarse grid (combined with a vorticity source), the location of the flow separation must be known in advance.

From the tests carried out, it appears that grid refinement gives better results in the simulation of the flow close to the wall. Results with grid refinement close to the right corner and to the left wall confirm this experience. Probably also the deviations found between the simulated and measured flow distributions close to the rounded pier and in the right canal can be improved in this way.

The eddy viscosity has an important influence onto the dimensions of the recirculating flow downstream of the right corner. The simulation with a constant eddy viscosity showed a too small recirculation area. The result was improved by using a turbulence model (e.g. the k-ε model), in which the eddy viscosity was computed. Now, the dimensions of the eddy were almost equal to the measurements.

The presented results emphasize the need for more practical applications, together with laboratory experiments and field measurements to obtain more knowledge about the physical understanding of the phenomena and the reliability of the simulation in mathematical models. In such a way these models will extend and improve their operational modelling system towards an efficient and user-friendly tool. These models can then be used as a base for a simulation system for specific situations from the available flow and sediment transport models. Subsequently, these models combined with userfriendly pre-and-post-

Advances in water modelling and measurement

processing facilities may develop into predictive tools in engineering and consultation practice.

9. Acknowledgements

We like to thank the Dutch Public Works for financing and supporting the present study, which was carried out in the framework of their BSW-Constructions and Water Research Programme.

10. References

1   Officier, M.J. Vreugdenhil, C.B. and Wind, H.G.: "Application in Hydraulics of Numerical Solutions of the Navier-Stokes Equations". In: Taylor, C.: "Recent Advances in Numerical Fluid Dynamics", Swansea (UK), Pineridge Press, 1984.
2   Rodi, W.: "Turbulence Models and their Application in Hydraulics". IAHR, Delft, The Netherlands, June 1980.
3   Struiksma, N., Olesen, K.W., Flokstra, C. and Vriend, H.J. de: "Bed Deformation in Curved Alluvial Channels.", Delft Hydraulics Publication no. 333, March 1985.
4   Beek, F.A. van, Bijker, R., Wind, H.G.: "Simulation of Erosion and Sedimentaion Around Marine Structures by Means of a Numerical Model". In: Proc. of the International Conference on Behaviour of Offshore Structures (Trondheim, Norway: June 1988), Volume 1, pp. 419-432.
5   Miller, D.S.: "Internal Flow Systems". BHRA Fluid Engineering, Series Engineering. Series; 4. 1987
6   Castro, I.P., Jones, J.M.: "Studies in Numerical Computations of Recirculating Flows". International Journal for Numerical Methods en Fluids, Vol. 7, 1987, pp. 793-823.

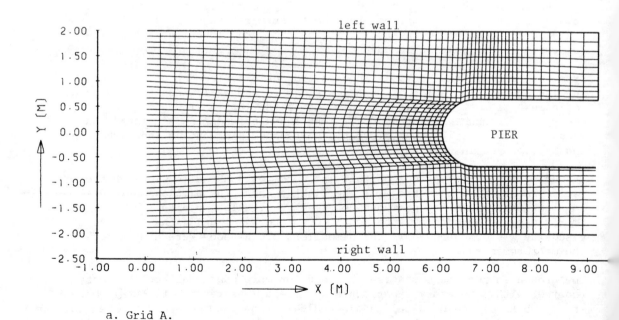

a. Grid A.

Figure 1. Tested flow geometries and computational grids.

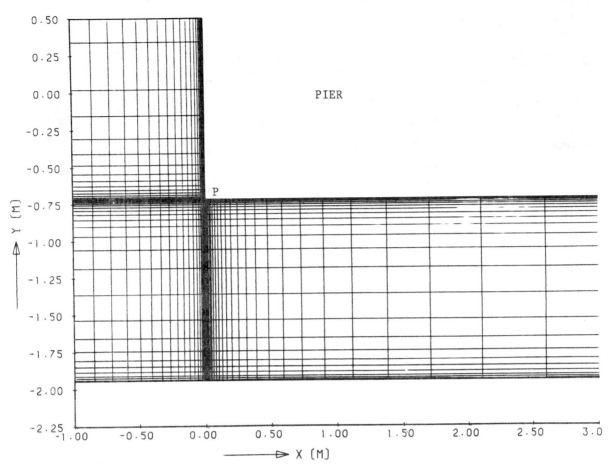

b. Grid B1.

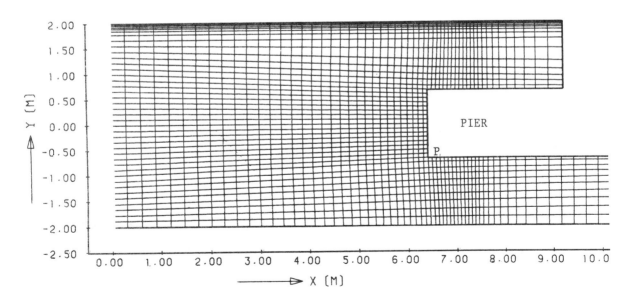

c. Grid B2.
Figure 1. Tested flow geometries and computational grids.

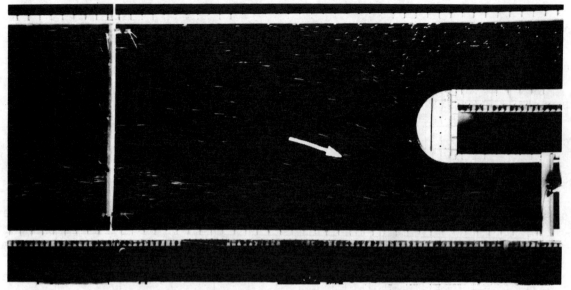

Figure 2. Flow pattern in physical model, test A (discharge: 0.1 m³/s, depth: 0.40 m)

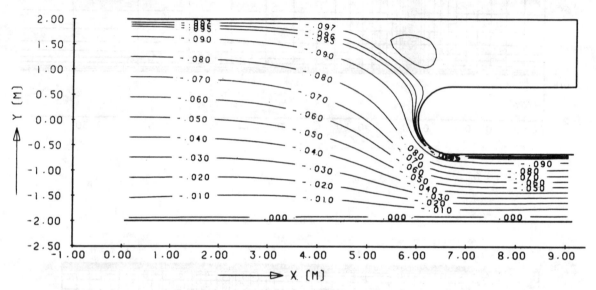

a. Streamlines, no-slip boundary condition (Grid A).

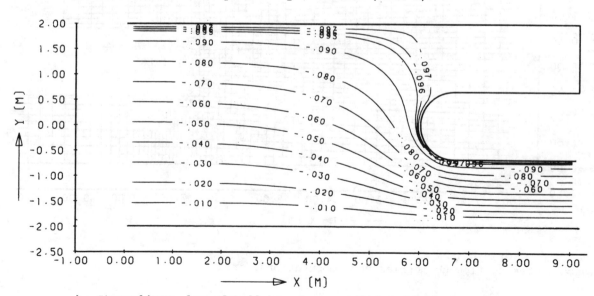

b. Streamlines, law-of-wall boundary condition (Grid A).
Figure 3. Flow simulations, test A (flow from left to right).

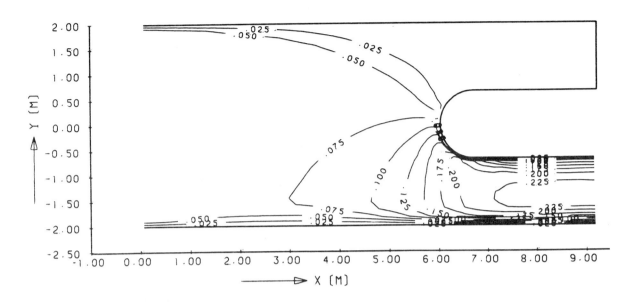

c. Magnitude of flow velocity, no-slip boundary condition (Grid A)
   (isolines: m/s).

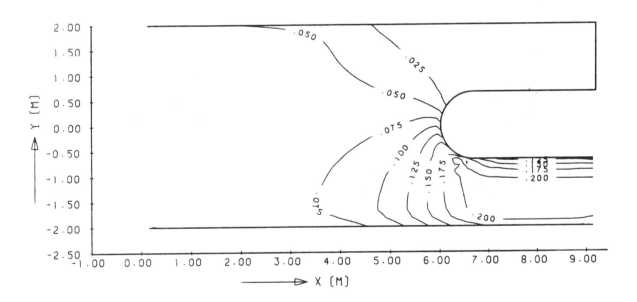

d. Magnitude of flow velocity, law-of-wall boundary condition (Grid A)
   (isolines: m/s).
Figure 3. Flow simulations, test A (flow from left to right).

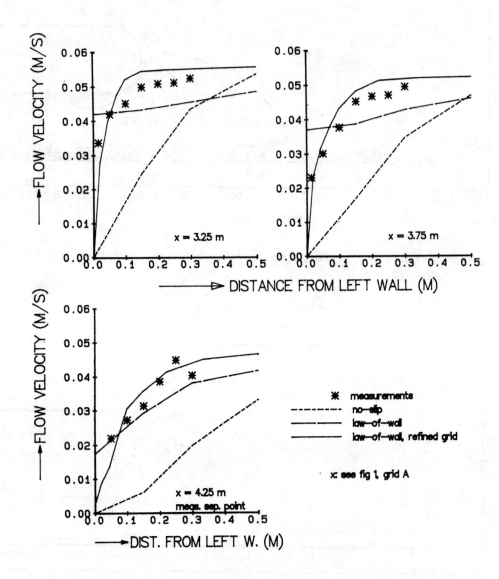

Figure 4. Flow velocity distribution near left wall, test A.

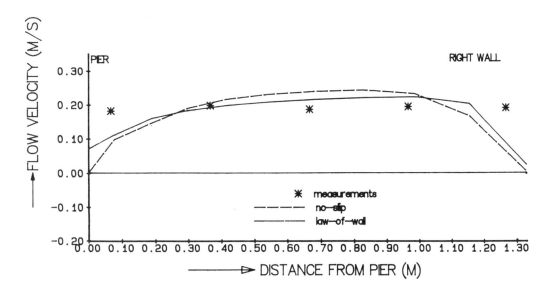

Figure 5. Velocity distribution in right canal, test A.

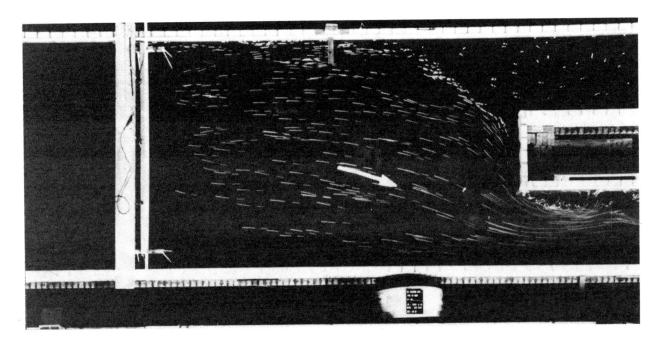

Figure 6. Flow pattern in physical model, test B (discharge: 0.1 m³/s, depth: 0.40 m)

213

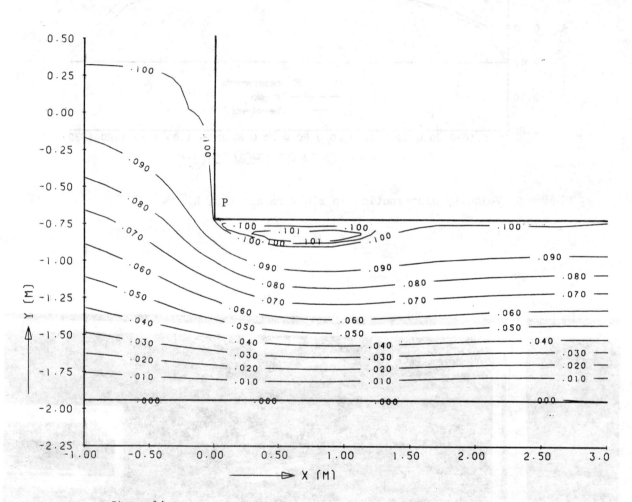

a. Streamlines
Figure 7. Flow simulation, test B1 (flow from left to right)

214

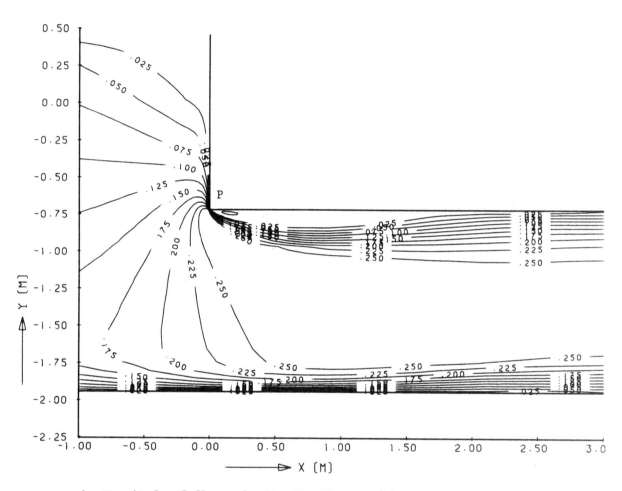

b. Magnitude of flow velocity (isolines: m/s).
Figure 7. Flow simulation, test B1 (flow from left to right)

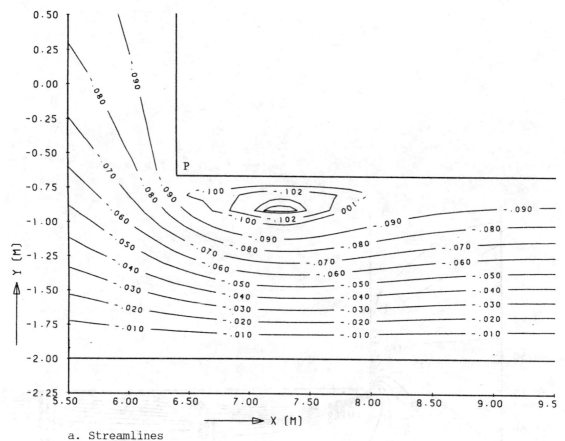

a. Streamlines
Figure 8. Flow simulation, test B2 (flow from left to right)

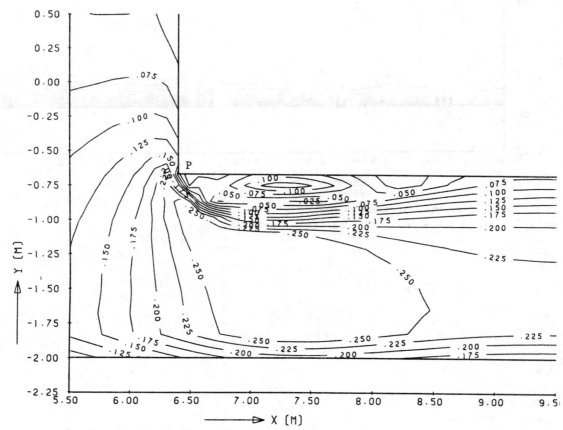

b. Magnitude of flow velocity (isolines: m/s).
Figure 8. Flow simulation, test B2 (flow from left to right)

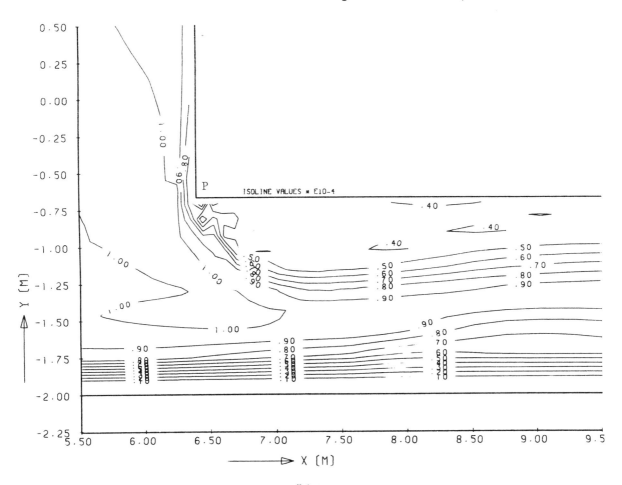

c. Eddy viscosity (isolines: m^2/s)
Figure 8. Flow simulation, test B2 (flow from left to right)

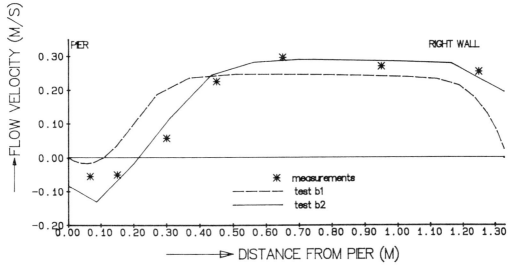

Figure 9. Velocity distribution in right canal (1.00 m downstream of P)

217

Chapter 15
TURBULENCE SIMULATION IN TIDAL FLOWS

G Vittori
University of Genoa, Italy

Summary

Recent field measurements in tidal flows have shown that the characteristics of turbulence during the accelerating and decelerating phases are markedly different. In the present paper tidal flow is studied by means of Saffman's turbulence model which has recently been proved to be successful in describing unsteady oscillatory and pulsatile flows. The present results are compared with experimental data of Knight & Ridgway (1) and a satisfactory agreement is found. The model is then systematically used to investigate how turbulence is affected by the oscillatory character of the mean flow. Unsteadiness is found to play an important role for values of the Keulegan-Carpenter number typical of tidal flows up to about 400.

1. Introduction

Many works have been recently devoted to the study of the unsteady turbulent flows induced by tides in the nearshore zone, due to the practical interest of such flows in relation to construction and transportation activities in coastal areas.

When studying tide propagation, it is usually assumed that the water depth is much smaller than the tide characteristic wavelength and consequently the long wave theory is commonly applied. In order to evaluate the friction term, it is common practise to assume the flow to be slowly varying and to relate it to the instantaneous depth averaged velocity. Turbulence is thus assumed in equilibrium and the production, dissipation and redistribution terms in the turbulence kinetic energy equation are supposed to balance each other.

However, recent field measurements and laboratory data in tidal flows (Anwar (2), Bohlem (3), Thorn (4), Anwar & Atkins (5)) have shown an increase in both the dispersion of tracer fluids and in the rate of sediment transport during the decelerating phase compared with the accelerating one. The

observed increase of transport phenomena is due to the stronger
turbulence intensity detected during the decelerating phase.
These experimental observations seem to indicate that turbulence
and all related quantities are affected by the time history of
the flow which cannot be accounted for by means of a quasi-steady
approach.

Recent works have dealt with turbulent tidal flows employing
turbulence models where time and spatial variation of turbulence
characteristics are taken into account. In particular Celik &
Rodi (6) used the standard K - ε model to study turbulent tidal
flow induced in an estuary. The standard k - ε model cannot go as
far as to describe the layer adjacent to the bottom where viscous
effects play an important role, so that the numerical solution is
to be matched with the analytical solution obtained in the
equilibrium layer where production and dissipation of turbulent
kinetic energy are supposed to balance each other and
redistribution term to be negligible. Results are thus affected
by the forcing of a logarithmic law close to the wall. The
applicability of the latter under unsteady conditions has been
recently questioned by many authors on the basis of experimental
data and theoretical considerations (Hino et al. (7), Blondeaux
(8)).

The present contribution is aimed at attempting to overcome
the above difficulties by employing Saffman's turbulence model
(Saffman (9)). The latter model has been recently proved to be
fairly successful in describing unsteady oscillatory and
pulsatile flows and has been shown to give a good description of
the flow near the wall where viscous effects become predominant
(Jacobs (10), Blondeaux & Colombini (11), Blondeaux (8)).

We will consider a fluid region of horizontal size of the
same order as the channel depth h, and assume the flow to be two
dimensional. Performing an analysis of the orders of magnitude of
the various terms of the governing equations, the problem of two
dimensional tidal flow in a region of order h is reduced to that
of an oscillatory flow in a closed duct with an oscillating
pressure gradient, provided the tide wavelength L is much larger
than h and its amplitude a is much smaller than h.

In the next section we formulate the problem giving also a
brief description of the turbulence model and of the numerical
method used to solve the problem. In section 3 we describe the
results and perform a comparison with the experimental results of
Knight & Ridgway (1). In particular, on the basis of the proposed
model, turbulence behaviour in tidal flows is analyzed and
differences between the accelerating and decelerating phases are
emphasized. The last section is devoted to some conclusions and
to discussing the need for further developments of the work.

2. Formulation of the problem

Let us consider a tidal wave of length L and amplitude a
propagating in a two-dimensional channel of depth h.

Let us introduce a cartesian coordinate system (x,y) with
the x axis lying on the bottom and the y axis normal to it.

If we assume that

$$\frac{a}{h} \ll 1 \quad ; \quad \frac{h}{L} \ll 1 \tag{1a,b}$$

at the leading order of approximation in h/L and a/h, the
momentum equation in the y-direction simply gives the hydrostatic
pressure distribution along the vertical, while the momentum
equation in the x-direction reads:

$$\frac{\partial u}{\partial t} = - gS_0 \cos \omega t + \nu \frac{\partial^2 u}{\partial y^2} - \frac{\partial}{\partial y} <u'v'> \qquad (2)$$

where t is time and the velocity vector has been split into an ensemble-averaged velocity (u,v) and a turbulent random velocity (u',v'), the symbol <> denoting ensemble averaging. In (2) g denotes gravity, ν kinematic fluid viscosity, S_0 and ω amplitude and angular frequency of surface slope oscillations respectively. The Reynolds stress tensor is then expressed in terms of an eddy viscosity ν_T and of an average strain tensor in the form:

$$- \frac{\partial}{\partial y} <u'v'> = \frac{\partial}{\partial y} \left[\nu_T \frac{\partial u}{\partial y} \right] \qquad (3)$$

In order to close the problem Saffman's (9) turbulence closure is employed. In the model above the eddy viscosity ν_T is assumed to be a function of turbulence local properties, namely a pseudo-energy e and a pseudo-vorticity Ω which are assumed to satisfy non-linear diffusion equations. In our case we find:

$$\frac{\partial e}{\partial t} = \alpha_e \, e \left| \frac{\partial u}{\partial y} \right| - \beta_e \, e \, \Omega + \frac{\partial}{\partial y} \left[(\nu + \nu_T \, \sigma_e) \frac{\partial e}{\partial y} \right] \qquad (4)$$

$$\frac{\partial \Omega^2}{\partial t} = \alpha_\omega \, \Omega^2 \left| \frac{\partial u}{\partial y} \right| - \beta \, \Omega^3 + \frac{\partial}{\partial y} \left[(\nu + \nu_T \, \sigma) \frac{\partial \Omega^2}{\partial y} \right] \qquad (5)$$

where α_e, α_ω, β_e, β_ω, σ_e, σ_ω are universal constants.
The quantity Ω is supposed to be proportional to the mean vorticity of the energy containing eddies and e to the kinetic energy of the motion induced by this vorticity. Equation (4) and (5) then follow by analogy with the vorticity and with kinetic energy equations of turbulence (Townsend (12)). From dimensional analysis it follows that $\nu_T = \gamma \, e/\Omega$. The values of the constants appearing in (4), (5) have been determined by Saffman & Wilcox (13) on the basis of theoretical arguments.
Equation (2-5) require appropriate boundary conditions. We assume the rigid lid approximation at the free surface. Furthermore the turbulence kinetic energy vanishes at the wall and the ensemble averaged velocity satisfies the no-slip condition. Saffman (9) postulated that at y=0, Ω depends only on a non-dimensional roughness parameter $k \, u_\tau/\nu$ through a universal function S

$$\Omega = \frac{u_\tau^2}{\nu \alpha_e} S \left(\frac{ku_\tau}{\nu} \right) \qquad (6)$$

where k is channel roughness and u_τ is friction velocity. A formula for S has recently been proposed by Blondeaux & Colombini (11).
The differential problem posed by (2-5) with the boundary conditions discussed above has been solved numerically by means of a finite difference method. The y-coordinate has been conveniently stretched in order to magnify the wall layer where strong gradients occur. The numerical approach has been extensively tested as discussed in the next section.

3. Discussion of the results

In order to test the performance of the proposed turbulence model, some comparisons have been performed with experimental data by Knight & Ridgway (1). The latter authors simulated tidal flow in an estuary, using a rectangular channel of length 20 m and width 0.15 m. One channel end was connected to a tank where the water level was made to oscillate, while the other end was closed. The test section was located at 7.5 m from the estuary mouth.

We consider three cases, which were also simulated by Celik & Rodi (6) by means of the $K - \epsilon$ turbulence model, namely experiments 1,3 and 5 of Knight & Ridgway (1). In the first two experiments the channel bottom was hydraulically smooth while in the third one an artificial two-dimensional roughness was introduced.

Because of the assumptions described in the previous section, we do not study tidal wave propagation in the estuary model but we only simulate the time development of turbulence in the test section feeding in the numerical model the value of surface slope detected experimentally.

In figure 1 the predicted distribution of the amplitude ($U(y)$) of velocity is compared with the experimental measurements and with Celik & Rodi's (6) predictions. It can be seen that the present model gives better results especially near the wall, where the standard $K - \epsilon$ model becomes inaccurate. It should be pointed out that in the experiments (particularly runs 1 and 3), due to the high values of the angular frequency, turbulent and viscous effects are confined within a thin layer adjacent to the channel bottom while the flow outside is nearly inviscid.

In figure 2 the present predictions for the time development of fluid speed at $y = 0.75$ h are compared with the experimental results of Knight & Ridgway (1). The forced and measured surface slope are also shown. The agreement is satisfactory and comparable with that obtained by Celik & Rodi (6).

On the basis of the above results and further comparisons not reported herein it can be stated that the present model gives accurate predictions of the vertical plane structure of tidal turbulent flows both in the smooth and rough regimes and can be confidently used to investigate turbulent tidal flows.

In figure 3 the turbulent shear stress, scaled with the fluid density ϱ and the amplitude $\bar{U}$ of depth averaged velocity, is plotted versus local velocity for different depths and for the values of the parameters of experiment number 5 of Knight & Ridgway (1). It can be seen that for a fixed value of velocity, shear stress attains different values depending on the sign of fluid acceleration. A similar trend is observed for the pseudo-energy e which is proportional to the turbulent kinetic energy (see figure 4). Indeed, figures 3 and 4 show that the values of the shear stress and of the turbulent kinetic energy are always larger when the fluid is decelerating than when it is accelerating. Also they are found to be different from the corresponding values for steady flows. The different behaviour of turbulence during the accelerating and decelerating phases was also observed experimentally by Anwar & Atkins (5) in their laboratory simulation of tidal flows (see figures 8 and 9 of the above authors). Similar results are obtained when simulating experiments number 1 and 3 of Knight & Ridgway (1). The increase in turbulence activity when the flow decelerates in most part of the channel clearly emerges from a comparison between figures 5 and 6 where the distributions of dimensionless pseudo-energy and speed are plotted respectively. Turbulence is generated explosively near the wall and reaches its maximum at the beginning of the decelerating phase; then it propagates quickly far from the wall where it is dissipated by viscous effects. This behaviour has been found both for smooth and rough walls.

In figure 7 the time development of the depth averaged velocity $(\bar{u}(t))$ and pseudo-energy $(\bar{e}(t))$ is plotted, to show how turbulence reaches its maximum intensity during decelerating phase.

The asimmetry between the distributions and intensities of turbulence energy characteristic of tide acceleration and deceleration respectively has a great influence on any transport phenomena. Assuming Reynolds analogy one can infer that sediment transport and pollutant dispersion are markedly affected by flow acceleration. Indeed figure 8 where ν_T is plotted versus y for different times in the cycle shows that eddy viscosity attains its maximum values during the decelerating part of the cycle for a large range of depths. Besides, figure 8 shows that ν_T can neither be assumed as time invariant, nor can be approximated by a simple function of the vertical distance as it was common practise in many previous works.

Of course the effects of unsteadiness decrease as the Keulegan-Carpenter number ($K_c = \bar{U}/\omega h$) characteristic of the tide becomes larger. In figure 9, the maximum values attained by the bed shear stress τ in a cycle is plotted versus the quantity $1/\omega h$ for fixed values of the other parameters. It can be seen that only for large values of $1/\omega h$, τ/ϱ approaches gS_0h which is the corresponding value for the steady case . From figure 9 and the values of h and $\bar{U}$ found in the numerical experiments, it can be stated that τ/ϱ differs from gS_0h by an amount larger than 10% for Keulegan-Carpenter numbers less than about 400. As in real situations Keulegan-Carpenter numbers less than 400 do occur, turbulence models, taking into account the time variation of turbulence characteristics, must be used.

4. Conclusions

Laboratory and field measurements show that turbulence is strongly affected by flow acceleration even in tidal flows which are characterized by very large time scale.

In describing the effects of turbulence in tidal wave propagation, the quasi-steady approach fails and the time hystory of the flow must be taken into account.

In the present paper a two-equations turbulence model has been used to predict the turbulence structure and flow behaviour in a two-dimensional channel where a tidal wave is propagating. Results have been obtained which agree satisfactorily with experimental observations.

We are currently investigating about the feasibility of simplified models where the eddy viscosity is a given function of space and time as Trowbridge & al. (14) assumed in their simulation of the turbulent boundary layer at the bottom of gravity waves. However these models need the eddy viscosity structure which can be provided only by experimental results or by comparison with more refined models like that considered in the present paper.

This work was supported by Italian Ministry of Education under grant MPI (60%) and is part of the writer Ph D thesis to be submitted.

References

1. Knight, D.W., Ridgway, M.A., 'Velocity distributions in unsteady open channel flow with different boundary roughness'. Proc. 17th IAHR Congress, Baden-Baden F.R.Germany, Paper No. A130, 2, pp. 437-444.

2. Anwar, H.O., 'Turbulent dispersion and meandering of a surface plume'. Proc. 16th Congress Int. Ass. for Hydraulic Research (Sao Paulo, Brasil July-Aug. 1975) Vol. 1, pp. 367-376.

3. Bohlen, W.F., 'Shear stress and sediment transport in unsteady turbulent flows'. Estuarine Processes, M.Wiley, ed., Vol. II, Academic Press, New York, N.Y. 1976, pp. 109-123.

4. Thorn, M.F.C., 'Hysteresis of fine sand suspension in a tidal estuary'. Note No. 17, Hydraulics Research Station, Wallingford, Oxfordshire, England, 1975, pp. 2-3.

5. Anwar, H.O., Atkins, E., 'Turbulence measurements in simulated tidal flow'. Journal of the Hydraulic Division, 106, Aug. 1980, pp. 1273-1289.

6. Celik, I., Rodi, W., 'Calculation of wave-induced turbulent flows in estuaries'. Ocean Engng., 12, 6, 1985, pp. 531-542.

7. Hino, M., Kashiwayanagi, M., 'Experiments on the turbulence statistics and the structure of a reciprocating oscillatory flow'. J. Fluid Mech., 131, 1983, pp. 363-400.

8. Blondeaux, P., 'Turbulent boundary layer at the bottom of gravity waves'. Journal of Hydraulic Research, 1987, 25, 4.

9. Saffman, P.G., 'A model for inhomogenous turbulent flow'. Proc. Roy. Soc., London, 1970, A317, 417.

10. Jacobs, S.J., 'Mass transport in a turbulent boundary layer under a progressive water wave'. J. Fluid Mech., 1984, 146, pp. 303-312.

11. Blondeaux, P. and Colombini, M., 'Pulsatile turbulent pipe flow'. 5th Symposium on Turbulent Shear Flows, Ithaca, 1985, New York.

12. Townsend, A.A., 'The structure of turbulent shear flow'. Cambridge University Press, 1956.

13. Saffman, P.G., Wilcox, P.C., 'Turbulence model predictions for turbulent boundary layers', A.I.A.A. Journal, 1974, 12, 541.

14. Trowbridge, J., Madsen, O.S., 'Turbulent wave boundary layers 2. second order theory and mass transport'. J. Geophysical Research, September, 1984, Vol. 89, No. C5, pp. 7999-8007.

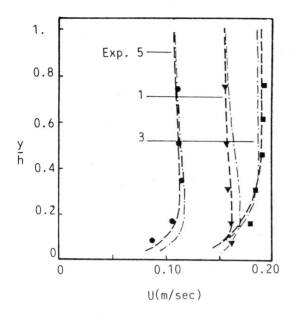

Figure 1:
Comparison of predicted and measured
(Knight & Ridgway (1)) velocity
amplitude profile.
(--- present predictions, -•-• Celik
& Rodi's predictions (6)).

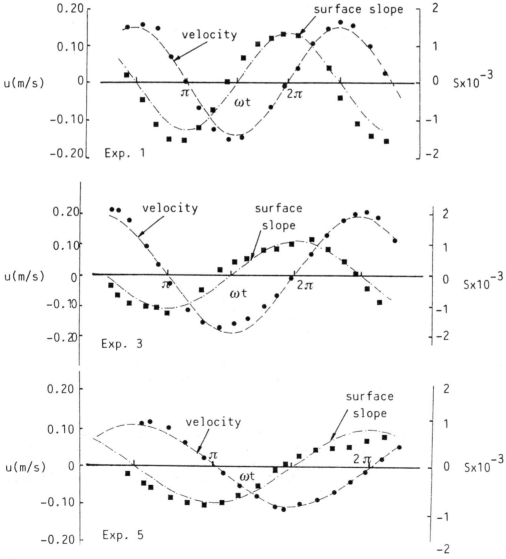

Figure 2: Comparison of predicted and measured (Knight &
Ridgway (1)) time development of fluid velocity at
y = 0.75 h and surface slope.

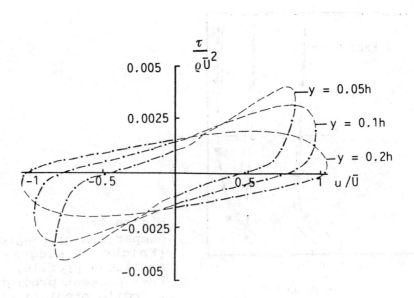

Figure 3: Shear stress versus local velocity at
different depths (Data as in exp. 5 of Knight
& Ridgway (1)).
(-•-•- accelerating phase, --- decelerating
phase).

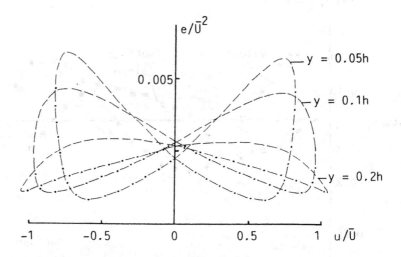

Figure 4: Pseudo-energy versus local velocity at
different depths (Data as in exp. 5 of Knight
& Ridgway (1)).
(-•-•- accelerating phase, --- decelerating
phase).

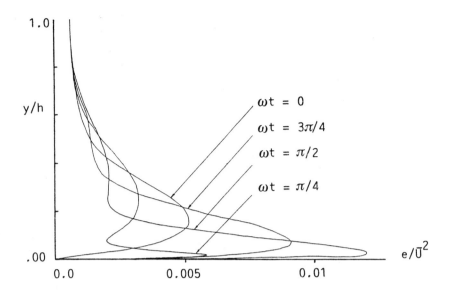

Figure 5: Pseudo-energy versus depth at different times in the cycle (Data as in exp. 5 of Knight & Ridgway (1)).

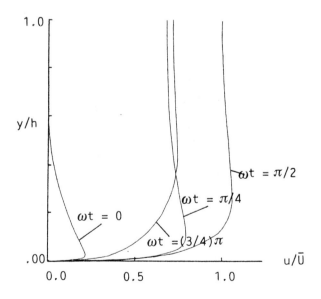

Figure 6: Velocity versus depth at different times in the cycle (Data as in exp. 5 of Knight & Ridgway (1)).

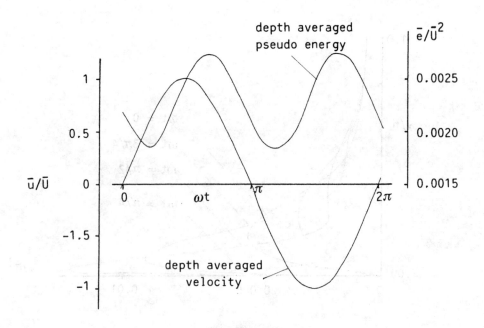

Figure 7: Depth-averaged velocity ($\bar{u}$) and pseudo-energy ($\bar{e}$) versus time (Data as in exp. 5 of Knight & Ridgway (1)).

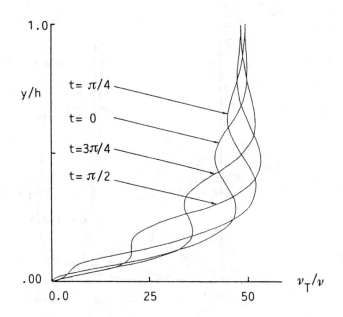

Figure 8: Eddy viscosity versus depth at different times in the cycle (Data as in exp. 5 of Knight & Ridgway (1)).

228

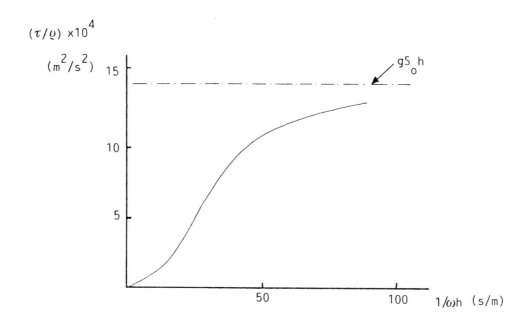

Figure 9: Maximum bed shear stress versus the quantity $1/\omega h$.

Chapter 16
BED FRICTION IN TIDAL FLOWS

B A O'Connor
University of Liverpool, UK
F J L Duckett
K Philpott Consulting Ltd, Canada

Summary

The paper describes a new, relatively simple approach to the specification of bed friction in tidal flows. In particular, equations are presented from which it is possible to predict average bedform dimensions and resulting friction factor coefficients and bed shear velocities. Such information may then be used to estimate sediment transport rates from suitable formulae. The method is tested on two field situations, one in the Ribble Estuary and the other off the Sizewell-Dunwich Bank area of East Anglia, with reasonable results.

Nomenclature

a, a_0	:	Constants in equations 17 and 7, respectively
f, f', f''	:	Darcy-Weisbach friction factors for total, dune, grain/ripple drag
g	:	Acceleration due to gravity (9.81 m/s^2)
h, h_*	:	Water depth and tidal-average water depth, respectively (m)
k_s	:	Bed roughness height (m)
q	:	Total sediment transport rate (kg/m-s)
t, t_*	:	Time and time available in a tidal cycle for erosion, respectively (s)
w_f	:	Particle fall velocity (m/s)
A	:	Exponent in equation 9
A_*	:	$\log_{10}(D_{50})$
B	:	Exponent in equation 17
D	:	Sediment grain size (m)
L	:	Bedform wavelength (m)
R_e	:	Fall velocity Reynolds Number (= $w_f D/\nu$)
S	:	Submerged relative density of sediment grains ($(\rho_s - \rho)/\rho$)
T	:	Water temperature (°C)
T_0	:	Tidal period (s)
U	:	Depth-mean tidal velocity (m/s)
U_*	:	Bed shear velocity (m/s)
W	:	Non-dimensional relative density (ρ_s/ρ)
X	:	Non-dimensional grain Reynolds Number (= $U_* D/\nu$)
Y	:	Non-dimensional grain Froude Number (= $U_*^2/(SgD)$)
Z	:	Non-dimensional depth (h/D)
α, β	:	Constants in equation 7
η	:	Y/Y_c
ν, ν_T, ν_{10}	:	Kinematic viscosity of water, at temperature T°C and 10°C, respectively (m^2/s)
ρ	:	Water density (kg/m^3)
ρ_s	:	Sediment grain density (kg/m^3)
ζ	:	Y/Z
ξ, ξ_{10}	:	X^2/Y, and with ν_{10}, respectively
Δ	:	Bedform height (m)
φ	:	$U_m t/h$
τ_b	:	Bed shear stress (N/m^2)

Subscripts

c	:	Threshold/critical value
d	:	Dune value
i,n	:	ith, nth tide value in Spring/Neap cycle
m	:	Maximum value
o	:	Steady flow wavelength ($2\pi h_*$)
r	:	Ripple value
50,90	:	% finer value for grain size

1. Introduction

The occurrence of large bedforms (sand waves) on the seabed has been amply demonstrated in recent years by the development of modern surveying equipment, such as side-scan sonar. Unfortunately, these large scale features pose serious problems for both the practical engineer concerned with laying pipelines or maintaining navigation depths in port approach channels, and the research engineer concerned with the design of estuary and coastal computer flow models, who has to include a proper representation of bed friction in his models.

While it is now becoming possible to model the unsteady, three-dimensional flow around bedforms, it is impractical to use such approaches over large estuary and coastal areas. Consequently, until computers develop further, a more practical simplified approach to the prediction of bed features and bed friction is required. The present paper provides such an approach, although it is recognised that further development of the ideas is likely to be required to make the method more universally applicable.

The method ignores stratification effects in the water column and assumes that the tidal motion produces gradually-varied flow so that acceleration and deceleration effects on bed friction are minor by comparison with surface drag effects. Highly rotating flows are also ignored for the present, so that the method is best suited to well-mixed estuaries and the nearshore coastal zone, in the absence of wave motion.

2. Bedforms in steady flow

It is generally recognised that bedforms develop from a flat bed of sediment once the environmental forces exceed some critical (or threshold) value. Since sediment transport of uniform-sized grains is usually taken as a function of four non-dimensional parameters (X,Y,Z,W) describing the fluid, sediment and flow properties, it follows that for a given water depth, grain size and relative density of sediment (Z,W fixed), a unique relationship exists between the Y and X parameters for a level bed of uniform-sized sediment grains, that is,

$$Y_c = f(X_c) \qquad\qquad 1.$$

where the subscript refers to 'critical' conditions.

Despite various attempts to determine improved functional forms for equation 1 so as to include non-uniform grains etc, the form produced by Shields, has generally been adopted by engineers for estuary and coastal situations of reasonable depth (Z $\geqslant$ 1000), see figure 1 and Yalin (1977).

At slightly greater flows then critical (X $\leqslant$ 8), it is generally suggested that small scale bedforms (ripples) occur with a wavelength (L_r) determined by the sediment grain size (D), that is:-

$$L_r \approx 1000 \, D \qquad\qquad 2.$$

and with a maximum steepness $(\Delta/L)_{rm}$ of between 0.09 - 0.20 for $Y/Y_c \approx 6$ and Z > 400, see Yalin (1977) and figure 2.

At higher flows (X > 24), ripples disappear and larger bedforms are produced (dunes) with a wavelength (L_d) determined by the flow depth (h), that is:-

$$L_d \approx 5\text{-}7h \qquad\qquad 3.$$

although a value of 2π is often used, based on theoretical studies, see Yalin (1977).

It has also been suggested that both ripples and dunes exist for large Z values for 8 $\leqslant$ X $\leqslant$ 24 and that dunes disappear when Y/Y_c > 65, see figure 1.

Experimental evidence, see Yalin (1977), indicates that dunes are less steep than ripples, with maximum values between 0.06 - 0.08 for $Y/Y_c \approx 13$, X > 31.6 and Z > 100. Later work by Yalin and Karahan (1979) suggested the relationship:-

$$(\Delta/L)_{dm} = 0.043 \, \log_{10} \, (h/D_{50}) - 0.052$$

$$\text{for } Z \leqslant 400 \qquad\qquad 4a.$$

$$= 0.06$$

$$\text{for } Z > 400 \qquad\qquad 4b.$$

and $\quad (\Delta/L)_d = 0.0127 \, x \, \exp(-x/\bar{x}) \qquad\qquad 5a.$

where $\quad \bar{x} = 214(\Delta/L)_{dm} \qquad\qquad 5b.$

and $\quad x = \eta - 1; \quad \eta = Y/Y_c \qquad\qquad 5c.$

which indicates zero dune height when $Y = Y_c$ and at 3-5% of the flow depth for $Y/Y_c = 65$.

An alternative equation has also been suggested by Fredsoe (1975), that is:-

$$(\Delta/L)_d = (1 - 0.06/Y - 0.4Y)^2/8.4 \qquad\qquad 6.$$

which can be re-written in the form:-

$$(\Delta/L)_d = (1 - \alpha/Y - \beta Y)^2/a_0 \qquad\qquad 7a.$$

where

$$a_0 = (1 - 2\sqrt{\alpha\beta})^2/(\Delta/L)_{dm} \qquad\qquad 7b.$$

$$\alpha = Y_1 Y_2/(Y_1 + Y_2) \qquad\qquad 7c.$$

$$\beta = 1/(Y_1 + Y_2) \qquad 7d.$$

and Y_1, Y_2 are the limiting values when $\Delta_d = 0$.

Using Yalin's values of $(\Delta/L)_{dm} = 0.06$; $Y_1 = Y_c$; $Y_2/Y_c = 65$; $Y_c = 0.05$, gives $a_0 = 9.52$; $\alpha = 0.0492$; $\beta = 0.303$ and provides similar results to equation 6.

Fredsoe's (1982) later theoretical work on dune geometry suggests that these empirical equations (4,6) provide realistic estimates of bedform steepness.

Equation 7 may also be used to describe ripple steepness. For example, if $(\Delta/L)_{dm} = 0.15$; $Y_1 = Y_c$; $Y_2/Y_c = 14$, 24; $Y_c = 0.05$ then $a_0 = 1.674$, 2.465; $\alpha = 0.0467$, 0.048; $\beta = 1.333$, 0.80, respectively, which provides a reasonable fit to experimental data, see figure 2.

Since ripples are thought to disappear for coarse sized sediment (> 600 μm), it is likely that maximum ripple steepness is also a function of particle grain size. Using very limited data, see Yalin (1977), a tentative relationship with fall velocity Reynolds Number (R_e) is suggested, that is:-

$$(\Delta/L)_{rm} = 0.147 - 0.0485 \log_{10} (R_e) \qquad\qquad 8.$$

where $R_e = w_f D/\mu$

and w_f, ν are given by the convenient expressions, see Muir Wood and Fleming (1981),

$$w_f = 10^A \text{ m/s} \qquad\qquad 9a.$$

where $A = 0.447 A_*^2 + 1.9611 A_* + 2.736 \qquad\qquad 9b.$

$$A_* = \log_{10} (D_{50}) \qquad\qquad 9c.$$

$$\nu = 1.9 \times 10^{-6} \exp(-0.042 T^{0.87}) \text{ m}^2/\text{s} \qquad\qquad 9d.$$

for $0 \leqslant T \leqslant 30°C$ and $D_{50} \leqslant 0.6$mm.

Equation 8 gives values of $(\Delta/L)_{rm}$ of 0.19, 0.144 and 0.082 for D_{50} values of 5.10^{-5}m, 1.10^{-4}m and 6.10^{-4}m, respectively, and may mean that modification is required for larger particles.

3. Time development of bedforms

Although considerable work has been done on the movement of sand waves, see for example Dyer (1986), there is little quantitative information on developing bedforms. Early work by Jain and Kennedy (1971), suggested that 'dunes' (the dunes may actually be ripples, based on Yalin's criteria of wavelength and X value (9.5)) approach a stable equilibrium (steady flow) wavelength (L_0) exponentially, that is:-

$$L/L_0 = 1 - \exp(- a\varphi) \qquad\qquad 10.$$

Yalin (1975) extended this earlier work and also suggested an exponential growth in wave height and length, although the developed equations are more complex than equation 10 and are difficult to use in practice, see Duckett (1984).

Most workers are agreed that ripples rapidly reach equilibrium values while dunes take many hours or days to reach equilibrium.

4. Bedforms in tidal conditions

Considerable observational work has been done in this area, see Duckett (1984) and Dyer (1986) and some empirical models exist, see for example, Allen (1982). Many authors confirm the discrepancies between laboratory-based equations and field data, see for example, Dalrymple et al (1978). Other authors, see for example, Terwindt (1971), suggest that ripples rapidly adjust to changing tidal conditions but that dunes persist relatively unchanged during a tidal cycle.

Fredsoe (1979) developed a model to predict initial change in dune height following a sudden change in river discharge in a river. He presented useful equations for dune height and length and tested his model with good results in a river situation with weakly varying flows. He also predicted bed friction by using relationships for grain and form drag, see below. The effect of waves has also recently been included (1986).

Further useful ideas have been contributed by the laboratory tests of Tsujimoto and Nakagawa (1984), who developed empirical equations to quantify the unsteady behaviour of dunes. Unfortunately, the equations did not provide such a good fit to experimental data and they are difficult to use in practice. However, the work suggested two points. Firstly, dune wavelength was found to be reasonably constant as dune height decayed and dune decay was more rapid than dune growth by a factor of two. These ideas have been incorporated in the present method of bedform prediction, described below.

5. Friction relationships

In steady flow situations, it has been usual to simply combine the effects of grain (surface) drag and form (shape) drag, ignoring the effect of grain impacts due to suspended load. In friction factor terms, the total friction factor (f) is divided into grain (f") and form components (f') components, that is:-

$$f = f' + f" \qquad \qquad \text{11a.}$$

where $f = 8Y.S.D_{50}.g/U^2$ 11b.

and $f" = 1.28 \left(\ell n (12h/k_s) \right)^{-2}$ 11c.

for fully turbulent flow with a logarithmic velocity variation over the flow depth for a bed of equivalent roughness height k_s. Since k_s and f' are likely to be related to the size and shape of the bedforms produced by flow with a depth-mean velocity of U, and bedform characteristics are related to Y, it follows that an iterative solution of equation 11 is required.

The value of k_s to use in equation 11c depends on the presence of ripples on the bed. For a flat bed, the effect of surface grain drag is usually related to some characteristic diameter of the bed sediment. For example, Engel and Lau (1980) suggest 2.5 D_{50} while Van Rijn (1982) suggests 3 D_{90}. Fortunately, equation 11c is not too sensitive to the precise choice of value. If ripples are present, k_s is usually related to ripple geometry. For example, Van Rijn (1982) suggests that:-

$$k_s = k_s \text{ (flat bed)} + 1.1 \, \Delta_r (1 - \exp(-25(\Delta/L)_r) \qquad \qquad \text{12.}$$

for $(\Delta/L)_r$ in the range 0.01 - 0.2; D_{50} in the range 0.1 - 2.4mm; U between 0.25 -1.1 m/sec; and h between 0.08 - 0.75m.

Van Rijn (1982) also noted that equation 12 gave somewhat lower values than the earlier work of Swart (1976) and Shinohara and Tsubaki (1959).

The effect of dunes is included through f' and numerous relationships exist. For example, Engelund (1966) suggested the equation:-

$$f' = 4(\Delta/h)_d(\Delta/L)_d \tag{13}$$

and later revised it to include a variable coefficient, that is:-

$$f' = [10 \exp(-2.5(\Delta/h)_d] . (\Delta/h)_d (\Delta/L)_d \tag{14}$$

Laboratory experiments by Engel and Lau (1980) also demonstrated the importance of dune steepness. For example, their results can be written in the form, see McDowell and O'Connor (1981):-

$$f = 1.31(\Delta/L)_d + 0.0165 \tag{15}$$

for Z > 1000.

Equation 15 also suggests that the form and surface drag should not be sub-divided. However, if the grain drag in Engel and Lau's experiments for Z > 1000 is examined, it is found to be of the same order as the constant (0.0175). Consequently, the first term in equation 15 approximates to f'.

Theoretical work by Haque and Mahmood (1983) using FEM techniques to solve inviscid flow equations for pressure variations along the bedform surface also tends to confirm the functional dependence of f' on Δ/h and Δ/L for $\Delta/h = 0.10$, although answers are less good for smaller and larger values, probably due to the inviscid nature of the equations, see Duckett (1984).

6.   Present model of tidal friction

Following Terwindt's (1971) field observations, it is assumed that ripples respond instantaneously to changes in flow condition but that dune dimensions change only slowly from one tidal cycle to another so that the flow and bedform conditions for most of the tidal cycle can be taken as in local equilibrium. It is also envisaged that dune dimensions of height and wavelength remain nearly constant during the tidal cycle except for a shift in directional orientation. Bed friction is thus taken as being represented by equation 11, as in steady flow, but with parameters controlled by the local depth-mean flow velocity (U), mean bed grain size (D_{50}) and the controlling influences of tidal period (T_0) and the Spring/Neap tidal variation of maximum tidal velocity (U_m), all of which contribute to the size of ripples and dunes found on the seabed during any particular local tide in the Spring/Neap cycle.

Since ripples respond quickly to local flow conditions, f" can be determined from equation 11c, with k_s determined from equation 12 and $(\Delta/L)_r$ determined from equations 7 and 8 with appropriate values of a_0, α and β; it is necessary to know values for the water temperature T, D_{50} and Y.

To determine f', equation 14 is used but with Δ_d and L_d values appropriate to the local tide. It is believed that the reversing nature of the tidal cycle limits the potential for dunes to grow to the large size found in rivers. The wavelength of tidal dunes is thus limited by the size of the maximum velocity during any particular tide and the time available during the flood or ebb phase of tide when velocities are in excess of some critical value for sediment movement. Since the Spring tide has the largest values of velocity and period available for sediment transport, it is believed that the Spring tide provides the wavelength of tidal dunes.

The height of any local tide dune is believed to remain more or less constant during the tidal cycle, as suggested by Terwindt's (1971) field observations. All that happens during the tidal cycle is a re-orientation to the flood or ebb flow direction. However, dunes are allowed to decay or grow in height slowly from one tidal cycle to another, if conditions allow, the latter being influenced by the location of the local tide in the Spring/Neap cycle.

The actual height of any local tide dune is believed to be the result of a balance between the potential of an individual tide to 'grow' a dune compatible with local maximum tidal conditions, and the destructive nature of each successive tide in destroying dune height created on earlier tides. The actual dune height is thus the maximum value of these 'growth' and 'destructive' processes.

Thus, on any local tide, the maximum size of dune that can be 'grown' by local conditions is determined by an iterative solution of equation 11 at maximum velocity (U_m) with f" determined as indicated earlier and f' determined by equation 14 with L_d specified as the maximum Spring tide value. The maximum size of 'destructive' dune that can exist on the local tide is found by allowing the local size of dune from all preceding tides, starting from the maximum Spring tide value, to decay at a specified rate, and then selecting the maximum value. By comparing the size of 'destructive' dune from the 1st, 2nd, 3rd etc preceding tide with the maximum locally 'grown' dune, it is possible to determine the overall maximum value. Clearly, if dunes from former tides decay rapidly, the overall maximum value will be the locally 'grown' dune. This probably tends to happen on the rising leg of the Spring/Neap cycle. However, if dunes decay slowly, the overall maximum value may be achieved by one of the decayed dunes rather than the locally 'grown' value and may tend to happen on the falling leg of the Spring/Neap cycle.

Once the maximum dune height is determined for the local tide, it is then possible to determine the friction factor at any tidal phase within the tide by an iterative solution of equation 11 to find f and Δ_r, since L_r, Δ_d and L_d (equal to the Spring tide value) are now known.

A check is also made during iteration on the values of X and Y produced by the calculation method and comparison made with their limiting values for ripple, ripple and dune, and dune occurrence, see Section 2 and figure 1.

Once f is determined, it is possible to calculate the local bed shear stress (τ_b) or shear velocity (U_*) from the equations

$$\tau = \rho f\ U^2/8 \qquad\qquad\qquad\qquad 16a.$$

and $\quad U_* = \sqrt{f/8}.\ U \qquad\qquad\qquad\qquad 16b.$

7. Dune wavelength in tidal flows

It is believed that the reversing nature of tidal flows limits the growth of dunes. Thus, the only period of time available (t_*) to cause dune growth is that available on the flood/ebb phase, once velocities exceed threshold values (U_c). By treating the flood or ebb phase as an equivalent steady flow with a characteristic velocity, taken as U_m, and depth, taken as the tidal average value (h_*), it is possible to describe the tidal dune wavelength using equation 10, that is:-

$$L_d/L_0 = 1 - \exp(- aU_m t_*/h_*) \qquad\qquad 17a.$$

where $\quad L_0 = 2\pi h_* \qquad\qquad\qquad\qquad\qquad 17b.$

t_* can be determined from a knowledge of U_c and the velocity variation with time during the tidal cycle. For example, for a sinusoidal variation of U, that is:-

$$U = U_m \sin (2\pi t/T_0) \qquad\qquad\qquad 18a.$$

$$t_* = T(1 - 4t_1/T_0)2 \qquad\qquad\qquad 18b.$$

$$t_1 = T_0 \sin^{-1}(U_c/U_m)/(2\pi) \qquad\qquad 18c.$$

By fitting equation 17 to field data, it has been possible to determine values for parameter a, see Duckett (1984) for details.

Since the tidal effect has been averaged out in equation 17, it is likely that L_d/L_0 is related to the sediment transport parameters (X,Y,Z,W) or to their combination ($\xi = X^2/Y$, $\eta = Y/Y_c$; $\zeta = Y/Z$) since W is likely to be constant in the field, see

Yalin (1977). Since equation 17 also contains U_m, which can be related to Y, it follows that a is likely to be a function of only (ξ, ζ), with ξ providing the major influence, since it has been used as a major parameter in recent sediment transport equations, see for example, Ackers and White (1973).

Correlation of a with ξ, see figure 3, produced the equation:-

$$a = 10^B \qquad\qquad 19a.$$

where $B = 1.29 \log (\xi) - 7.13$ 19b.

based on a ν value for $T = 20°C$, since temperature data was not readily available.

By varying T in the range $0°-30°C$, it was found possible to encompass much of the scatter in the correlation, see figure 3, and suggests that the influence of ζ may be of secondary importance. Given that $10°C$ is more likely as an average water temperature for the field observations, it may be more appropriate to calculate ξ_{10} (for $T = 10°C$) and to use a reduced value for the constant in equation 19, that is, 6.84 in place of 7.13.

The value of B for any other temperatures $(T°C)$ may then be obtained from the equation:-

$$B = 1.29 \log (\xi_{10}) + 2.58 \log (\nu_T/\nu_{10}) - 6.84 \qquad\qquad 19c.$$

where ν_T is the value of ν at $T°C$ and ν_{10} is the value of ν at $10°C$.

A measure of the accuracy of fit of equation 19 to field data is provided by figure 4 (correlation coefficient = 0.92).

8. Field applications

Unfortunately, there is little comprehensive field data available to fully test the present ideas. However, two applications have been chosen to illustrate separate aspects of the present approach. Firstly, data from the Ribble Estuary (UK) is used to predict dune dimensions and to illustrate frictional variations during the Spring/Neap cycle, and secondly, shear velocity predictions are compared with values derived from field velocity profiles for a coastal site near the Sizewell/Dunwich Banks, East Anglia, UK.

For ease of application, the various equations have been drawn together into a computer programme, with the addition of an equation to describe the decay of dunes generated on earlier tides in the Spring/Neap cycle. For example, if dune decay occurs at the same rate as dune growth:-

$$\Delta_n = \Delta_i \exp(- a \sum_{i+1}^{n} U_{m(i+1)} \cdot t_{*(i+1)}/h_*) \qquad\qquad 20.$$

where n is the local tide in the Spring/Neap cycle (e.g. tide 4) and i is an earlier tide value (e.g. tide 1, tide 2 or tide 3).

For the Ribble conditions, equation 20 indicates a 50% decay over some 5-6 tides. More rapid decay can be obtained by scaling up the exponential function.

For both test cases, it has also been necessary to make various assumptions due to lack of information. In particular, it has been assumed that tides are semi-diurnal (T = 12.42 hrs); tidal ranges vary linearly between Spring and Neap tide values over a 28 tide cycle; maximum tidal velocities vary linearly with tidal range; the tidal velocities vary sinusoidally over each tidal cycle; the water depth (h) varies about the tidal-mean value (h_*) as a cosine function; and the critical velocity (U_c) for sediment motion on a rippled/duned covered bed is 0.54 m/sec, based on field observations by Dyer (1980).

8.1 Ribble Estuary, UK (18)

For the Ribble tests, use was made of the known parameters h_* (= 8.6m); D_{50} (= 230 μm); and U_m (= 1.5 m/s). In addition, it was assumed that $(\Delta/L)_{dm}$ = 0.06; Y_c = 0.05; the tidal range varies from 6.4-8.2m from Neap to Spring tides; and T = 20°C (to produce the correct value of a from equation 19).

The results of the computation for dune characteristics is shown in table 1 for a selection of tides in the 28 tide cycle. It can be seen that dune wavelength is fixed by the Spring tide at 8.3m, while dune height varies from 0.3-0.4m and may be compared with observed values of 6.9m and 0.46m, respectively (18). Table 1 also shows larger bedforms on neap tides, suggesting that any reduction in maximum tidal velocity, as for example by deeper dredging of the estuary, would lead to the formation of larger dunes. The lack of development of the Ribble dunes is also evidence from table 1, since L_0 = 54m, indicating only 15% growth.

The variation of total friction factor during a particular tide, the largest Spring tide, is shown in figure 5. It can be seen that the largest values occur towards low water when both ripples and dunes are present.

By examining identical tides on the falling (tide 5) and rising (tide 25) legs of the Spring/Neap cycle, it is possible to see the effect of previous flow history on bed roughness. Table 2 shows that differences of some 5-10% are found in total friction factor and 2-4% in shear velocity. From an engineering point of view these differences would not be considered as significant. However, larger differences may be possible for other situations.

8.2 Sizewell/Dunwich, East Anglia, UK (20)

The second application was to a nearshore coastal site in East Anglia, UK, were data on tidal velocities and suspended sediment transport rates were collected by Lees (1981). Unfortunately, a complete set of data was unavailable but shear velocities were estimated from the measured velocity profiles.

The computer programme was run for known values of h_* (= 14m), D_{50} (= 144 μm) and assumed values of U_m (= 1.5m/s), Y_c (= 0.06), $(\Delta/\lambda)_d$ = 0.06; and tidal ranges of 1.1-1.9m.

The results indicated that no dunes were present on the seabed, probably due to the small size of the bed sediment (X < 8), but a rippled bed was predicted for most of the tidal cycle. A comparison of the predicted U_* values with those of Lees is shown in figure 6. It is apparent that quite a good comparison is obtained.

8.3 Maximum sediment transport rate

Given that the computer programme gives a reasonable estimate of the maximum shear stress during the tidal cycle, it may also be possible to predict maximum sediment transport rates by use of a suitable equation. For example, the Engelund and Hansen (1967) formula for total load has the form:-

$$q = \frac{0.4\ \rho_s}{f} (g\ S\ D_{50}{}^3)^{0.5} . Y^{2.5} \quad \text{kg/m-s} \qquad 21.$$

and although produced for dune-covered beds of somewhat coarser material than that at Sizewell, it produces values (0.8 kg/m-s) at maximum flow of the same order as measured by Lees (1.3-3.1 kg/m-s). Clearly, the prediction of sediment transport is an area requiring future study.

9. Sensitivity testing

A limited amount of sensitivity testing has been done with the present approach, see Duckett (1984). Tests on equation 19, using the Ribble test data suggest that errors in D_{50} and ν are the most important ones. However, this data is readily obtained from field information.

Other tests used a reduced dune decay equation, increased maximum dune steepness (0.065 and 0.080) and Swart's (1976) equation for ripple roughness. The results suggest that only the change in dune steepness is significant, producing a 25% increase in dune height for the largest steepness value tested (0.080).

10. Conclusions

The simple model presented herein for the prediction of bedform dimensions and friction factors in well-mixed estuaries and coastal zones seems to give realistic answers for the situations tested. The method uses ideas on bedform steepness from work in rivers, which is known to be realistic, and a new relationship for the wavelength of tidal dunes, which seems to give a reasonable fit to a range of field data. Further testing of the approach on more comprehensive field data would enable the present ideas to be developed further and make them more universally applicable.

11. References

1. Ackers, P. and White, W.R.: "Sediment transport: new approach and analysis". Proc. ASCE, Hydraulics Div., 99, HY11, Nov. 1973, pp. 2041-2060.

2. Allen, J.R.L.: "Simple model for the shape and symmetry of tidal sand waves". Marine Geology, 48, 1982, pp. 31-73.

3. Bouma, A.H., Hampton, M.A., Rappeport, M.L., Whitney, J.W., Teleki, P.G., Orlando, R.C. and Torreson, M.E.: "Movement of sand waves in Lower Cook Inlet, Alaska". In: Proc. Offshore Technol. Conference (Houston, USA, 1978), 1978, Paper OTC 3311, pp. 2271-2284.

4. Dalrymple, R.W., Knight, R.J. and Lambiase, J.J.: "Bedforms and their hydraulic stability relationships in a tidal environment, Bay of Fundy, Canada". Nature, 275, 1978, p. 100-104.

5. D'Anglejan, B.F.: "Submarine sand dunes in the St. Lawrence Estuary". Canadian Journal of Earth Sciences, 8, 1971, pp. 1480-1486.

6. Diegaard, R. and Fredsoe, J.: In: Proc. Coastal Engineering (Taipei, Taiwan, November 9-14, 1986), II, 1986, paper 78, pp. 1047-1061.

7. Dyer, K.R.: "Velocity profiles over a rippled bed and the threshold of movement of sand". Estuarine and Coastal Marine Science, 10, 1980, pp. 181-199.

8. Dyer, K.R.: "Coastal and estuarine sediment dynamics". Wiley & Sons Ltd, UK, 1986.

9. Duckett, F.: "A study of friction in unsteady flows". M.Sc Thesis, University of Manchester, 1984.

10. Engel, P. and Lau, Y.L.: "Friction factor for two-dimensional dune roughness". J. Hydraulic Research, 18, 3, 1980, pp. 213-225.

11. Engelund, F.: "Hydraulic resistance of alluvial streams". Proc. ASCE, Hydraulics Div., 92, 1966, pp. 315-326.

12. Engelund, F.: "Hydraulic resistance for flow over dunes". I.S.V.A., Tech. Univ. Denmark, Prog. Rep. No. 44, 1978.

13. Fredsoe, J.: "The friction factor and height-length relations in flow over a dune-covered bed". Inst. Hydrodyn. and Hydraulic Engrg, Tech. Univ. Denmark, Prog. Rep. 37, Dec. 1975, pp. 31-36.

14. Fredsoe, J.: "Unsteady flow in straight alluvial streams: modification of individual dunes". J. Fluid Mech., 91, 3, 1979, pp. 497-512.

15. Fredsoe, J.: "Shape and dimensions of stationary dunes in rivers", Proc. ASCE, Hydraulics Div., 108, No. HY8, Aug. 1982, pp. 932-947.

16. Haque, M.I. and Mahmood, K.: "Analytical determination of form friction factor". Proc. ASCE, Hydraulics Div., 109, 1, 1983, pp. 590-610.

17. Harvey, J.G.: "Large sand waves in the Irish Sea". Marine Geology, 4, 1966, pp. 19-55.

18. Hydraulics Research Station.: "Investigation of siltation in the estuary of the river Ribble". Report No. EX 281, 1965.

19. Jain, S.C. and J.F. Kennedy.: "The growth of sand wave", In: Proc. Intern. Symposium on Stochastic Hydraulics (Pittsburgh, USA, 1971), 1971, pp. 449-471.

20. Lees, B.J.: "Sediment transport measurements in the Sizewell-Dunwich Banks area, East Anglia, U.K.". Spec. Publs. Int. Ass. Sediment., 5, 1981, pp. 269-281.

21. McDowell, D.M. and O'Connor, B.A.: "Numerical analysis of sediment transport in coastal engineering problems". Terra et Aqua, 23/24, 1981, pp. 11-16.

22. Muir Wood, A.M. and Fleming, C.A.: "Coastal hydraulics". London, UK, McMillan, 1981.

23. Nasner, H.: "Prediction of the height of tidal dunes in estuaries". Proc. ASCE Conf. on Coastal Engineering (Copenhagen, Denmark, June 24-28, 1974), II, 1974, pp. 1036-1050.

24. Shinohara, K. and Tsubaki, T.: "On the characteristics of sand waves formed upon beds of the open channels and rivers". Research Institute for Applied Mechanics, VII, 25, Kyushu University, Kyushu, Japan.

25. Swart, D.H.: "Predictive equations regarding coastal transport". In: Proc. 15th Coastal Eng. Congress (Honolulu, Hawaii, 1976), II, paper H1, 1976, pp. 1113-1133.

26. Terwindt, J.H.J.: "Sand waves in the southern bight of the North Sea". Mar. Geol., 10, 1971, pp. 51-67.

27. Tsujimoto, T. and Nakagawa, H.: "Unsteady behaviour of dunes". In: Proc. 1st Internat. Conf. on Hydraulic Design in Water Res. Eng. (Southampton, UK, 1984), paper 5, pp. 85-99.

28. van Rijn, L.C.: "Equivalent roughness of alluvial bed", Proc. ASCE, Hydraulics Div., 108, HY10, 1982, pp. 1215-1218.

29. Yalin, M.S.: "Mechanics of sediment transport". London, UK, Pergamon Press, 1977.

30. Yalin, M.S. and Karahan, E.: "Steepness of sedimentary dunes". Proc. ASCE, Hydraulics Div., 105, HY4, April 1979, pp. 381-392.

31. Zarillo, G.A.: "Stability of bedforms in a tidal environment". Marine Geology, 48, 1982, pp. 337-351.

Table 1. Dune characteristics, Ribble Estuary, UK

Tide	$(\Delta/L)_d$	$L_d(m)$	$\Delta_d(m)$
1(Sp)	0.037	8.3	0.30
5	0.042	8.3	0.35
10	0.043	8.3	0.36
15(N)	0.047	8.3	0.39
20	0.043	8.3	0.36
25	0.039	8.3	0.33
28(s)	0.038	8.3	0.32

Table 2. Tidal variation of friction parameters for idential tides, Ribble Estuary, UK

Time (hrs)	h (m)	U (m/s)	f'	f"	f	U_* (m/s)	Tide
0	12.4	0	0.011	0.0083	0.0192	0	5
	12.4	0	0.0096	0.0083	0.0179	0	25
1.01	11.9	0.69	0.0114	0.0147	0.0261	0.0394	5
	11.9	0.69	0.010	0.0153	0.0253	0.0388	25
3.04	8.7	1.40	0.0152	0.0088	0.0240	0.0769	5
	8.7	1.40	0.0133	0.0088	0.0221	0.0739	25
5.32	5.1	0.61	0.0241	0.0145	0.0386	0.0424	5
	5.1	0.61	0.0212	0.0156	0.0368	0.0414	25

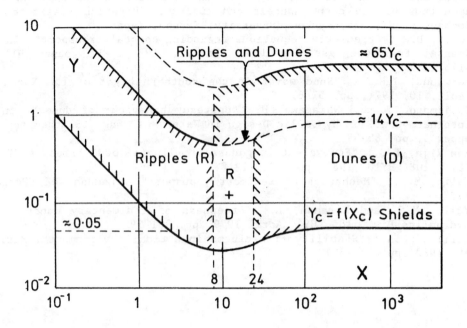

1.   Ripple and dune domains, after Yalin (29).

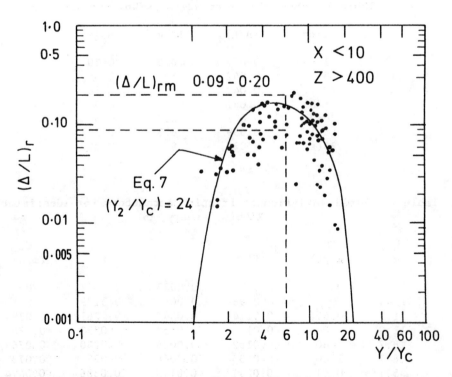

2.   Comparison of ripple steepness from eq. 7 with observations, Yalin (29)

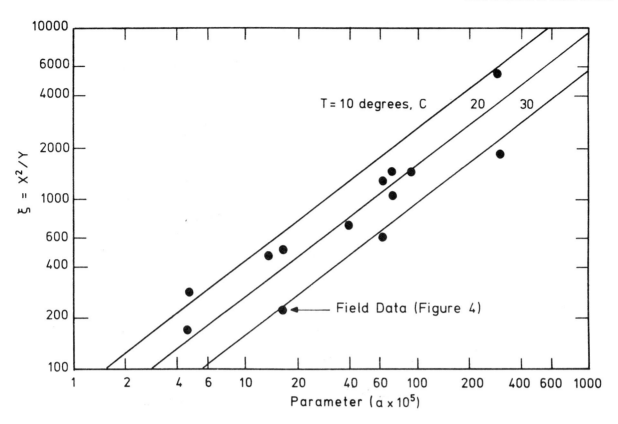

3. Variation of bedform growth parameter a with ξ.

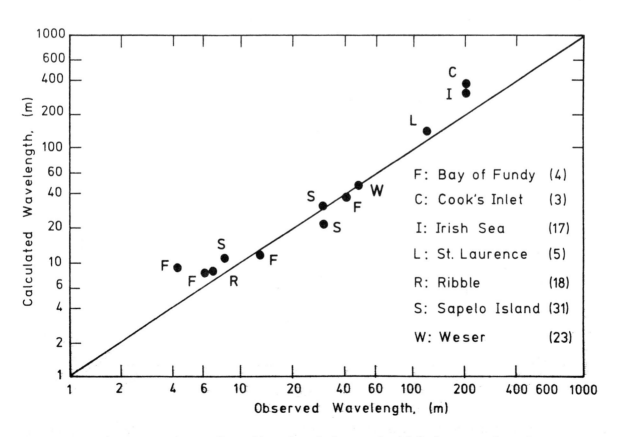

4. Comparison of predicted and observed tidal dune wavelengths.

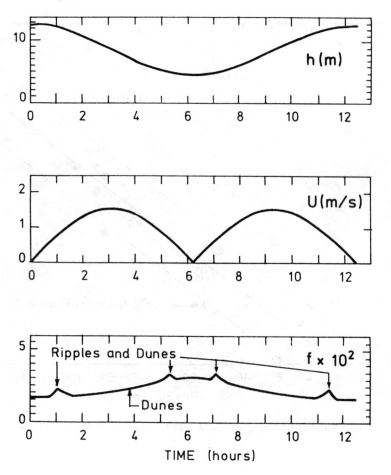

5. Variation of water depth, depth-mean velocity and predicted total friction factor for a typical Spring tide, Ribble Estuary, UK.

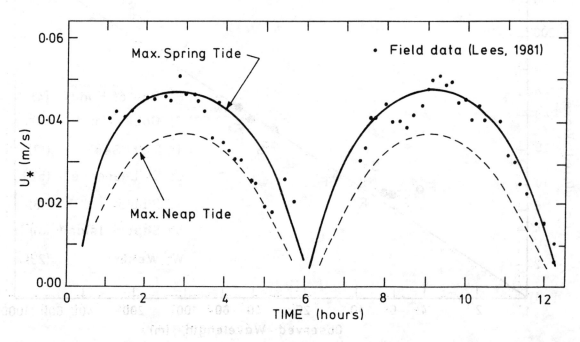

6. Variation of calculated (Lees) and predicted Spring and Neap tide U_* values for Sizewell/Dunwich area, East Anglia, UK.

PART 2
Coastal Waters

Chapter 17

A REVIEW OF THE USE OF HF RADAR (OSCR) FOR MEASURING NEAR-SHORE SURFACE CURRENTS

D Prandle
Proudman Oceanographic Laboratory, UK

ABSTRACT

By examining results from a number of OSCR deployments it is shown that the system can operate under a wide variety of conditions and measure the predominant tidal currents to within an accuracy of 5cm s^{-1}, with phase and direction within 5°, from as little as 7-10 days of half-hourly recordings. Theoretical calculations indicate more accurate measurements but, in reality, significant spatial variability of the flow field may limit such accuracy. These values can be significantly improved by more frequent observations over longer sampling intervals extending for a longer period (a month or longer) and with careful attention to the alignment of the beams relative to the predominant current directions. Useful information has also been obtained regarding both the net residual surface drift and the wind-driven components. In three deployments in the Irish Sea, the magnitude (>10cm s^{-1}) of these residual drifts is of first order importance for the spread of buoyant contaminants.

Present development of the OSCR system will provide a miniaturised, automated system with reduced deployment costs and improved temporal and spatial coverage.

1. INTRODUCTION

The Proudman Oceanographic Laboratory has been involved in more than 10 deployments of the OSCR (Ocean Surface Current Radar) H.F. Radar system starting with the preliminary trials in 1983 and continuing with two deployments planned in 1988. This paper represents a summary of general experience gained and is intended to provide guidance to prospective users of HF Radar systems. In particular, information is presented pertinent to a number of commonly encountered questions concerning the reliability and accuracy of the measurements, the choice of sites and necessary duration of observations.

2.   THE OSCR HF RADAR

Measurements of high frequency radar echoes back-scattered from the sea surface have been used to derive information on both waves, surface winds and surface currents (Barrick 1978, Shearman 1983 and Leise 1984). The OSCR system was developed at the Rutherford Appleton Laboratory (King et al 1984), it uses a beam-forming receive antenna comprising 16 separate elements equi-spaced over a line 90m long oriented orthogonally to the centre line of the radar beam coverage.   Signals from 16 beams each 6° wide can be differentiated, along each beam signals can be separated into 'bin' lengths of 1.2 or 2.4km. Operating at a frequency of 27mHz measurements can be made over distances up to 40km offshore.

The measured spectra exhibit two peaks corresponding to a resonance between the radio wavelength λ_R = 10m and both advancing and receding surface waves of length λ_W = $\lambda_R/2$, (figure 1). In still water, these 'Bragg' peaks occur at frequency shifts of $\pm C/\lambda_W$ where C is the wave celerity. Where the waves propagate over a background surface current U, the Bragg peaks are offset by $\Delta f = U/\lambda_W$.

The observational period (or dwell time τ) used is in multiples of 67 seconds providing spectrum resolution δf of $1/\tau$. Thus the resolution δU of U is given by

$$\delta U = \lambda_w . \delta f = 500/67_n$$

(1)

i.e. 3.7cm s^{-1} for n=2, 1.9cm s^{-1} for n=4 etc.
The surface current U represents a scalar component of the actual surface vector R resolved along the radar beam direction. To obtain the spatial distribution of surface current vectors, two radar units are used and values of R derived at beam-crossing points within the region of overlap.
While some uncertainty exists in defining the depth of the surface layer to which the radar current measurements pertain, a value of the order of 1m (for λ_R = 10m) is generally assumed (Stewart and Joy 1974, Ha 1979).

3.   RANGE OF OBSERVATIONAL CONDITIONS

Ideally it seems the OSCR units should be located in close proximity to the shore.   However successful deployments have been made from a wide range of locations including a rock shelf 100m above sea level (Red Wharf Bay) and from a location 10km inland (Cumbria).

The optimum sea-state for OSCR operation is a near flat-calm. As wave heights increase the level of the background spectra rises thereby reducing the distinction of the Bragg peaks. Since the magnitude of this backgrtound level relative to the Bragg peaks is larger for more distant bins, increasing sea-state eventually leads to an effective reduction in the range of OSCR. Precise quantification of the relationship between range and sea-state has not been attempted but, in broad terms, slight data loss may be expected for wind speeds of 10m s^{-1} while successful observations have continued in winds exceeding 20m s^{-1} . In those deployments using 20 1.2km bins, i.e. a range of up to 25kms, observational losses rarely exceed 5%, with minor logistical difficulties probably accounting for half of these losses. The only instance in which the sea-state was apparently insufficient to identify Bragg peaks occurred in measurements within a pool of sea water at the head of an estuary cut off from the open-sea on the ebbing tide.

The sampling rate for conventional moored current meters is generally at 10 minute intervals, one hourly intervals may be regarded as the maximum to resolve tidal motions while a smaller value is required for typical wind-driven currents.   The present OSCR system can only sample one beam at a time and, moreover, only 1 of the two units is operational at any one time (to avoid

246

possible signal interference). Thus where say 12 beams from each site are to be sampled, a sampling interval of only 2 x 67s must be adopted to provide hourly values. Even in those deployments where only 6 beams are monitored it is usual to retain this same sampling duration providing values at half-hourly intervals.

4. ACCURACY OF MEASUREMENTS OF TIDAL CONSTITUENTS

Godin (1972) showed that the accuracy of resolution of well-separated tidal constituents is given by

$$\Delta A = \nu \sqrt{2/N} \tag{2}$$

$$\Delta g = \frac{\nu}{A} \sqrt{\frac{2}{N}} \tag{3}$$

where ΔA is the standard error in the tidal amplitude A, Δg the standard error in the phase g, ν the standard error of the (white) noise and N the number of observations.

By examining the spatial variability of A and g along a beam aligned with the major axis of the (predominant) M_2 tidal ellipse, Prandle (1987) deduced a value of $\nu \approx 4 cm\ s^{-1}$. These data were taken with $T = 124s$ and hence this value of ν is in close accordance with the theoretical value from (1) of $3.7 cm\ s^{-1}$.

Tidal Ellipses

The above results refer to the accuracy of the separate radial current components R_1 and R_2. In combining these to calculate the magnitude of the major axis of the tidal ellipses, R, an equation of the form

$$R = \frac{R_1 \sin(\alpha + \beta/2) - R_2 \sin(\alpha - \beta/2)}{\sin \beta} \tag{4}$$

is used, where α is the inclination of the bisector of the two beams to the current vector direction and β is the angle between the two beams (see figure 2). By assuming ν is a constant independent of either α or β, the resulting standard error in R can be calculated from (4). Figure 2 shows the result as an amplification factor of ν expressed as a function of both α and β. This figure confirms the criterion, adapted in earlier POL studies, that beam crossing angles (β) should exceed 30°. However the figure also emphasises the importance of aligning the beams as closely as possible to the major current direction (whenever this is known beforehand). These two conditions can be conveniently combined by noting that for $\beta > 1.3\alpha$ the error amplification factor is always less than 1.5.

Two important conclusions may be drawn from these latter results. First, for tidal current ellipses with large ellipticities (i.e. currents rotate progressively rather than simply oscillate between opposing flood and ebb directions) the error amplification factor cannot easily be reduced since the alignment angle α will always be large during a significant part of the tidal cycle. Second, the possibility of extrapolating the current vector results beyond the overlap region using results from a single site is shown to have potentially large errors since (except when $\alpha \rightarrow 0$) any combination of data from adjacent beams is akin to using data from crossing beams in which β is only 6°.

5. M_2 TIDAL ELLIPSE PARAMETERS MEASURED BY (i) OSCR AND (ii) MOORED CURRENT METERS

Table 1 summarises comparisons of M_2 tidal ellipse parameters as measured by OSCR and by adjacent moored current meters in four separate deployments. In the

Advances in water modelling and measurement

first comparison (Prandle and Ryder 1985), two current meter rigs were used located 500m apart. The difference between the ellipse parameters for these two moorings was of similar magnitude to the difference between the OSCR data and each mooring. Similarly in the second comparison (Prandle 1987) the much larger OSCR amplitude at the surface can be shown to be part of a pronounced vertical current structure at that location - an additional bottom-mounted current meter at this same location indicates a well-ordered change in all four ellipse parameters between surface and bed.

In summary, while table 1 indicates encouraging agreement between tidal ellipse parameters measured by OSCR and moored current meters, the differences exhibited are of the same order as the corresponding vertical and horizontal variability in these parameters and hence no direct conclusions regarding the accuracy of the OSCR measurements can be made.

6. SELF-CONSISTENCY OF TIDAL PARAMETERS MEASURED BY OSCR

A recent OSCR experiment, off the Fylde Coast (Prandle 1988), presented an interesting opportunity to check for self-consistency. Measurements were first made at $\frac{1}{2}$-hourly intervals (134 second observational period) for 10 days from Walney Island (figure 3) with a bin length of 2.4kms concurrently with measurements from Blackpool. About a week later the measurements from Walney Island were repeated but with a bin length of 1.2km and with the second unit now located at Fleetwood.

Here we examine the self-consistency of the M_2 tidal constituent as measured (i) from 1.2 and 2.4km bins along the same beam from Walney (radial components) and (ii) from combinations of radial values of beams from Walney, Blackpool and Fleetwood that intersect in close proximity.

6.1 M_2 radial components along beam 186° from Walney Island for 1.2 and 2.4km bins

Figures 4(a) and 4(b) show, respectively, M_2 amplitude and phase values along beam 186° for both 1.2 and 2.4km bins. Table 2 quantifies this comparison and extends the comparison to include S_2, M_4 and Z_0 (residual). M_2 results for the two bin lengths show averaged discrepancies of 5.2cm s^{-1} in amplitude and 9.5° in phase. Much of this can be attributed to the pronounced spatial variability in tidal currents, although some systematic difference appears in the more off-shore bins. The latter difference may be partly related to the short duration of the observations, 15, or preferably 29, days of observations will significantly reduce the errors in resolving close-period tidal constituents.

Similar results were found from comparisons of data from other beams at Walney Island.

6.2 M_2 tidal ellipses calculated from combinations of beams from 3 sites that intersect in close proximity.

Figures 5(a) and 5(b) show the M_2 tidal ellipse distributions obtained by combining radial data from Walney Island, Blackpool and Fleetwood. The separate tidal data for 1.2 and 2.4km bins used from Walney Island were vector-averaged. Particular attention is focussed on the locations A to F at each of which 3 separate beams intersect in close proximity.

While there is generally good broad agreement between the separate sets of results, close examination of the distributions (figure 5(b)) indicates systematic differences. In particular, the high level of spatial consistency found between results from any two sites (with values almost always lying appropriately between adjacent values thereby yielding smooth spatial distributions) is not maintained between results from different sites. At all seven positions A to F, the highest amplitude is derived from the combinations of beams from Walney Island and Blackpool while the smallest amplitude is derived from combinations of beams from Blackpool and Fleetwood. The magnitude of these discrepancies is shown in table 3, amounting, on average, to $\pm$5cm s^{-1} in amplitude, $\pm$ 5° in direction, $\pm$3° in phase and $\pm$0.05 in eccentricity. From the theory of (2) and (3), for an estimated noise level of 4cm s^{-1} with approximately 500 observations, the standard error in amplitude should be reduced to 0.25cm s^{-1} and in phase to 0.24° (with A $\simeq$ 60cm s^{-1}). The much higher discrepancies found at locations A to F may be attributed to (i) short observation periods, (ii) error amplification due to relative alignment of beams and current vector (figure 1) and (iii) spatial variability in the flow field. From the results examined in section 6.1, the latter factor seems pronounced in this region which lies close to the outflow of a major bay.

7. MEASUREMENT OF NON-TIDAL CURRENT COMPONENTS WITH OSCR

Despite the discrepancies discussed in earlier sections, the measurement of surface tidal currents provided by OSCR is sufficiently accurate (and is particularly well-suited) for the requirements of coastal engineers engaged in the design of sea outfalls etc. Indeed, agreement between OSCR measurements and tidal currents obtained from numerical models has generally been so close that some modellers have concluded that OSCR measurements are unnecessary where good models exist. Without pursuing this latter circuitous argument, it is important to recognise the usefulness of OSCR in measuring non-tidal components, in particular the net steady surface drift and the wind-driven component.

In examining the latter, it must be recognised that longer-term deployments are necessary to accurately resolve these components, generally at least 30 days. Independent observational verification of these components is much more difficult than for the tidal components.

7.1 Wind-driven component

Collar and Howarth (1987) describe earlier attempts to evaluate the accuracy of HF Radar measurements of surface currents. In two experiments, OSCR measurements were compared directly with freely drifting drogues and also with moored current meters. Despite the inherent differences in the sensitivity and response of these instruments over the range of time and space scales of water movements, the overall agreement was encouraging.

These comparisons were specifically aimed at developing techniques to measure near-surface wind driven currents and, thus, unlike other OSCR deployments reported here, locations with small tidal currents were chosen. In light to moderate wind conditions, time-averaged agreement between OSCR measurements and floats (with centre of drag 0.5m below the surface) was close to 1cm s^{-1} with a standard deviation of 3cm s^{-1}. In stronger winds, OSCR measurements of the wind-driven component sometimes exceeded the other observations by approximately 10cm s^{-1} with the difference being related to the relative direction of surface wave propagation and current.

Prandle (1987) showed that for the Red Wharf Bay experiment an average of 66% of the variance of the non-tidal current could be accounted for by a single slab-like wind driven mode . Likewise in both the Cumbrian and Fylde deployments, spatially consistent maps of wind-driven surface currents associated respectively with orthogonal wind-stress components have been drawn.

7.2 Net surface drift

Though lacking direct independent corraboration, large (>10cms), spatially coherent net surface drift patterns have been constructued for OSCR deployments at Red Wharf Bay, Cumbria and the Fylde (figure 6). The exact nature of these is presently being examined, they are almost certainly associated with horizontal density gradients arising from the outflow of less-dense river water. Their magnitudes in these three cases is such that they may rapidly (less than one day) become more important than tidal flow in determining the mixing of buoyant contaminants. Moreover, accurate simulation of these currents by three-dimensional numerical models poses an outstanding challenge.

8. SUMMARY

It has been shown that the major tidal currents can be measured to an acceptable accuracy by OSCR (amplitude within 5cm s^{-1}, direction and phase within 5°) from as little as 7-10 days of measurements made under a wide variety of conditions. While this accuracy is sufficient for most engineering applications, significant improvement is required to permit more rigorous scientific analyses of the generation of higher harmonic tidal constituents etc. Such improved accuracy can be obtained by (i) longer observations, typically of 30 days, (ii) careful alignment of beams in relation to current directions, (iii) more frequent observations for longer sampling intervals.

The latter condition should be satisfied by the on-going development of the OSCR system which will allow the entire field of 16 beams to be monitored simultaneously. This development will be of even greater advantage in examining the non-tidal current components. The technical development of the system will also provide miniaturised, automated hardware that will both reduce the cost and ease logistical operations.

Research is also underway to produce a system with much smaller bin size suitable for near-shore coastal research. Likewise research is continuing into other HF Radar systems, including longer-range (several hundred kilometres) versions especially suited for remotely measuring sea-state as well as surface currents and surface winds (Wyatt 1985).

ACKNOWLEDGEMENTS

A major contributory factor to the success of the oceanographic experiments described has been the continuing high-quality technical support of the OSCR system. This support was originally provided by the system's developers, namely Dr. J. King and D. Eccles of the Rutherford Appleton Laboratory and more recently by Marex, Isle of Wight. We are also indebted to the North West Water Authority for permission to reproduce results from commissioned reports.

REFERENCES

Barrick, D.E. 1978 H.F. radio oceanography - a review.
    Boundary Layer Meteorology. 13. 23-43.
Collar, P.G. & Howarth, M.J. 1987 A comparison of three methods of measuring
    surface currents in the sea: Radar, current meters and surface drifters.
    H.M.S.O. Dept. of Energy. Offshore Technology Report. 87-272 46pp.
Godin, G. 1972 The analysis of tides.
    Liverpool University Press. 264pp.
Ha, E.C. 1979 Remote sensing of ocean surface current and current shear by
    H.F. backscatter radar.
    Ph.D Thesis. Stanford University.
King, J.W. et al. 1984 OSCR (Ocean Surface Current Radar) observations of
    currents off the coasts of Northern Ireland, England, Wales and
    Scotland. Current measurements offshore.
    Society of Underwater Technology. London. 38pp.
Leise, J.A. 1984 The analysis and digital signal processing of NOAA's surface
    current mapping system.
    IEEE J. Oceanic Eng. OE-9, 106-113.
Prandle, D. 1987 The fine-structure of nearshore tidal and residual
    circulations revealed by H.F. radar surface current measurements.
    J. Physical oceanography. 17(2), 231-245.
Prandle, D. 1988 Analysis of surface current data measured by H.F. Radar
    (OSCR) off the Fylde Coast.
    POL Internal Document No. 2. 17pp. (unpublished).
Prandle, D. & Ryder, D.K. 1985 Measurement of surface currents in Liverpool
    Bay by high-frequency radar.
    Nature, 315(6015), 128-131.

Prandle, D. & Eldridge, R. 1987 Use of surface currents measured by H.F. radar
    in planning coastal discharges.
    Marine Pollution Bulletin, 18(5), 223-229.
Shearman, E.D.R. 1983 Propagation and scattering in m.f/h.f groundwave radar.
    IEE Proc. (F), 130, 579-590.
Stewart, R.H. & Joy, J.W. 1974 H.F. radio measurements of surface currents.
    Deep-Sea Research. 21.12, 1039-1049.
Wyatt, L. 1985 Netherlands/UK, radar, wave-buoy experimental comparison
    (NURWEC) October-December 1983. Report on the radar results.
    Dept. Electrical and Electronic Engineering, University of Birmingham.

Table 1. M_2 tidal constituent, amplitude semi-major axis.

Comparison of OSCR and current meter measurements.

	OSCR	Current Meters (2)	OSCR	Current Meter	OSCR	Current Meter	OSCR	Current Meters (3)
Amplitude $(cm\ s^{-1})$	61	52	80	58	51	46	33	33
Direction	342°	353°	99°	100°	15°	18°	71°	60°
Phase	232°	231°	226°	226°	245°	249°	45°	42°
Eccentricity (+ve a.c)	+0.03	-0.03	-0.13	-0.01	0.05	0.19	0.32	0.33
Record length (days)	1	47	28	29	9	30	$6\frac{1}{2}$	
Reference	Prandle & Ryder (1985)		Prandle (1987)		Prandle (1988)		Collar & Howarth (1987)	

Table 2.   Amplitude and phase differences along Beam 186°
from Walney Island as measured from (i) 1.2km
bins and (ii) overlapping 2.4km bins.

	Mean Value		Root Mean Square Difference
	Amplitude		Phase
M_2	37.6	5.2	9.5°
S_2	13.8	2.7	11.7°
M_4	5.7	2.3	8.6°
Z_0	2.5	1.9	

$$\text{cm s}^{-1}$$

Table 3.   Root Mean Square variations in M_2 tidal current ellipse parameters
Variations are between three values obtained from combinations
of radial values along 3 beams that intersect in close proximity.

Location	Amplitude	Direction	Phase	Eccentricity
A	1.7	1.0°	1.1°	0.01
B	3.1	3.9°	3.4°	0.06
C	2.3	3.1°	5.2°	0.08
D	6.0	6.7°	1.7°	0.02
E	8.2	6.0°	1.3°	0.04
F	6.8	5.2°	3.9°	0.06
G	5.8	6.1°	6.8°	0.07
Mean Values	4.8	4.6°	3.3°	0.05

$$(\text{cm s}^{-1})$$

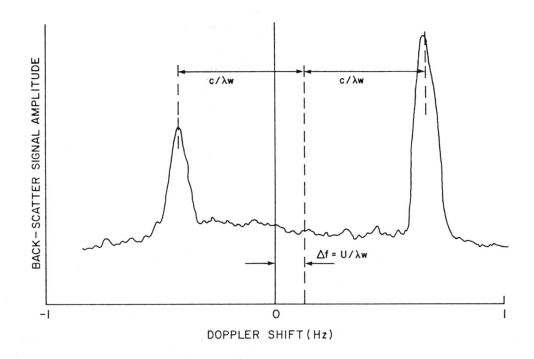

1.   Shift of Doppler Spectrum Δf due to sea surface current U.

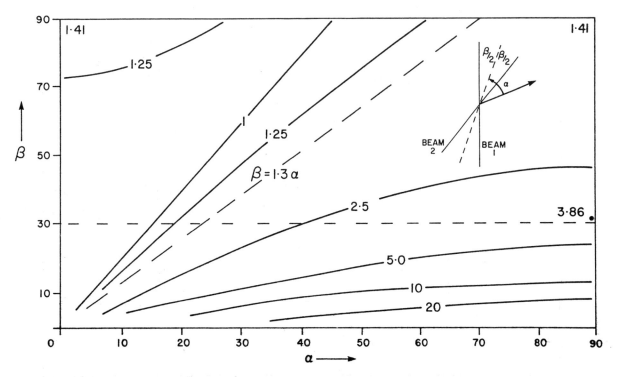

2.   Amplification factor of standard error as a function of α and β.
          α angle between the bisector of the two beams and the current
          vector direction.
          β angle between the two beams.

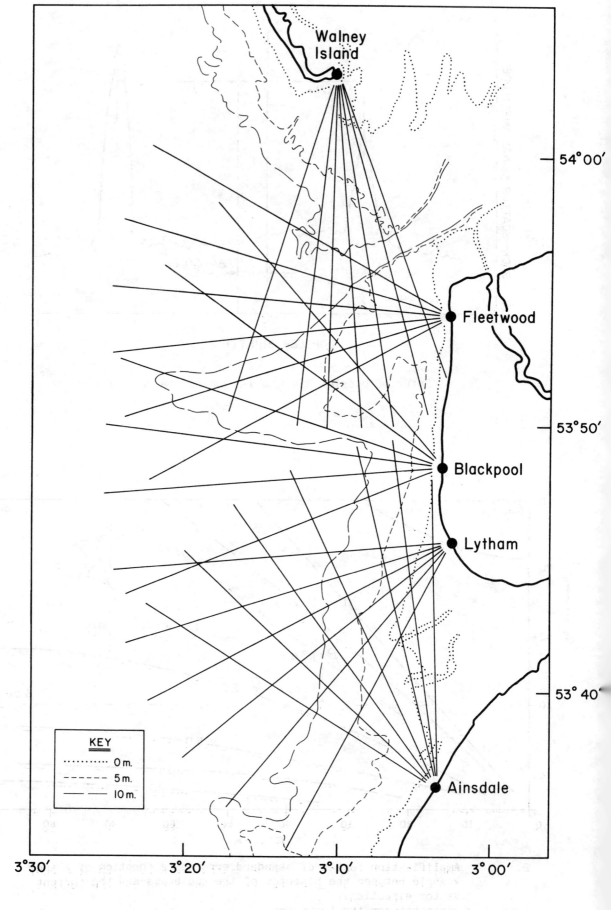

3.   Location of OSCR units and beam directions in Fylde deployment.

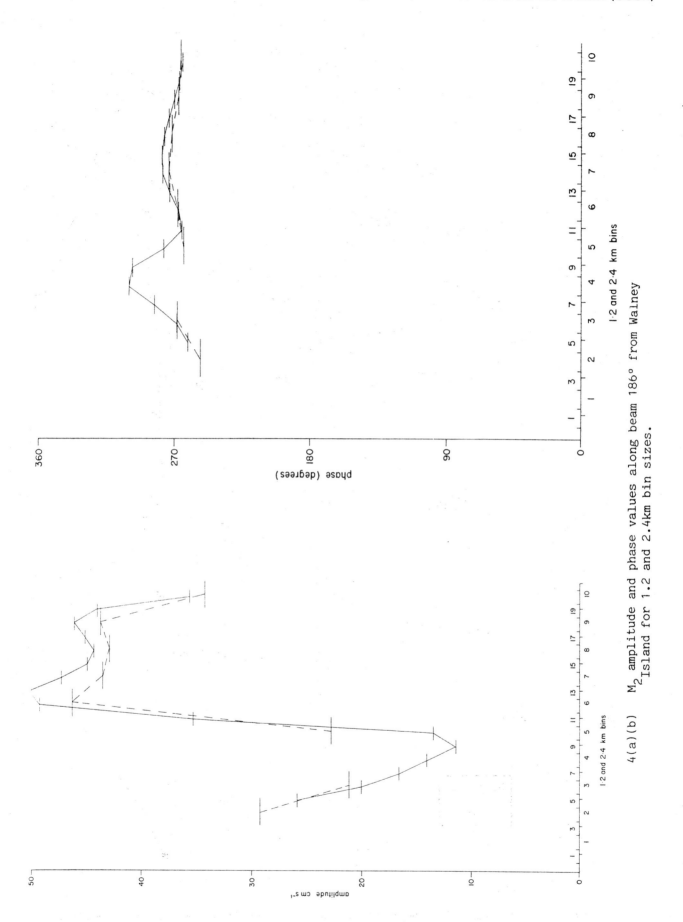

4(a)(b) M_2 amplitude and phase values along beam 186° from Walney
Island for 1.2 and 2.4km bin sizes.

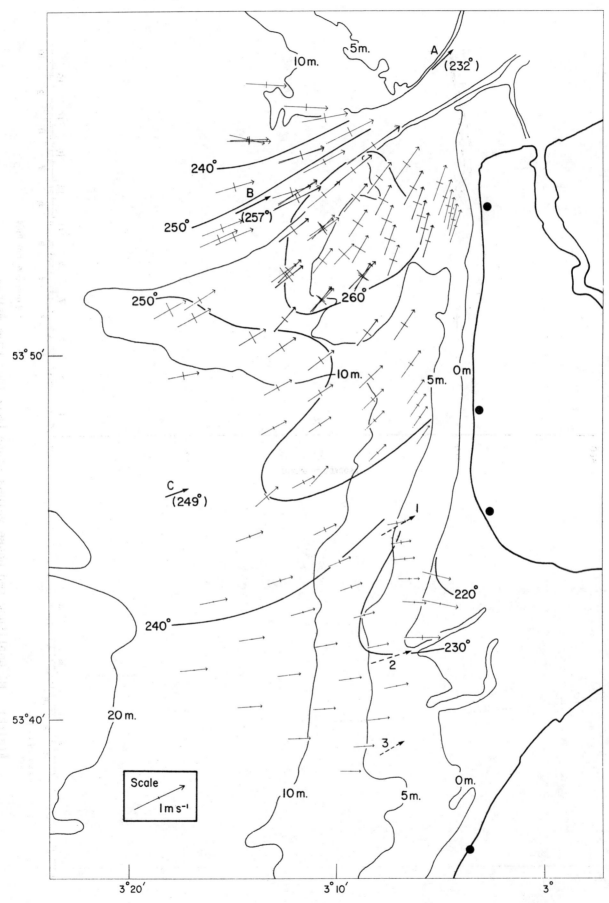

5(a) M_2 tidal ellipse distributions obtained from combination of radial
         values.

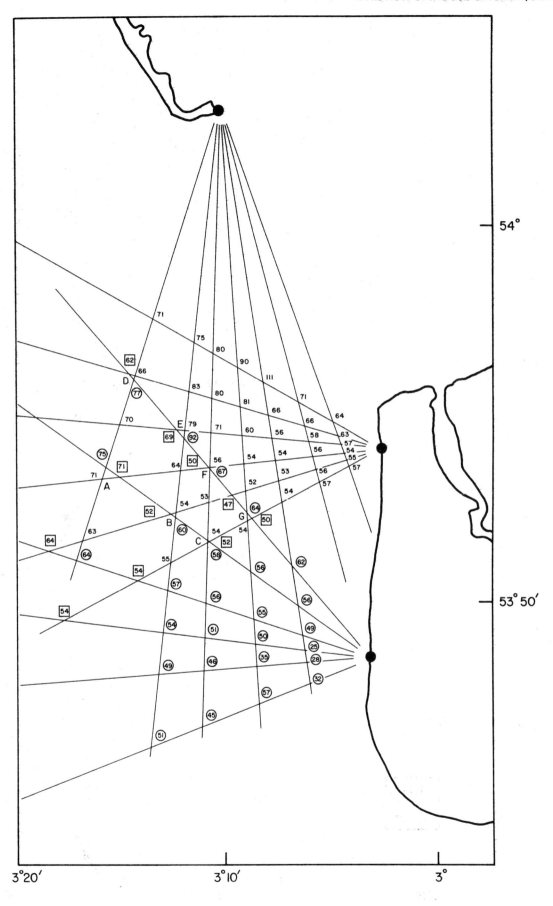

5 (b)   Amplitudes of M_2 semi-major ellipse axis in region of overlap.

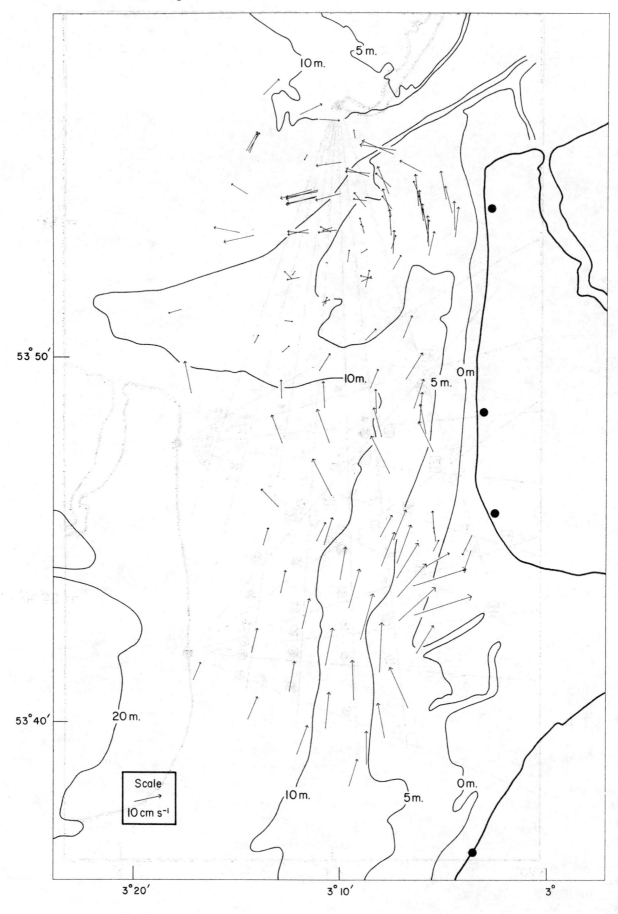

6.   Net surface residual drifts.

Chapter 18
ESTIMATES OF EDDY DIFFUSION IN THE UPPER OCEAN BASED ON THE BACKSCATTER STRENGTH MEASUREMENTS OF AN ACOUSTIC DOPPLER CURRENT PROFILER

G Griffiths

Institute of Oceanographic Sciences Deacon Laboratory, UK

Summary

This paper shows that estimates of K_v, the coefficient of vertical eddy diffusion, can be made near the sea surface using measurements of the acoustic backscatter from bubble clouds acquired as ancillary data by an Acoustic Doppler Current Profiler. The data shows K_v to be dependent on wind and wave conditions and suggests that the near-surface current shear decreases with increasing wind speed.

1. Introduction

The development of three dimensional numerical models for the prediction of shelf sea currents is hampered by uncertainty over the magnitude and variation with depth of a major frictional parameter, K_v, the vertical eddy diffusion coefficient. As K_v is a property of the flow, unlike molecular diffusion which is a property of the fluid, it cannot be measured in isolation at a laboratory. At sea, factors including current magnitude, water depth, wind speed, stratification, surface waves and Langmuir circulation are likely to affect K_v. Hence it is clearly a bulk parameter incorporating the effects of many physical processes. This was identified by Heaps (1) as a main difficulty in the modelling of both wind driven and tidal flow. The dramatic effect of the choice of K_v on the form of the current profile can be seen in Figure 1.

Heaps and Jones (2) formulated a two layer model in which K_v for the two layers were determined using observed near-surface current profiles. Griffiths (3) has shown that a 1 MHz Acoustic Doppler Current Profiler (ADCP), developed at IOSDL, can provide near-surface current profiles that could be used as data input to such inverse models. This paper shows that the same instrument, by measurement of the acoustic backscatter strength of near-surface bubble clouds, can provide an independent estimate of K_v. The potential therefore exists to determine K_v with one instrument using two independent methods yet having common time and spatial scales. This should aid the understanding of the limitations inherent in using a parameter such as K_v to model the effects of many physical processes.

The 1 MHz ADCP was designed primarily as an instrument to measure current profiles. This is achieved by calculating the Doppler shift, in discrete cells, of the backscatter from a sound pulse propagating away from the instrument. In order to ensure linear operation given a wide range of received signal strength variations (both temporal and spatial) the instrument varies its gain on an ensemble to ensemble basis. The signal strength and the digital gain code are logged with the current data and form the basis of the technique to be described. Whereas Thorpe (4,5) used a specialized instrument to obtain near-surface acoustic backscatter estimates more data would be available if suitable measurements could be made using the ancillary data from ADCPs - an instrument likely to see increasing use as a survey tool.

Thorpe (4,5) described in detail the dynamics of bubble clouds in the ocean. These bubble clouds are produced when wind waves break and the small bubbles diffuse downwards under the influence of turbulence. It was shown that K_v could be estimated by comparing the observed scattering strength with that obtained from a model of the steady state bubble distribution.

2. Determining K_v from backscatter measurements

2.1 The model

The underlying assumption of the bubble distribution model is that the sub-surface bubbles produced by breaking waves have the same vertical turbulent diffusion coefficient as that of momentum, K_v. Thorpe (4) gives a simple analytical model for a steady bubble population with depth as:

$$\frac{d}{dz}\left(K_v \frac{dN}{dz}\right) + W_b \left(\frac{dN}{dz}\right) - \sigma.N = 0 \qquad (1)$$

where N is the number of bubbles per unit volume at depth z, tending to zero as z tends to infinity, W_b is the rise speed of the bubbles and σ an inverse time-scale representing the change in radius with depth and the diffusion of the gasses from the bubbles to the water. If we assume that K_v, σ and W_b are independent of depth, Thorpe (4) model A - a reasonable assumption given the limited depth range considered in this paper - then the solution to equation 1 is:

$$N = No \exp(-\beta z) \qquad (2)$$

with

$$K_v.\beta^2 - W_b.\beta - \sigma = 0 \qquad (3)$$

where $\beta = \dfrac{-1.dN}{N\ dz} = d^{-1}$, say $\qquad (4)$

Therefore if σ and W_b are known or can be estimated we can obtain an estimate for K_v based on the measurement of N:

$$K_v = W_b.d + \sigma.d^2 \qquad (5)$$

Following the assumption of Thorpe (4) we take W_b as 5.4 mm/s and σ as $0.018s^{-1}$. The assumed rise speed corresponds to that at the peak of the bubble size distribution (50μm) observed by Johnson and Cooke (6), and σ corresponds to a bubble lifetime of about 1 minute corresponding with the observations of Thorpe (7). Estimates of K_v obtained using this method are completely dependent on these two observed parameters.

The 1 MHz ADCP obtains a measurement of the mean amplitude of the backscattered signal within a number of cells over an ensemble (typically 50 pings acquired over 12 minutes). This can be converted to a figure representing the scattering strength per unit volume by correcting for the two way spherical spreading and absorption and the change of cell volume with range. If we assume that the resulting scattering strength is proportional to the area of bubbles per unit volume, then $M_v \propto N$, where M_v is the scattering strength. Hence equation 4 can be re-written as:

$$d^{-1} = \frac{-d}{dz}(\ln M_v) \qquad (6)$$

Using equations 5 and 6, K_v can be estimated from the ADCP data as:

$$K_v = \frac{W_b . h}{m} + \frac{\sigma . h^2}{m^2} \tag{7}$$

where h is the ADCP cell depth increment and m is the natural log of the ratio of the scattering strengths in adjacent cells.

2.2 Using the model with ADCP data

There are two requirements that need to be met before the method can be used with ADCP data:

(a) at least two ADCP cells within the bubble clouds,
(b) a way of determining the presence of the bubble clouds.

The first requirement is presently achieved by the upward looking 1 MHz ADCP on a sub-surface mooring, alternatively a downward looking ADCP could be mounted on a surface buoy.

The second requirement can be achieved by examination of the scattering strengths in the uppermost cells as scattering due to bubble clouds increases rapidly towards the surface, a signature which is clearly discernible from biological or particulate scattering. In order to remove the contribution to the received signal due to scattering from biological material, the scattering strength at cell 5 (12.8 m depth) was removed from cells 6-10 (11.2 m to 4.6 m), as in wind speeds of <15 m/s very few bubble clouds penetrated down to cell 5. Figure 2, a contour map of the scattering strength over a two day period, shows that below ~12 m the scattering was generally uniform. Except for a few short intervals, bubble clouds were present in the uppermost cell throughout the record. Adjacent cell pairs were used in the calculation of K_v if the signal levels in both cells were above that of cell 5 and increased monotonically towards the surface. The number of observations meeting the above criteria during the period considered in Section 3.1 were:

Cells 9-10	167	5.3 m (depth of centre of cell pair)
8-9	92	7.0 m
7-8	54	8.7 m
6-7	29	10.4 m

out of a possible 240. This shows in a qualitative manner the decay of the bubble clouds with depth.

3. Estimates of K_v from our observations

3.1 Celtic Sea - March 1987

As part of an experiment to measure current profiles in a well mixed region the 1 MHz ADCP was deployed on a mooring (51°N 7°W) to measure the current profile in the uppermost 20 m of the water column. Details of the mooring and the results for the tidal current measurements are given in Flatt et al. (8). Figure 3 shows the time series of K_v in the form of hourly averages at 5.3 m whenever there were more than three estimates (out of a possible five) within the hour. The error bars are the standard error of the mean. Also shown is the magnitude of the hourly mean RMS current variation obtained from a VAESAT buoy, equipped with an electromagnetic current sensor at 1 m depth, Collar et al. (9). Previous observations have shown that the VAESAT RMS current variation is related to the local sea-state and in this case is a better estimator of the sea-state than the six-hourly wind speed interpolated from the coastal stations at Valencia, Aberporth, Roches Point and Scilly. The gaps in the estimates of K_v in Figure 3 were due to the wind speed being below the threshold for bubble cloud formation in the sampling region (6.2-4.6 metres). This was especially noticeable after Day 78.7 when the RMS current variation dropped below 0.35 m/s. Figure 4 shows a scatter diagram of the RMS current variation and K_v, the correlation is encouraging given the error bars on K_v and the apparent phase difference between the RMS and K_v for the first major peak.

Let us assume that K_v is of the form, Thorpe (4):

$$K_v = k \, u^* \, z \qquad (8)$$

where k is the Von Karman constant taken as 0.4 and u* is the friction velocity. When Thorpe (5) compared his observed estimates of K_v, transformed into an equivalent u* using (8) he found that, in general, the ratio of the observed to calculated u* varied between 1 and 3. Table 1 shows the values obtained using the ADCP data and the interpolated wind speed, assuming that:

$$u^* = \sqrt{\frac{\rho_a \, C_d \, W^2}{\rho_w}} \qquad (9)$$

with the drag coefficient suggested by Garratt (10):

$$C_d = 7.5E-4 + 6.7E-5 \, W \qquad (10)$$

ρ_a, ρ_w are the air and water densities and W the wind speed in m/s.

Figure 5 shows a scatter plot of the magnitude of the current shear against K_v, both obtained from the data in cells 9 and 10 at 6.1 m and 4.5 m respectively. The scatter in the shear for $K_v > 0.1$ m²/s is less than for $K_v < 0.1$ m²/s, and there is certainly no evidence for shear increasing with K_v, even though K_v was correlated with sea-state (Figure 4). The negative bias to the shear was due to a sub-surface maximum in the tidal current as shown in Flatt et al. (8).

3.2 South-West Approaches - July/August 1986

The 1 MHz ADCP was moored beneath a spar buoy, near the databuoy DB2 (operated by Thorn EMI on behalf of UKOOA) in the South-West Approaches, in a water depth of 168 m. As in the Celtic Sea experiment the ADCP cells covered the upper 20 m of the water column although during this experiment the water column was stratified. Prior to the first storm the mixed layer depth was between 20 and 30 m, whereas after the storm it was deeper than 35 m. Wind and wave data were obtained from DB2 as hourly values, the height of the wind measurement being 6.8 m. A correction to the standard height of 10 m was not deemed necessary.

Figure 6 shows the time series of K_v for the period Day 211 to 214 with a storm during day 213. Sea surface slope given by:

$$S = \frac{\pi \, H_s}{\lambda} \qquad (11)$$

where H_s is the significant waveheight is also shown, together with the wind speed, $W_{6.8}$ and also $W_{6.8}$ divided by the wave speed, c, obtained from the zero crossing period, T_z:

$$c = \frac{g \, T_z}{2\pi} \qquad (12)$$

Energy transfer from the wind to the waves increases as a function of $W_{6.8}/c$. When $W_{6.8}/c > 1$ the sea surface would be dominated by wind waves accompanied by wave breaking whereas for $W_{6.8}/c < 1$ swell would be expected to dominate.

Hatched areas on the bottom of the figure show times when $W_{6.8}/c > 1$ and when the mean depth of the estimate of K_v, 5.3 m, was within 10 times the rms waveheight, ζ, of the surface, assuming $\zeta = H_s/4$. Kitaigorodski et al. (11) postulate this layer of thickness 10ζ to be dominated by turbulence generated by breaking waves. The range of K_v, 0.020 m²/s to 0.208 m²/s in wind speeds of 10.3 m/s to 17.5 m/s is comparable to the Celtic Sea data. Figure 7 shows the ratio of observed K_v to that calculated using (8) as a function of $W_{6.8}/c$. Data points when 10ζ was less than 5.3 m are indicated. Note that, in general, they have a lower K_v ratio for a given $W_{6.8}/c$ than points where 10ζ exceeds 5.3 m, the difference between the means of the two sets being 0.51 with a 90% confidence limit of ±0.47. This lends some support to the postulate of Kitaigorodski et al. (11).

262

Figure 8 shows the linear regression of wind speed and K_v for the period of the storm. Points with $K_v > 0.05$ m²/s are marked with the change in wind speed (in units of ½ m/s) and wind direction (in units of 10 degrees) that took place during the hour over which K_v was calculated. The figure shows that, for a given wind speed, K_v was likely to be higher if the wind was increasing in strength rather than decreasing. There was no statistically significant effect on K_v of the changes in wind direction, though it should be noted that the direction changes during the storm were rather limited (rms $\approx 12°$).

Let us assume that the stress in the near-surface layer of the ocean is linearly related to the applied wind stress:

$$\rho_w K_v \frac{\partial u}{\partial z} \propto \rho_a C_d W^2 \qquad (13)$$

If we drop the densities and substitute for C_d from equation (10) then:

$$K_v \frac{\partial u}{\partial z} \propto W^2 + 0.089W^3 \qquad (14)$$

Over the range of wind speeds 10 m/s to 20 m/s the right hand side of equation 14 can be approximated by the single power term $\frac{1}{2}W^{2.56}$. If $K_v \propto W^n$ and $n > 2.56$ (over 10 m/s to 20 m/s) then $\partial u/\partial z$ will be a decreasing function of wind speed. Using a log-log regression of K_v upon W on the complete data set shown in Figure 8 a slope $n = 2.76$ (± 0.45) was obtained. It was noted above that the values of K_v were significantly higher when the wind speed was increasing therefore the slope, n, was calculated for the two cases of (a) wind speed increasing and (b) wind speed steady or decreasing. Case (a) gave $n = 2.97$ (± 0.76) and case (b) gave $n = 2.55$ (± 0.53). Although the standard errors on the slopes are rather large there is some indication, during periods of increasing wind especially, that the near-surface shear may be a decreasing function of wind speed. This supports the direct current shear measurements as shown in Figure 5.

4. Discussion

As to the magnitude of our estimates of K_v, we can compare them to the observations of Thorpe (5) using (5) and $d(0.1\lambda)$ as listed in his table 1. His values of K_v vary from 0.0152 m²/s to 0.1194 m²/s over a range of wind speeds from 6.9 m/s to 14.4 m/s (note that the depth 0.1λ varied from 5.6 m to 15.6 m). Our Celtic Sea observations of K_v covering a range of 0.025 m²/s to 0.16 m²/s at a depth of 5.3 m in winds of 9.5 to 14.6 m/s are certainly of the same order of magnitude. Both our observations and those of Thorpe give eddy diffusion coefficients generally larger than those listed in Pollard (12), who does however quote the findings of Hoeber (13), in the upper 12 m of the tropical Atlantic, which indicated an enormous range of K_v, correlated with the wind variations, varying on short time scales. Such high values of K_v (say > 0.1 m²/s) are necessary in the model of Heaps and Jones (2) which predicts low current magnitude shear in a near-surface layer of very high K_v. During calmer periods, with $K_v < 0.05$ m²/s, the current shear predicted by the model would be relatively greater, in keeping with our general observations that current shear tends to reduce as wind speed and K_v increase.

5. Conclusions

It has been shown that estimates of K_v in the upper ocean can be made with an ADCP using a simple analytical model of the sub-surface bubble distribution. Estimates of K_v derived using this method may then be used to check the reasonableness of the value of K_v obtained from the current profiles using the inverse method of Heaps and Jones (2).

The data collected in the Celtic Sea and the South-West Approaches strongly suggests that K_v increases with wind speed. At a 95% confidence level the data also showed that, at a given wind speed, K_v was likely to be higher if the wind speed was increasing rather than decreasing. Somewhat higher values of K_v were found when the measurement region was within 10ζ of the sea surface. The above results suggest a near-surface region of intense turbulence with high values of K_v rapidly distributing downwards the energy input of the wind. As a result of the rapid diffusion of energy at times of high K_v the current profiles, measured below 4.5 m, showed less shear than during times of lower K_v.

An extension of the technique to give the depth variation of K_v may be possible using the numerical models of bubble distribution outlined in Thorpe (4). However, due to the exponential decay of the number of bubbles per unit volume it is unlikely that the method could be used at depths of greater than about 10 m.

6.   Acknowledgements

I would like to thank Mr M.J. Howarth and Professor S.A. Thorpe for their very useful comments and suggestions on the draft manuscript of this paper.

7.   References

1.   Heaps, N.S.: "Development of numerical models for the prediction of currents". In: Proc. Conference on Current Measurements Offshore (London, UK: May 17, 1984). London, UK, SUT.

2.   Heaps, N.S. and Jones, J.E.: "Two-layer Ekman model for wind-induced shear flow near the sea surface". Continental Shelf Research, 6, 6, 1986, pp.741-763.

3.   Griffiths, G.: "Intercomparison of an Acoustic Doppler Current Profiler with conventional instruments and a tidal flow model". In: Proc. 3rd IEEE Working Conference on Current Measurement (Airlie, USA: Jan 22-24, 1986). New York, USA, IEEE.

4.   Thorpe, S.A.: "On the determination of K_v in the near surface ocean from acoustic measurements of bubbles". Journal of Physical Oceanography, 14, 1984, pp.855-863.

5.   Thorpe, S.A.: "Measurements with an Automatically Recording Inverted Echo Sounder: ARIES and the bubble clouds". Journal of Physical Oceanography, 16, 1986, pp.1462-1478.

6.   Johnson, B.D. and Cooke, R.C.: "Bubble populations and spectra in coastal waters; a photographic method". Journal of Geophysical Research, 84, 1979, pp.3761-3766.

7.   Thorpe, S.A. and Hall, A.J.: "The characteristics of breaking waves, bubble clouds, and near-surface currents observed using a side-scan sonar". Continental Shelf Research, 1, 1983, pp.353-384.

8.   Flatt, D.F., Griffiths, G. and Howarth, M.J.: "Measurement of current profiles". In: Proc. Conference Oceanology International '88 (Brighton, UK: Mar 8-11, 1988). London, UK, Graham and Trotman.

9.   Collar, P.G., Hunter, C.A., Perrett, J.R. and Braithwaite, A.C.: "Measurement of near-surface currents using a satellite telemetering buoy". In: Proc. 5th International Conference on Electronics for Ocean Technology (Edinburgh, UK: Mar 24-26, 1987). London, UK, IERE, pp.49-56.

10.   Garratt, J.R.: "Review of drag coefficients over oceans and continents". Monthly Weather Review, 105, 1977, pp.915-929.

11.   Kitaigorodski, S.A., Donelan, M.A., Lumley, J.L. and Terray, E.A.: "Part 2: Statistical characteristics of wave and turbulent components of the random velocity field in the marine surface layer". Journal of Physical Oceanography, 13, 1983, pp.1988-1999.

12.   Pollard, R.T.: "Observations and models of the structure of the upper layers of the ocean". Chapter 8 in "Modelling and prediction of the upper layers of the ocean". Ed. E. Kraus, Oxford, UK, Pergamon Press, 1977.

13.   Hoeber, H.: "Eddy thermal conductivity in the upper 12 m of the tropical Atlantic". Journal of Physical Oceanography, 2, 1972, pp.303-304.

Table 1: Comparison of K_V derived from (8) and observations at 5.3 m

Day	W	u*	K_V(8)	K_V(obs)	Ratio
77.0	9.5	.012	.025	.050	2.0
77.25	11.7	.016	.034	–	–
77.5	14.4	.020	.042	.070	1.7
77.75	13.0	.018	.038	.080	2.1
78.0	14.6	.021	.045	.150	3.3
78.25	11.1	.015	.032	.040	1.3
78.5	12.2	.016	.034	.060	1.8

This rather limited data set seems to confirm the observations of Thorpe
that the estimates of K_V are larger than those given by (8). A better
comparison would have been possible if local hourly wind measurements had
been available.

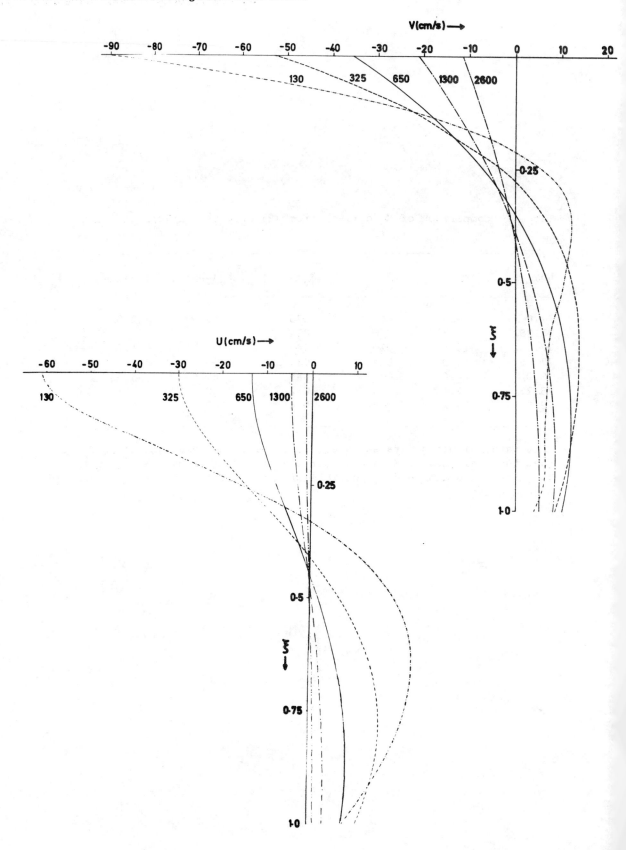

Figure 1: Vertical distribution of u- and v-current at the central position of
a rectangular basin after the establishment of a steady wind-driven
circulation - showing the effect of the choice of K_v (in cm²/s).
From Heaps (1).

266

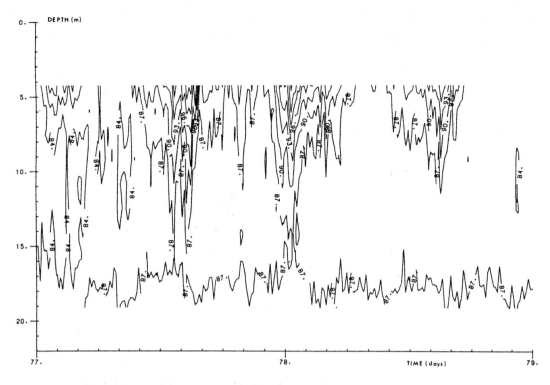

Figure 2: Contour diagram of the scattering strength in 3 dB increments, Celtic
Sea, March 1987.

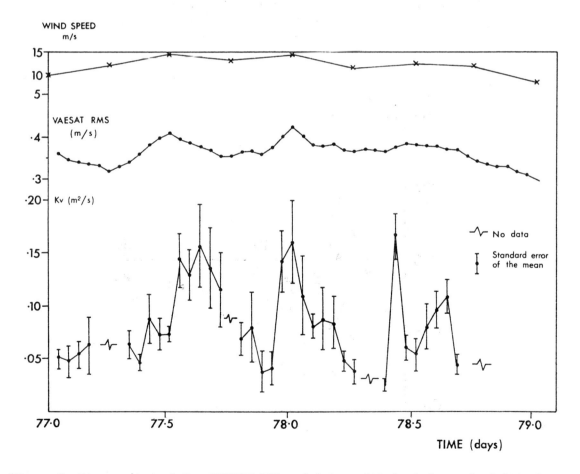

Figure 3: Time series of K_v, VAESAT RMS and interpolated wind speed, Celtic Sea,
March 1987.

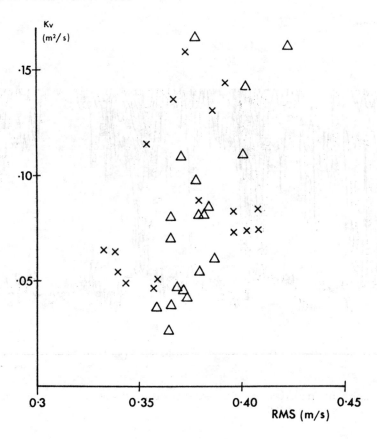

Figure 4: Scatter plot of K$_v$ and VAESAT RMS.
X = first two sections shown in figure 3.
Δ = second two sections shown in figure 3.

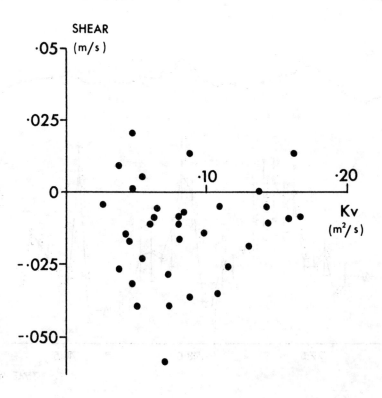

Figure 5: Scatter plot of the current magnitude shear against K$_v$, Celtic Sea, March 1987.

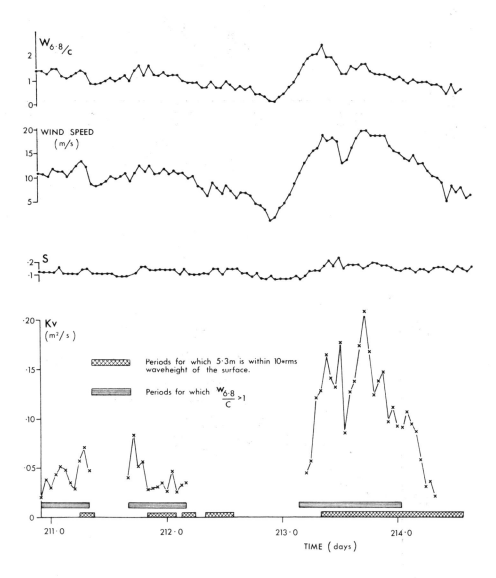

Figure 6: Time series of K_v, sea surface slope, wind speed and wind speed divided by wave speed for the South West Approaches.

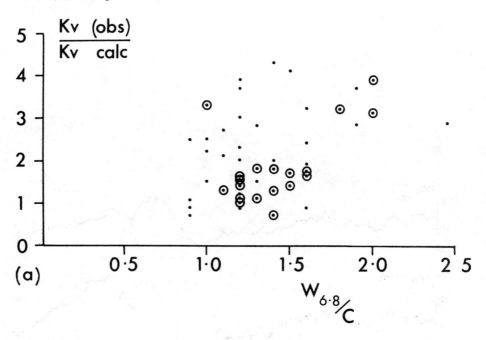

Figure 7: Scatter plot of K_v(obs)/K_v(calc) against w/c.
• 10ζ >5.3 m, ⊙ 10ζ <5.3 m.

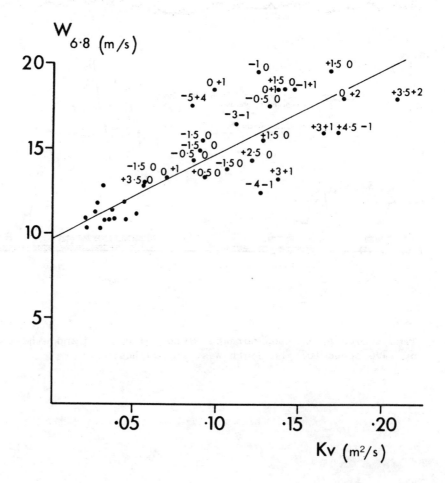

Figure 8: Regression of wind speed and K_v for the South West Approaches. Points with K_v >0.05 m²/s are marked with changes in wind speed (in units of ½ m/s) followed by changes in wind direction (in units of 10 degrees).

Chapter 19
LABORATORY AND SITE MEASUREMENTS OF NEARBED CURRENTS AND SEDIMENT MOVEMENT BENEATH WAVES: A SPECTRAL APPROACH

J Hardisty
The University of Hull, UK

Abstract

One of the most important processes in coastal engineering is the erosion and transport of sediment by wave action in shallow water. Although theoretical models for nearbed wave induced currents were proposed in the last century, and a number of studies have investigated the resulting sediment movements under symmetrical sinusoidal flows in laboratories, it is not clear which theories are applicable to prototype beaches.

Experiments are reported which involved the laboratory and field measurement of surface elevation and nearbed wave induced current spectra which support the asymmetry predicted by second order Stokes wave theory although the amplitude of the harmonics is less than predicted by theory. Results are also reported from laboratory sediment transport experiments in an asymmetrically oscillating flow using acoustic techniques which, for the first time, reveal the spectrum of wave induced bedload transport. This data is combined with the hydrodynamic theory to develop a more realistic spectral model for sediment transport beneath shallow water waves.

Introduction

The problem of explaining and predicting the evolution and stability of the two and three dimensional form of the nearshore seabed bathymetry under the action of waves is of considerable importance to marine geomorphologists, coastal engineers, sedimentologists and oceanographers. The form is the result of net sand and gravel transport and the problem is usually separated into investigations of longshore sediment transport for models of coastal plan shape and onshore/offshore sediment transport for the shorenormal seabed profile. Attempts may then be made to combine the two to examine the full three dimensional bathymetry.

The former set of processes which constitute the longshore transport of sediment are now relatively well understood and the recent reviews by, for example, Komar (1983) provide the theoretical and empirical background as well as functional formulae for the computation of time averaged mass transport rates. Research in this area is concerned with higher frequency events and with the effects of more realistic wave height, period and direction distributions on the essentially monochromatic formulae so that the rate process can be better understood over shorter time scales.

In contrast the shorenormal, or more correctly the orthogonal transport of sediment in the onshore/offshore direction is a poorly understood phenomena and these orthogonal processes are the subject of the present paper. In general the objective is to drive a model with some characteristic of the deep water wave conditions. Processes known as the shoaling wave transformations will affect these characteristics as the waves approach the shoreline, resulting in general in a reduction in wave length and an increase in wave height. Hydrodynamic processes then transform the local surface wave characteristics into a velocity field close to the sediment surface. Sediment dynamic processes which are powered by these flows then move seabed sand and gravels along the orthogonal profile, leading through erosion and deposition to the shape of the orthogonal profile. These four sets of processes, and the associated parameters are listed in Table 1. In practise a series of feedback loops exist between the profile and each of these processes, but these are discussed in Hardisty (1985, 1986) and are not of interest here.

The object of the present paper is to examine some of the theoretical analyses which have been presented for each of these processes, and to assess their predictive capabilities with laboratory and field data. In particular, earlier work assumed that the surface elevations, the nearbed flows and the resulting sediment transport are monochromatic phenomena which are essentially sinusoidal and can be adequately described by a single frequency and amplitude. Recent research has begun to accept the more realistic demand that time series of the surface profile or nearbed velocity field are randomly distributed about some more regular variations, and that analysis in the frequency rather than the temporal domain is necessary.

There are of course, different types of waves. In the present experiments the plate extends from the surface to the bed and remains vertical throughout the wave cycle. The amplitude of the oscillations varies between about 0.08m and 0.20m, and small amplitude low frequency oscillations generate closely sinusoidal waves as discussed below.

Theory

Regular Waves

Waves (Figure 1) are traditionally assumed to be monochromatic and to be adequately described by the height (H) which is equal to twice the amplitude (a), and the period (T). The following sections present first order, monochromatic and second order solutions for the three sets of processes of interest here. The surface profile of the wave is given by linear, first order Airy wave theory as:

$$\eta(x,t) = \frac{H}{2}\cos(kx - \sigma t) \qquad \text{Eq.1}$$

and by second order, Stokes wave theory as:

$$\eta(x,t) = \frac{H}{2}\cos(kx - \sigma t) + \frac{\pi}{8}\frac{H^2}{L}\frac{\cosh(kh)[2 + \cosh(2kh)]}{\sinh^3(kh)}\cos[2(kx - \sigma t)] \qquad \text{Eq.2}$$

where η is the surface elevation

H is the wave height (equal to twice the wave amplitude, a)

k is the wave number defined by $k = 2\pi/L$

s is the wave (radian) frequency defined by $\sigma = 2\pi/T$

L is the wave length

T is the wave period

h is the local water depth

and the hyperbolic functions are defined by $\cosh[(k(z+h)] = [e^{k(z+h)} + e^{-k(z+h)}]/2$

and $\sinh(kh) = [e^{kh} - e^{-kh}]/2$

We note that the Stokes wave profile represents a contribution at the fundamental frequency, with a second contribution at the harmonic, which is twice the fundamental wave frequency which is positive under the crest and the trough resulting in a steepening and shortening of the crest profile and flattening and lengthening of the trough profile, so that the wave surface elevation becomes trochoidal. The magnitude of the harmonic component increases as the wave shoals, and is given by the ratio of the two terms in Eq.2 as shown below.

Regular Shoaling Transformations

The wave length in all water depths is taken as the linear solution to the dispersion equation:

$$L = \frac{g}{2\pi}T^2\tanh(kh) \qquad \text{Eq.3}$$

where the hyperbolic tangent is defined by $\tanh(kh) = \sinh(kh)/\cosh(kh) = [e^{kh} - e^{-kh}]/[e^{kh} + e^{-kh}]$

This equation is not directly soluble because L appears both on the left hand side, and within the hyperbolic tangent on the right hand side. Simpler, deep and shallow water approximations are frequently employed giving $L_\infty = 1.56T^2$ and $L = T\sqrt{(gh)}$ respectively, but in practise the full Eq.3 can be solved iteratively with a short computer routine. The local wave height is also taken from linear theory which gives:

$$H = H_\infty \sqrt{\left(\frac{1}{2n}\frac{C_\infty}{C}\right)} \qquad\qquad \text{Eq.4}$$

where H_∞ is the deep water wave height
  C is the local wave celerity ($C = L/T$)
  C_∞ is the deep water wave celerity ($C_\infty = L_\infty/T$) and:

$$n = \frac{1}{2}\left(1 + \frac{2kh}{\sinh(2kh)}\right) \qquad\qquad \text{Eq.5}$$

A common substitution in Eq.4 is to use the shoaling coefficient, $K_s = \sqrt{C_\infty/C}$ so that the local wave height is given by:

$$H = H_\infty \sqrt{\left(\frac{K_s}{2n}\right)} \qquad\qquad \text{Eq.6}$$

The wave breaks when water reaches a depth approaching the wave height, and thereafter the H is taken to be given by Dyer (1986):

$$H = 0.72\ h(1 + 6.4\tan\beta) \qquad\qquad \text{Eq.7}$$

where β is the seabed gradient

Regular Orthogonal Currents

In principle the flow field can be described by the Laplace Equations, but in practise these cannot be solved analytically. A number of approximations were proposed by 19th century mathematicians to overcome these difficulties and again the simplest is the first order theory which gives:

$$u_{in} = u_{ex} = \frac{\pi H}{T}\frac{\cosh[k(z+h)]}{\sinh(kh)}$$

where u_{in} is the amplitude of the onshore velocity
  u_{ex} is the amplitude of the offshore velocity
  z is measured positive upwards from the still water surface

since $z = -h$ at the seabed then the hyperbolic cosine becomes unity and the flow is given by:

$$u_{in} = u_{ex} = \frac{\pi H}{T\sinh(kh)} \qquad\qquad \text{Eq.8}$$

The second order, Stokes wave equations give:

$$u_{in} = \frac{\pi H}{T\sinh(kh)}\left(1 + \frac{3\pi H}{4L\sinh^3(kh)}\right) \qquad\qquad \text{Eq.9}$$

$$u_{ex} = \frac{\pi H}{T\sinh(kh)}\left(1 - \frac{3\pi H}{4L\sinh^3(kh)}\right) \qquad\qquad \text{Eq.10}$$

which again reveals the presence of the primary component with the first harmonic given by the second term in the brackets resulting in flow asymmetry, a stronger shorter duration shorewards flow is followed by a weaker but longer duration seawards flow. It is now recognised that this asymmetry is necessary for dynamic equilibrium on a non-horizontal beach surface (Bowen, 1980; Hardisty 1985, 1986). Other solutions include the Gerstner Wave Theory, Solitary Wave Theory and Cnoidal Wave Theory but are not discussed here.

Regular Bedload Transport

Many sediment transport formulae exist (cf Horikawa, 1988, p196-199), the simplest of which are the Bagnold stream power type which for bedload can be written (Bowen, 1980):

$$j = \frac{e_b\ C_d\ \rho\ (u^2 - Bu_{cr}^2)u}{\cos\beta(\tan\phi - u\tan\beta/|u|)} \qquad\qquad \text{Eq.11}$$

273

where e_b is an efficiency coafficient

        C_d is the drag coefficient

        ρ is the density of water

        ϕ is the angle of dynamic friction of the sediment

        u is the flow velocity

        u_{cr} is the threshold flow velocity

$$B = \sqrt{\frac{\tan\phi - \tan\beta}{\tan\phi}} \cos\beta \quad \text{corrects the threshold for the bedslope}$$

This type of equation is particularly appropriate because it includes the effect of bedslope on the sediment transport rate, and it is the bedslope which controls some aspects of the non-linear behaviuour of the beach system (Hardisty, 1986). The efficacy of this type of formula was recently examined by Whitehouse and Hardisty (1988), and Hardisty and Whitehouse (1988).

Irregular Waves

Real beach waves are not regular. Instead, a wave train exhibits a random distribution of wave heights, and the presence of a range of wave periods which are distributed about at least one mean period due either to the incidence of more than one wave train or to non-linear near shore processes.. Longuet-Higgins (1952) showed that the wave height closely approximates to a Rayleigh distribution:

$$P(H) = \frac{2H}{H_{rms}^2} \exp\left[-(H/H_{rms})^2\right] \qquad \text{Eq.12}$$

where $P(H)$ is the probability of occurence of a wave height H

        H_{rms} is the root mean squared wave height

The most frequent waves are at $H = 0.707 H_{rms}$ and the significant wave height, $H_{1/3}$ is $1.42 H_{rms}$ as shown in Figure 2. Again the period is seldom constant. The randomness produces a spread in periods, and the wave energy is better represented by a spectrum. A commonly used equatiuon for a fully developed sea state was given by Pierson-Moskowitz (1964) as:

$$E(f) = \frac{2\pi\alpha g^2}{\sigma^5} \exp[-B(\sigma/\sigma_o)^4] \qquad \text{Eq.13}$$

where $E(f)$ is the spectral density function in the frequency, f, domain

        σ is the wave frequency ($\sigma = 2\pi/T$)

        α is an empirical constant = 0.0081

        B is an empirical constant = 0.74

        σ_o is related to the wind speed 19.5 m above the sea-surface, $U_{19.5}$, by $\sigma_o = g/U_{19.5}$

A range of P-M spectra are shown in Figure 3.

Irregular Shoaling Transformations

The problem of height transformation and breaking for a realistic wave distribution is particularly complex because, in contrast to monochromatic waves, there is no well-defined breakpoint for random waves. The largest waves tend to break farthest offshore and small waves close to shore. This section is based upon the review by Thornton and Guza (1983) which identifies two classes of random wave shoaling and breaking models. All of the models assume that the waves can be described by a joint distribution of height and frequency as above, but that they are very narrow banded such that all wave heights of the distribution are associated with an average frequency f and thus the deep water waves are described by the single-parameter Rayleigh probability distribution function. In the earlier "local models" the shoaling is dependent only on the local water depth so that the shoaled, unbroken height distribution is calculated by modifying form of Eq.12 with the shoaling coefficient:

$$P(H) = \frac{2H}{K_s H_{\infty rms}^2} \exp\left[-(H/K_s H_{\infty rms})^2\right] \qquad \text{Eq.14}$$

Thornton and Guza (1983) and Horikawa (1988, Fig.3.12) show that transforming wave height distributions can be adequately described in this way and discuss the "integral path" models which cumulate the shoaling transformations from deep water to the local depth and also the effect of breaking on the distribution. The surface elevation $\eta(x,t)$ is taken as the superposition of an infinite number of independent sinusoids:

$$\eta(x,t) = \sum_{n=1}^{\infty} a_n \cos(k_n x + \sigma_n t + \varepsilon_n) \qquad\qquad\qquad Eq.15$$

where a_n is the amplitude

k_n is the horizontal vector wave number

σ_n is the frequency

ε_n is the phase angle

Irregular Orthogonal Currents

Relationships between the local sea surface elevation and the velocity field at depth are generally assumed to be given by flat bottom linear Airy theory both inside and outside the surf zone (e.g. Bowen, 1969; Longuet-Higgins, 1970; Komar and Inman, 1970). Comparisons between the spectra of sea surface elevation and velocity, measured at the same horizontal location in relatively shallow water, have been made by various authors.

Guza and Thornton (1980) give the horizontal velocity as:

$$u(x,t) = \sum_{n=1}^{\infty} \left(\frac{\sigma_n \cosh[k_n(h+z)]}{\sinh(k_n h)} \right) \eta_n$$

The bracketed term is the spectral transfer function relating the velocity spectrum to the surface elevation spectrum. Again at the seabed where z=-h this expression becomes

$$u(x,t) = \sum_{n=1}^{\infty} \left(\frac{\sigma_n}{\sinh(k_n h)} \right) \eta_n \qquad\qquad\qquad Eq.16$$

Thus the transfer function involves simply applying first order theory (Eq.8) to the height spectrum at each frequency interval. Guza and Thornton (1980) and Horikawa (1988 Fig.3.15) show that nearbed flow spectra can be adequately described in this way.

Irregular Bedload Transport

There are presently few theoretical or experimental analyses of the orthogonal transport of sediment under irregular waves. This is largely due to the lack of instrumentation which is capable of providing sufficient temporal resolution for the transport parameters. Recent technical developments are however proving promising. Hanes and Huntley (1986) describe results with an optical back scatter sensor which was used to measure the suspended load concentration beneath beach waves. Similarly Hardisty (1987) describes the acoustic detection of bedload with the sensors detailed below. Existing models for sediment transport would suggest a transfer function similar to Eq.11 and would therefore predict a direct correlation between the transport rate and the flow and would not predict any cross frequency energy transfer.

Experimental Methods

The Instrument System

The system (Figure 4) consists of a pressure transducer, a three component ultrasonic current meter, and a hydrophone mounted on a tubular steel frame frame (Figure 5), which was pinned to the beach face close to the waters edge at low tide. Signals from the sensors were amplified and transmitted along cables to a four channel, analogue tape recorder for storage, and to display dials. The components of the system will be detailed, before describing the laboratory and field recording procedure and analytical techniques.

The pressure transducer was a Druck PDCR 820 Series 2 bar sensor and amplifier. The output from the amplifier was recorded on the first channel of a Racal Fourstore magnetic tape recorder with FM cards on all channels. In situ calibration was achieved by noting the mean water depth on a surveying staff which was fixed vertically close to the frame. The depth was recorded manually at fifteen minute intervals throughout the deployment, and calibrated with the transducer output from the recorder playback.

The shorenormal flows were measured with a Sensordata SD12 Ultrasonic Current Meter, which operates by exciting a transducer at 30 kHz, and recording the transmission time for the sonic pulse over a known path length. The signals are amplified in the underwater unit mounted on the frame, and converted to flow speeds in a deck readout unit. This unit displays the results and presents them to an RS232 connector and to three analogue output sockets. The analogue voltage from the z axis was, in the present experiment, recorded on the second channel on the tape recorder, and the output from the x axis was recorded on the fourth channel on the tape recorder. The manufacturer's calibration of the instrument and analogue outputs was assumed, but the tape recorder playback signal was calibrated against the current meter offsets in still water before the deployment.

The hydrophone was designed to provide an estimate of the seabed sediment transport rate. Details of this technique, which is based on measuring the self generated noise (SGN) which results from the collisions of moving grains are given in Hardisty (1987), and in a series of papers by Heathershaw and Thorne (1985) and Thorne (1985, 1986). The hydrophone employed in the present experiments consisted of a small condenser and preamplifier, encapsulated in silicone rubber and mounted at the focus of a brass, parabolic reflector. The reflector was pointed vertically downwards and was designed to detect SGN and to be less sensitive to the general ambient noise in the submarine environment. Power for the pre-amplifier and the acoustic signal were cabled to a Dawe Instruments Sound Level Meter, which provided second stage amplification and the signal was recorded on the third channel of the tape recorder. The relationship between the intensity of SGN and the bedload transport rate has been demonstrated qualitatively by Hardisty (1987) and Thorne (1986), but the theoretical basis for the correlation is the subject of ongoing research. We concentrate on the hydrodynamics of the processes in the present paper, and the discussion of sediment transport is therefore simply based upon an analysis of the voltage playback signal from the hydrophone records.

In the field an R.T.Labs TVC10UB underwater video camera was mounted adjacent to the frame, and displayed and recorded the sensors and the seabed throughout the deployment. A voice over facility on the fourth channel was used during the calibrations, and to provide time marks throughout the deployment. This interrupts the shoreparallel flow record, and therefore these results are not analysed in detail in the present paper. The whole system is mounted in a specially constructed trailer which is towed to the site, and is capable of providing both DC and, through inverters, AC power for all electrical systems.

Laboratory Experiments

A series of laboratory experiments were conducted during early 1988 to develop the system, and those described here were carried out on 10:7:88 and 12:7:88 in an Armfield 10m working section wave channel. The channel is 0.3 m wide, and 0.4 m deep and waves were generated from a variable speed oscillating vertical plate. The sensors described above were mounted about 6.25 m from the wave generator, and a slatted and bristled wooden beach absorbed wave energy at the distal end of the channel.

It is well known that suspended sediment affects the boundary layer characteristics in tidal currents (Hori Kawa, 1988) but insufficient data is presently available with which to assess this process within wave generated flow fields.

Field Experiments

The instrument system was deployed on the low tide terrace of the intertidal profile close to the entrance to Chichester Harbour on the Channel coast of southern England during the flood and ebb of the daytime tide on 15:7:88. The beach profile exhibited a steep, gravel and cobble berm backing a flat, fine sand lower section. The instruments were deployed close to low water, and the signals were cabled back to the tape and video recorders ashore. A surveying staff was attached to a nearby groyne to provide the calibration of the pressure transducer. The rig submerged at about 1145 GMT, and was recovered in the swash zone at 1610 GMT. The pressure transducer and the acoustic system operated satisfactorily throughout the deployment, but the billowing, suspension transport of the fine sediment which was observed on the video camera was inaudible to the hydrophone and acoustic data is not therefore discussed here.

The analysis of the results from the American "Nearshore Sediment Transport Study" are reported in a large number of recent papers and are developing aspects of the spectral approach, particularly in the description of the near shore flow field. The present paper reviews that work whilst extending the discussion to bedload transport processes, although reference will be made to the parallel developments on the sediment dynamics of the suspended load.

Maximum water depth of 1.65 m occured between 1300 and 1400, but the ultrasonic current meter fouled at 1250. The data analysed therefore concentrated on the first sixty minutes of the deployment.

Results and Analysis

Laboratory Results

The laboratory experiments were designed to test the regular equations for monochromatic waves. The first set of experiments are summarised by Figure 6 which presents results obtained on 10:7:88 to examine the hydrodynamic predictions of linear and second order wave theory. Figure 6(a) show the surface profile spectrum, and Figure 6(b) the nearbed flow spectrum for 4.4 cm wave heights in 31.5 cm of water with a period of 0.59 s. Figures 6(c) and (d) and Figures 6(e) and (f) show corresponding records for wave with the same period and heights of 3.9cm and 2.1cm in water depths of 14 cm and 6 cm respectively. Considering firstly the surface profiles, the three water depths evidence the change from linear to second order wave theory through the growth of the second harmonic. Figure 7 tests Eq.2 with this data by plotting the amplitude of the first harmonic normalised by the wave height, and divided by the wave height because of the second order term. The local wave length was here calculated from the wave period and water depth using Eq.3. The results give observed values of 0.005, 0.032 and 0.075 against the theoretical values of 0.022, 0.040 and 0.115 for Figure 6(a), (b) and (c) respectively. It appears then that the form of the surface spectra resembles a second order Stokes wave with the appearance of the first harmonic in all three datasets, even in deep water (kh>1.6 for h>L/4). The magnitude of the first harmonic is however somewhat smaller than predicted by theory. Additionally the three flow spectra show that, in these water depths, there is little evidence of the harmonic at the bed so that the flow is sufficiently described by Eq.8 rather than Eqs. 9 and 10.

The second set of laboratory experiments were performed on 12:7:88 to again investigate the hydrodynamics of a near monochromatic wave train, but now to also study the sediment dynamic spectra. The hydrophone was placed with its focus 4 cm above the steel bed of the flume and waves with a period of 1.48 s, a mean height of 9.8 cm were run in a water depth of 21.4 cm. The flume channel is itself an acoustically noisy environment, and therefore the spectra of the amplified microphone signal was taken whilst waves were running, but prior to the emplacement of any sediment. The results are shown in Figure 8(a) showing acoustic noise being generated at the wave frequency (0.675 Hz), very little at the first harmonic (1.35 Hz), and then decaying amplitudes of the higher harmonics (2.05 Hz etc.). A carpet of 1cm glass Ballotini balls was then spread in a single particle layer along the bed of the flume beneath the sensors, and the spectrum was again measured. The results are shown in Figure 8(b) revealing little increase in energy at the fundamental frequency, but a considerable increase at the harmonic. This is to be expected since the sediment transport is occuring in both the onshore and the offshore directions, that is at twice the frequency of the wave motions. The difference between the ambient and total spectra is the energy produced by SGN, and is shown in Figure 10(c) and discussed below.

Figure 9 illustrates contiguous measurements of the surface elevation, nearbed flow and acoustic intensity with Ballotini moving on the bed. It illustrates that the sediment transport does not correlate directly with the flow speed, but instead and in this instance the transport appears to lag the flow by about 50° or about 0.25s. This is a sensible finding because observations confirm that there is an inertia in the mass movement of the sediment, so that it lags the flow, but is contrary to all of the existing sediment transport functions of the type illustrated by Eq.11.

Measurements were taken of the surface elevation spectra, and nearbed flow spectra under the same conditions as the acoustic measurements and the results are shown in Figure 10. The upper trace shows the surface elevation spectrum revealing a decaying series of harmonics down from the wave frequency (0.675, 1,35, 2.025 Hz etc.). Note that the frequency axis is here different from that employed in Figure 6, and the waves are larger, which explains why higher harmonics are visible. Figure 10(b) shows the nearbed flow spectrum, with the presence of the first harmonic, and a small second harmonic

Field Results

Examples of the surface elevation and shorenormal flow component records are shown in Figure 11 for the period from 1230 in a water depth of about 105 cm. Full spectra for a 2000s time series were obtained beginning at this point over the frequency range 0 to 1 Hz and are shown in Figure 12, which clearly demonstrates the irregular nature of the phenomenon and it is difficult to envisage the application of the theory to such complexity. However a more detailed examination of the data does permit some progress to be achieved. Firstly the vertical distances between wave crests and subsequent troughs were measured manually for one hundred waves beginning with the sequence shown in Figure 11 and the resulting distribution is shown in Figure 13. It is apparent that two Rayleigh type of distributions are present and this is in accord with the results reported by Thornton and Guza (1983) who found a good fit for a single Rayleigh distribution at a number of locations inside and outside the breaker zone.

Secondly, time series analyses of shorter segments of the datasets revealed a far simpler picture which was amenable to the theoretical analyses discussed earlier. For example Figure 14 (b) shows the surface elevation record with (a) the spectrum of the surface elevations and (c) the spectrum of the corresponding nearbed, shorenormal flows for a 62.5 s interval at 1248. Both records show the presence of primary frequencies at 0.19 Hz (period 5.2 s) and appear to show well developed harmonics at n=2, 3 etc..

Discussion

The preliminary experiments reported here appear to support some aspects of a spectral approach to the modelling of beach processes. In summary the orthogonal model would consists of:

1. Characterising the deep water wave parameters by the sum of one or more Rayleigh height distributions about $H_{\infty rms}$ with a Pierson-Moskowitz (or similar) frequency distribution and appropriate directional information.
2. Shoaling the $H_{\infty rms}$ using a form of K_s, with allowance for frictional losses and breaking etc.
3. Transforming the energy spectrum to develop a local, nearbed flow field.
4. Transforming the local, nearbed flow field to develop a bedload transport rate.
5. Utilise the results with continuity considerations to model the evolution and stability of the orthogonal bathymetry.

A theoretical basis is now available for each of these five stages and all are now amenable to experimental testing using the equipment and procedures detailed above. The present experiments have particularly examined stages 1, 3 and 4. It appears that Stokes wave theory provides an appropriate and asymmetric description of the surface elevation spectra for monochromatic wave trains in shallow water, and that a linear, first order spectral transformation predicts the nearbed flow field. Flow attenuation with depth is however rather greater than predicted by theory. Finally a bedload spectrum has been demonstrated which, at least qualitatively, is related to the velocity field, although the transformation is complex and, for oscillatory currents, appears to require some allowance for the acceleration of the flow field and the inertia of the particles. These results are of course preliminary but the present dataset together with Hardisty (1987) and similar results from suspension measurements under waves by Hanes and Huntley (1986) do suggest that models of nearshore sediment transport which ignore the unsteady nature of the flow field are missing an important and possibly dominant factor in the dynamics of sediment transport. The only analysis of which the author is aware and which explicitly includes allowance for the unsteady terms is due to Hinze (1975, p463) which gives a momentum equation for turbulent flow:

$$m_s a_s = m_f a_f - C_m m_f(a_s - a_f) - C_d \frac{6}{8d} m_f (u_s - u_f)|u_s - u_f| + F_B + (m_s - m_f)g + F \qquad \text{Eq.17}$$

where m_s is the particle mass
m_f is the mass of an equivalent fluid volume
d is the particle diameter
C_m and C_d are added mass and drag coefficients
a_s and a_f are the particle and fluid accelerations
u_s and u_f are the particle and fluid velocities
F_B and F are the Bassett Force and an external potential force.

The left hand side represents the force to accelerate the particle and the right hand terms represent the pressure gradient, the force on the added fluid mass, the viscous force, the Bassett term, gravity and the external potential force respectively. Research is being carried out under the N.E.R.C. grant listed below to evaluate these terms and hence to provide a better understanding of the transfer function between the hydrodynamic and sediment dynamic spectra.

Acknowledgements
The author wishes to acknowledge the assistance of Jeremy Lowe during both the experimental and analytical stages of this work, which was partially supported by the Natural Environment Research Council Grant GR3/6645 *Higher Frequency Acoustic Measurements of Bedload Transport Processes in Turbulent Flow*.

BIBLIOGRAPHY

Bowen,A.J. (1969) The generation of longshore currents on a plane beach. *J.Mar.Res.*, **27**, 206-215.

Bowen,A.J. (1980) Simple models of nearshore sedimentation: beach profiles and longshore bars. In: The Coastline of Canada, S.B.McCann (Ed), *Geol. Surv. Can.*, **80-10**, 1-11.

Guza,R.T. and E.B.Thornton (1980) Local and shoaled comparisons of sea surface elevations, pressures and velocities. *J.Geophys.Res.*, **85**, 1524-1530.

Hardisty,J. (1985) A note on negative beach slopes and flow asymmetry. *Mar. Geol.*, **69**, 203-206.

Hardisty,J.(1986) A morphodynamic model for beach gradients. *Earth Surf. Processes and Landforms*, **11**, 327-333.

Hardisty,J. (1987) Higher frequency acoustic measurements of swash processes. *Physical Geography*, **8**, 169-178.

Hardisty,J. (1988) An Introduction to Wave Recording for Geomorphological Applications. *Tech. Bull. Brit. Geomorphological Res. Group* (in press)

Hardisty,J. and R.J.S.Whitehouse (1988) Evidence for a new sand transport process from experiments on Saharan dunes. *Nature*, **332**, 532-534.

Hanes,D.M. and D.A.Huntley (1986) Continuous measurements of suspended sand concentration in a wave dominated nearshore environment. *Continental Shelf Research*, **6**, 585-596.

Heathershaw,A.D. and P.D.Thorne (1985) Seabed noises reveal role of turbulent bursting phenomenon in sediment transport by tidal currents. *Nature*, **316**, 339-342.

Hinze,J.O. (1975) *Turbulence*, McGraw-Hill, New York, 796pp

Horikawa,K. (1988) *Nearshore Dynamics and Coastal Processes*, University of Tokyo Press, Tokyo, 522pp

Komar,P.D. (1983) C.R.C. Handbook of Coastal Processes and Erosion. C.R.C. Press, Boca Raton, Florida, 305pp.

Komar,P.D. and D.L.Inman (1970) Longshore sand transport on beaches. *J.Geophys.Res.*, **75**, 5914-5927.

Longuet-Higgins,M.S. (1952). On the statistical distribution of the heights of sea waves. *J.Marine Res.*, **2**, 245-266.

Longet-Higgins,M.S. (1970) Longshore currents generated by obliquely incident sea waves. *J.Geophys.Res.*, **75**, 6778-6789.

Pierson,W.J. and L.Moskowitz (1964) A proposed spectral form for fully developed wind seas based on the similarity theory of S.A.Kitagorodskii. *J.Geophys.Res.*, **69**,5181-5189.

Thorne,P.D. (1985) The measurement of acoustic noise generated by moving artificial sediments. *J.Acoustic Soc.Am.*, **78**, 1013-1023.

Thorne,P.D. (1986) An intercomparison between visual and acoustic detection of seabed gravel movement in turbulent tidal currents. *Mar.Geol.*, **54**, M43-M48.

Thornton,E.B. and R.T.Guza (1983) Transformation of wave height distribution. *J.Geophys.Res.*, **88**, 5925-5938.

Whitehouse,R.J.S. and J.Hardisty (1988) Experimental assessment of two theories for the effect of bedslope on the threshold of bedload transport. *Mar.Geol.*, **75**, 135-139.

Table I

Beach Parameters and Processes (after Hardisty 1988)

	Regular	Irregular
Deep Water Surface Wave Parameters	Eq.1-2	Eq.12-13
Transformation Processes		
T1. Shoaling length reduction	Eq.3	
T2. Shoaling height increase	Eq.6	Eq.14
T3. Frictional height reduction		
T4. Percolation height reduction		
T5. Refraction		
T6. Reflection		
T7. Breaking	Eq.7	
Local Surface Wave Parameters		
Hydrodynamic Processes		
H1. Wave induced orthogonal currents	Eq.8-10	Eq.16
H2. Wave induced longshore currents		
Local Nearbed Flow Parameters		
Sediment Dynamic Processes		
S1. Orthogonal transport	Eq.11	Eq.17
S2. Longshore transport		
Local Sediment Transport Parameters		
Morphodynamic Processes		
M1. Orthogonal Profile Continuity		
M2. Longshore Continuity		
M3. Transformation feedback		
M4. Hydrodynamic feedback		
M5. Sediment dynamic feedback		
Geomorphological Parameters		

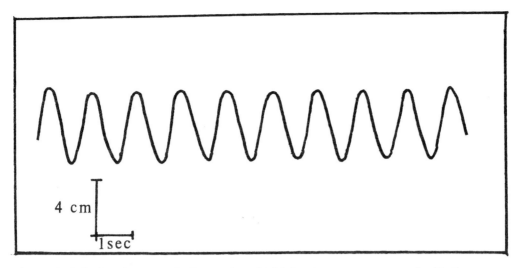

Figure 1. Full size chart record of near sinusoidal laboratory waves measured with the pressure transducer on 4:7:88. Chart speed was 1 cm/s, wave period was 1.35 s, mean wave height was 7.06 cm and the water depth was 20.3 cm. The sensor was mounted on the bed of the channel about 5m from the wave generator.

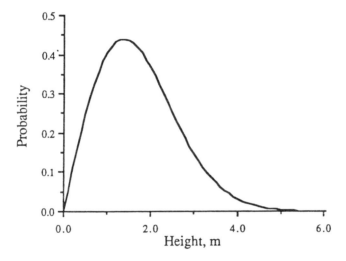

Figure 2. Example of the Rayleigh wave height distribution given by Eq.2 for H_{rms} = 1.95 m.

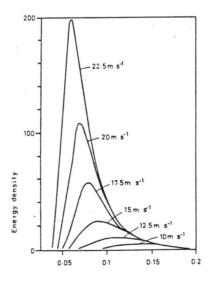

Figure 3. Example of the Pierson-Moskovitch spectrum for a fully developed sea and various wind speeds.

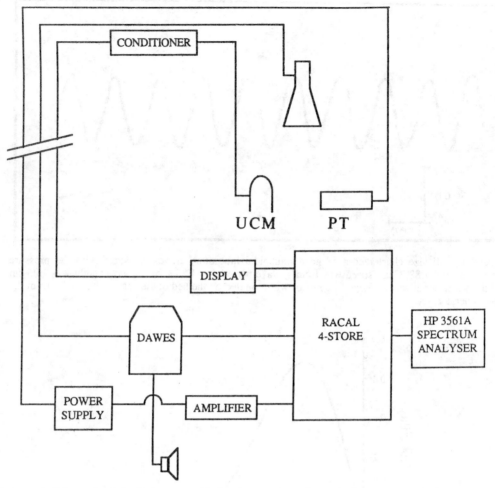

Figure 4. Schematic of the full system with three component ultrasonic current meter, hydrophone and pressure transducer.

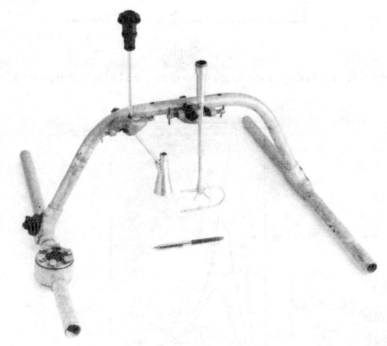

Figure 5. Photograph of the sensors mounted on the galvanised steel frame prior to beach deployment with the pressure transducer (PT in Figure 4), in the casing on the left hand base and the ultrasonic current meter (UCM in Figure 4) mounted to the right of the hydrophone on the horizontal bar.

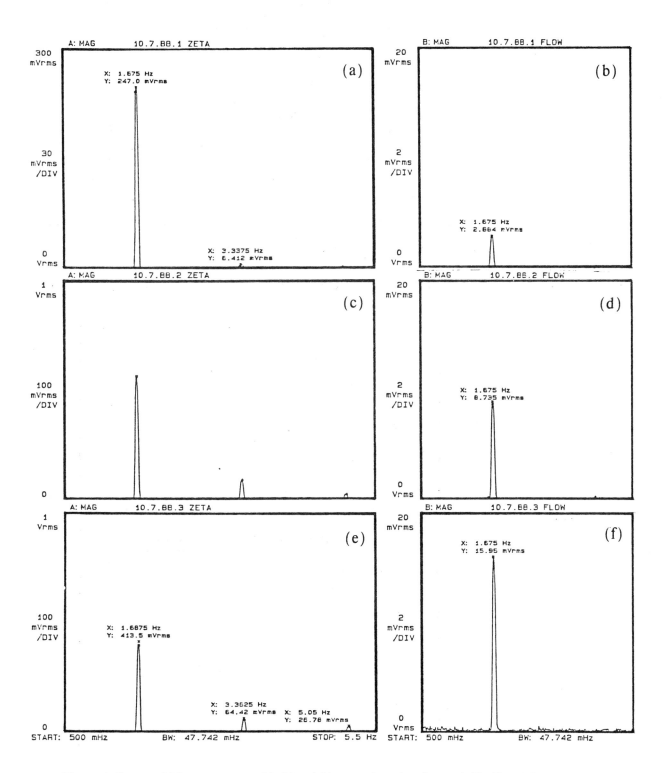

Figure 6. Spectra of laboratory waves: (a), (c) and (e) are surface elevations and (b), (d) and (f) are nearbed velocity fields. Wave heights and water depths for the upper, middle and lower pairs were 4.4 cm in 31 cm, 3.9 cm in 14 cm and 2.1 cm in 6 cm respectively and the flow calibration was 1 cm/s/mV.

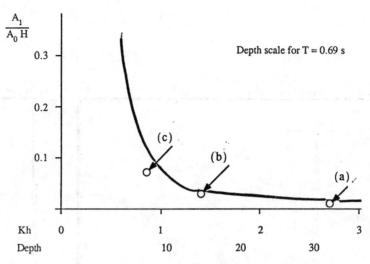

Figure 7. Test of Eq.2 using the relative magnitude of the principal (A_0) and harmonic (A_1) in Figure 6(a), (c) and (e). H is the wave height and the graph is plotted on kh and calibrated for the absolute depth with the wave period of 0.69 s by using the local wavelength calculated from Eq.3.

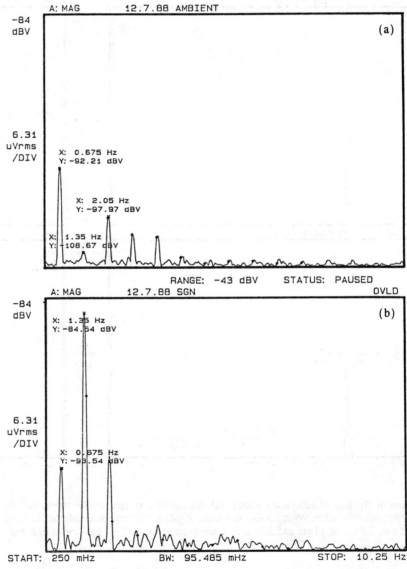

Figure 8. Energy spectra of the hydrophone outputs beneath the waves shown in Figure 9: (a) with the wave channel empty, (b) with 1 cm Ballotini glass spheres being transported on the channel bed beneath the hydrophone.

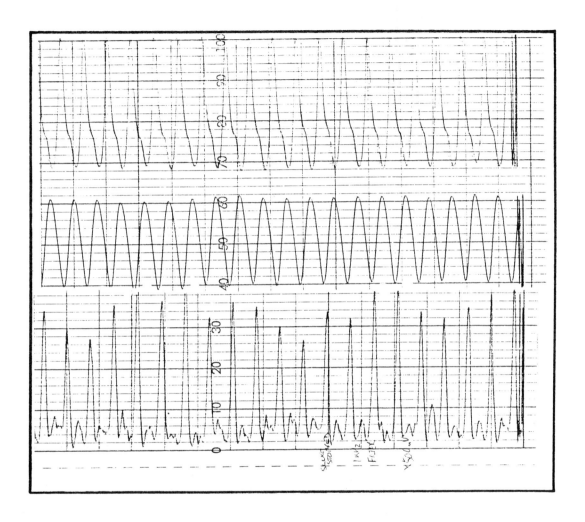

Figure 9. Example of contiguous measurements of (upper) the surface elevation, (middle) the nearbed orthogonal flow and (lower) the filtered acoustic intensity. The mean wave height was 6.5 cm, wave period was 1.43 s and the water depth was 15 cm. The acoustic signal was high pass filtered at 1 k Hz to remove ambient noise, principally from the wave generator. Note that the SGN lags the flow.

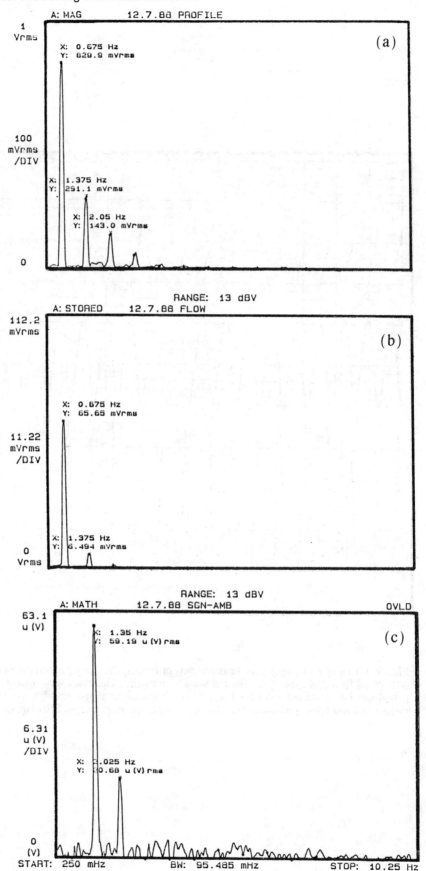

Figure 10. A preliminary summary of the spectral approach to beach processes showing the elevation, near-bed flow and SGN spectra in the wave channel. The flow calibration is 500mV=1.0 m/s with a +1m/s offset.

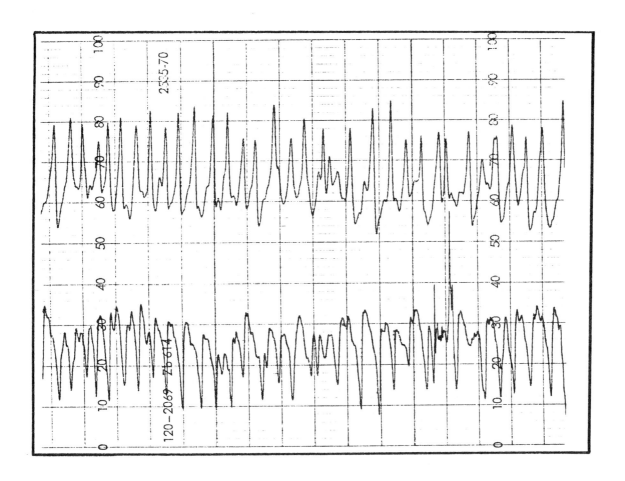

Figure 11. Time series of the surface elevations and orthogonal flow component at the field site.

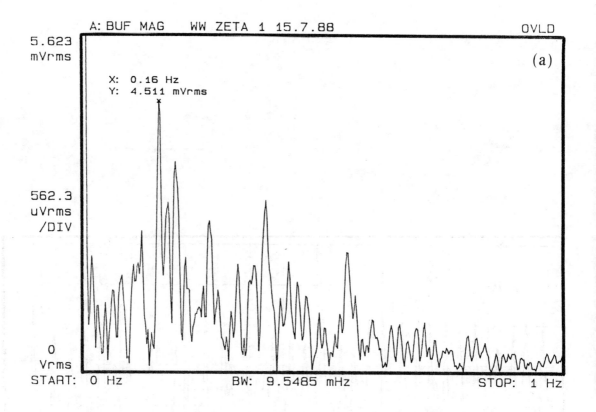

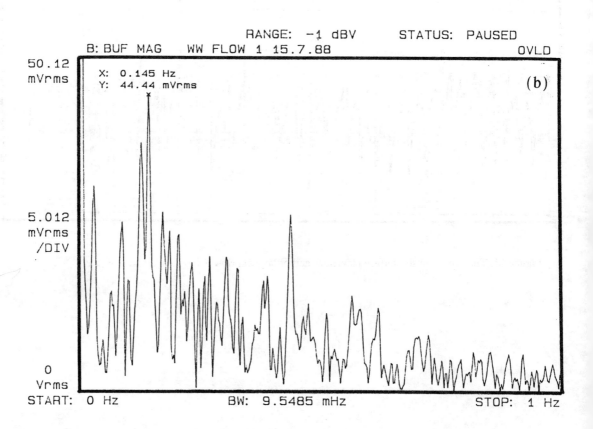

Figure 12. Long duration (2000 s), narrow bandwidth spectra for the surface elevations and nearbed flows at the field site. The flow calibration is 500mV=1.0 m/s and the elevation calibration is 10mV=1m.

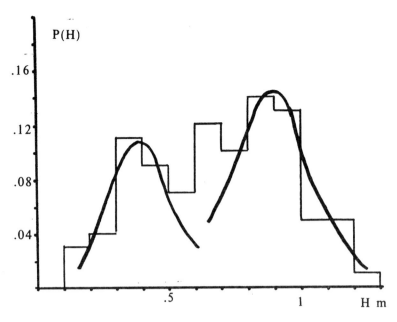

Figure 13. Wave height distributions for 100 waves suggesting the presence of two Rayleigh type of distributions.

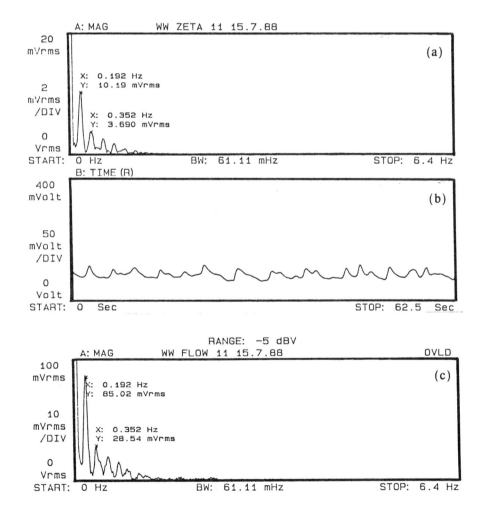

Figure 14. Short duration (62.5 s), wide bandwidth spectra of (a) the surface elevations, (c) the orthogonal flows and (b) time series of the surface elevations at the field site. Calibrations as in Figure 12.

Chapter 20
STEADY WAVE THEORY AND THE SURF ZONE

R J Sobey and K Bando
University of California, USA

Summary

An analysis of the utility of steady wave theory in the surf zone is presented. With an appropriate formulation of the shoaling problem, it is shown that Fourier wave theory provides mathematically tolerable solutions beyond the break point that have some physical credibility. Relevant characteristics such as wave height, wave number, setup, undertow and asymmetry about the MWL follow physically reasonable trends. Asymmetry about the crest and the influence of the sloping bed cannot be accommodated by a steady wave theory but there are at least qualitative predictions of dissipation.

1.   Introduction

There is an understandable reluctance to give any serious consideration to the utility of steady wave theory in studies of surf zone processes. Nevertheless, oscillatory flow remains an essential feature of the flow throughout the surf zone and, at any location, steady wave theory describes the gross features of the local flow pattern. Conditional acceptance of the local validity of steady wave theory immediately provides an almost complete description of the flow field through the surf zone and a wealth of qualitative detail on the hydrodynamics of the surf zone.

Consideration of slowly-varying bottom topography in steady wave theory is the shoaling problem, where the bed is assumed to be locally horizontal. Formulation of the linear shoaling problem and predictions from linear theory are well known but have led to incomplete formulations of the shoaling problem at higher order. The context of the present study is Fourier wave theory, at truncation order eighteen to accommodate the transition from deep to shallow water waves in a consistent manner.

2.   Fourier Steady Wave Theory

The selection of an appropriate steady wave theory (1) is of some considerable relevance to the present analysis. Stokes wave theory is valid only in deep water but gives hopelessly spurious solutions or no solution at all in shallow water. Cnoidal wave theory is valid in shallow water but gives no solution in deep water and is not convergent for even moderately high waves in shallow water. Only Fourier wave theory provides uniform validity from deep to shallow water and from small to large waves, as long as the truncation order is sufficiently high to adequately represent the steeper waves in shallow water (2). A truncation order of eighteen is adequate for these purposes. The remaining constraint on the selection of a wave theory is the definition of phase speed; Stokes first definition is common but it is necessary to use Stokes second definition for the shoaling problem because of the zero mass flux constraint imposed by the beach.

There are various formulations of Fourier wave theory in the literature (3,4,5), of which the Rienecker and Fenton (5) formulation is preferable as it accommodates both definitions of phase speed and the problem formulation is superior in the normalization of the hyperbolic functions in deep water. The present formulation (6) is a variation on the Rienecker and Fenton formulation that scales the equations in terms of the known wave frequency rather than the unknown wave number. It also permits overspecification but this is not a major consideration (2).

A steady reference frame is adopted, which is located at and moves with the wave crest. The solution is represented by a truncated Fourier series expansion for the stream function

$$\psi(x,z) = -\bar{u}z + \frac{g^2}{\omega^3} \sum_{j=1}^{N} B_j \frac{\sinh jk(h+z)}{\cosh jkh} \cos jkx$$

which automatically satisfies the kinematic bottom boundary condition and the periodic lateral boundary conditions. The given parameters defining a steady wave solution are

$$H \qquad h \qquad T \qquad C_E \ \text{ or } \ C_S$$

where H is the wave height, h is the water depth to the still water level (SWL), T is the wave period ($= 2\pi/\omega$), C_E is the coflowing Eulerian current and C_s is the Stokes drift. The unknowns are

$$k \qquad \bar{u} \qquad C_E \ \text{ or } \ C_S \qquad Q \qquad R$$

$$\eta_m, \quad m = 0(1)M$$

$$B_j, \quad j = 1(1)N$$

of which there are M+N+6. k is the wave number, $\bar{u}$ is the mean fluid speed at any z wholly within the fluid, -Q is the constant volume flow rate per unit width under the steady wave, R is the Bernoulli constant in the steady frame, $\eta_m = \eta(x_m)$ are the water surface nodes where $x_m = (m-1)\pi/kM$, and B_j are the dimensionless Fourier coefficients.

The problem formulation provides 2M+6 implicit algebraic equations in these M+N+6 unknowns, each equation being cast in the form

$$f_i(k, \ \bar{u}, \ C_E \ \text{ or } \ C_S, \ Q, \ R, \ \eta_m, \ B_j) = 0$$

The equations define the wave height

$$f_1 = \eta_0 - \eta_M - H$$

the mean water level (MWL)

$$f_2 = \frac{1}{2M} \left(\eta_0 + 2 \sum_{m=1}^{M-1} \eta_m + \eta_M \right)$$

the Eulerian current

$$f_3 = \frac{2\pi/k}{T} - \bar{u} - C_E$$

the Stokes drift

$$f_4 = \frac{2\pi/k}{T} - \frac{Q}{h} - C_S$$

the kinematic free surface boundary condition (KFSBC) at each of the M+1 free surface nodes

$$f_{5+2m} = \psi(x_m, \eta_m) + Q$$

and the dynamic free surface boundary condition (DFSBC) also at each of the free surface nodes

$$f_{6+2m} = \frac{1}{2} [\frac{\partial \psi(x_m, \eta_m)}{\partial x}]^2 + \frac{1}{2} [\frac{\partial \psi(x_m, \eta_m)}{\partial z}]^2 + g\eta_m - R$$

This system of equations is recast in dimensionless form in terms of ω and g and can be solved in a least squares sense for $M \geq N$, which was accomplished using the IMSL subroutine ZXSSQ, a finite difference Levenberg-Marquardt algorithm with strict descent, in double precision. This algorithm seeks the minimum value of

$$\Sigma = f_1^2 + f_2^2 + \dots + f_{2M+6}^2$$

In the present analysis M=N=18 throughout. This solution provides the initial conditions in deep water for the following shoaling computation.

3.   Fourier Shoaling

There are a number of alternative higher order shoaling formulations in the literature but all are arguably incomplete, either in formulation or wave theory selection or both (7). The Stiassnie and Peregrine (8) formulation is appropriate, which imposes zero mass flux I, constant wave action flux N and constant Bernoulli constant $\overline{B}$, all in the fixed, unsteady reference frame and not the moving, steady frame of the wave theory. The given parameters defining each new steady wave solution become

$$h \qquad T \qquad I_0 \qquad N_0 \qquad \overline{B}_0$$

where $I_0, N_0, \overline{B}_0$ are defined by the deep water initial conditions solution from the classical steady wave solution. The M+N+6 unknowns become

$$k \qquad \overline{u} \qquad C_E \qquad Q \qquad R$$

$$\eta_m, \quad m = 0(1)M$$

$$B_j, \quad j = 1(1)N$$

where the wave height is determined from the solution as
$$H = \eta_0 - \eta_M$$

together with the setup of the local MWL from the global SWL
$$\overline{\eta} = \frac{1}{2M}\left(\eta_0 + 2\sum_{m=1}^{M-1}\eta_m + \eta_M\right)$$

C_E is identified as the undertow current which opposes the direction of wave propagation. Less complete formulations will exclude $\overline{\eta}$ and/or C_E.

The 2M+6 implicit algebraic equations now define the undertow
$$f_1 = \frac{2\pi/k}{T} - \overline{u} - C_E$$

the mass flux in the fixed frame where $I_0 = 0$ is imposed by the beach,
$$f_2 = I$$

the Bernoulli constant in the fixed frame
$$f_3 = \overline{B} - \overline{B}_0$$

the wave action flux in the fixed frame
$$f_4 = N - N_0$$

and the KFSBC and DFSBC equations remain unchanged. The evaluation of mass flux, Bernoulli constant and wave action flux utilizes various integral relationships among the parameters (1,8,9), although considerable care must be taken to translate these relationships to a fixed frame located at the global SWL. The least squares solution algorithm remains otherwise unchanged with an appropriate initial estimate of the solution provided by the converged solution at the previous water depth.

4.   Solution Characteristics

Incident wave conditions of wave height 2 m, water depth 100 m, wave period 10 s, zero setup and zero Stokes drift have been adopted as an illustration of the capability of the method. A Fourier steady wave solution following Section 2 defines mass flux (at zero), the wave action flux and the Bernoulli constant, all in the fixed frame, that are conserved in the Fourier shoaling computation of Section 3. Shoaling computations were continued from water depth 100 m through to depth 3.8 m. Figure 1 shows the evolution of wave height, wave number, setup and undertow with water depth. Also included is the Airy (linear) wave shoaling solution for wave height and wave number, the Airy solutions for setup and undertow being zero.

This figure clearly illustrates the major features of the shoaling evolution which involves a gradual increase in wave number and the familiar (from linear theory) initial decrease in wave height followed by a rapid increase to values considerably above the deep water wave height. The associated evolution of the setup of the MWL and the undertow current are also available from the present formulation of Fourier shoaling. Both are negative. The setdown of the MWL from the global SWL increases steadily in magnitude towards shallower water. The magnitudes are not large but the dynamic influence is known to be significant in the near shore zone (10). The magnitude of the reverse undertow also increases in magnitude, in accord with observations in shoaling and breaking waves.

The sharp discontinuity in the evolution traces at a depth of about 4.3 m is indicative of wave breaking. It is not and can not be predictive, as the Fourier shoaling computation excludes relevant influences such as the sloping bottom, wave asymmetry and dissipation. It is however strongly suggestive of the physical processes that determine wave breaking and provides a useful description of the dependent variable interactions that lead to wave breaking.

A specific advantage of the present Fourier formulation is the ability to continue the computation beyond the 'break point'. The computations were continued and the solutions remained visually successful and physically reasonable. The evolution of the dependent variables remains equally reasonable and follows the general trends of available observations. Again there can be no expectation of a predictive capability near and beyond breaking but there is potential value in the descriptive nature of these shallow water computations. Figure 2 shows shoaling and 'surf zone' evolution of the mass, momentum(J), energy(F) and wave action fluxes together with the Bernoulli constant, all in the fixed frame. These provide complementary details of the dependent variable interactions. Note in particular that conservation of mass flux, wave action flux and Bernoulli constant implies conservation of energy flux but not momentum flux.

The occurrence and consequences of breaking appear to have much in common with a choke condition in open channel flow. The initial conditions in deep water define specific values of the wave action flux and the Bernoulli constant. These values are conserved as the water depth progressively decreases by adjustments principally in the wave height but also in wave number. Changes in setup and undertow are somewhat smaller and have a lesser influence on the conservation equations (7). The rapid fall in wave action flux, momentum flux and energy flux initiated at the 'break point' suggests the choke analogy for wave breaking. The water depth at the 'break point' is the smallest depth at which there exist a viable steady wave solution with the same wave action flux and Bernoulli constant as in deep water. At shallower depths, viable solutions do not exist with the same wave action flux and Bernoulli constant (and energy flux).

Wave action flux is clearly not conserved beyond the 'break point' but this does not necessarily destroy the potential relevance of the numerical solution. The least squares algorithm finds the minimum of the sum Σ. For solutions in water deeper than or equal to the water depth at the 'break point', this sum at convergence is not zero but neither is it especially large. The finite sum is largely attributable to the f_4 equation and to a lesser extent the f_2 and f_3 equations (see Figure 2), which are not satisfied, and the other $2M+3$ equations are essentially zero. Has the problem formulation been invalidated ? Arguably, no ! In principle a least squares solution is a weak solution in that it seeks a minimum of a sum of squares, not a zero for each of the $2M+6$ equations. In water depths up to the 'break point', the sum of squares is essentially zero and the solution must be characterized as strong. Beyond the 'break point' a characterization as moderately weak is perhaps appropriate. The solutions continue to follow physically reasonable trends and are arguably neither invalid nor especially weak. A strong solution is clearly not achievable with the present formulation in the surf zone, as it is a steady wave theory that excludes bed slope, asymmetry and dissipation.

The predicted solution trends beyond the 'break point' follow the trends of observational data in the surf zone. There is a continuous dissipation of both momentum and energy flux through the surf zone (Figure 2) which leads to a rapid decrease in wave height and reductions in the setdown and the magnitude of the undertow. The trend in wave height is seen to parallel the water depth, which is consistent with wave theories in shallow water where the limit wave is (11) approximately $H/h = 0.83$.

If it is necessary to use a steady wave theory in the surf zone, Fourier shoaling (at truncation order 18) is seen to provide tolerably realistic predictions, including predictions of quantities such as setup, undertow, wave height decay, momentum flux dissipation and energy flux dissipation that cannot be provided by linear theory.

5.   Conclusions

An analysis of the potential of steady wave theory in the surf zone has resulted in a moderately positive perspective. Providing that the context is Fourier wave at an adequate truncation order and that the shoaling formulation is complete, the nature of the least squares solution algorithm is sufficiently flexible to provide tolerable predictions of trends throughout the surf zone. The predictive capability will clearly not extend beyond trends but the results suggest that the potential value of steady wave theory in the surf zone should not be dismissed without consideration. Oscillatory flow remains the dominant characteristic of the flow in the surf zone, which must account for the relative success of steady wave theory.

6.   References

1.   Sobey, R.J., Goodwin, P., Thieke, R.J. and Westberg, R.J.jr.:"Application of Stokes, Cnoidal and Fourier wave theories". Journal of Waterway, Port, Coastal, and Ocean Engineering, ASCE, 113, 1987, 565-587.
2.   Sobey, R.J.: "Truncation order of Fourier wave theory". In: Procs 21st. International Conference on Coastal Engineering (Costa del Sol - Malaga, Spain: June, 1988) ASCE, to appear.
3.   Dean, R.G.:"Stream function representation of nonlinear ocean waves". Journal of Geophysical Research, 70, 1965, 4561-4572.
4.   Chaplin, J.R.:"Developments of stream-function wave theory". Coastal Engineering, 3, 1980, 179-205.
5.   Rienecker, M.M. and Fenton, J.D.:"A Fourier approximation method for steady water waves". Journal of Fluid Mechanics, 104, 1981, 119-137.
6.   Sobey, R.J.:"Apparent stress in shoaling waves". Shore and Beach, 55, 1987, 113-117.
7.   Bando, K. and Sobey, R.J.:"Variations on higher-order shoaling". Manuscript under review, 1988.
8.   Stiassnie, M. and Peregrine, D.H.:"Shoaling of finite-amplitude surface waves on water of slowly-varying depth". Journal of Fluid Mechanics, 97, 1980, 783-805.
9.   Longuet-Higgins, M.S.:"Integral properties of periodic gravity waves of finite amplitude". Procs, Royal Society, London, A342, 1975, 157-174.
10.   Battjes, J.A.:"Surf-zone dynamics". Annual Review of Fluid Mechanics, 20, 1988, 257-293.
11.   Longuet-Higgins, M.S.:"The unsolved problem of breaking waves". In: Procs 17th International Conference on Coastal Engineering (Sydney, Australia: March, 1980) ASCE, 1980, 1, 1-28.

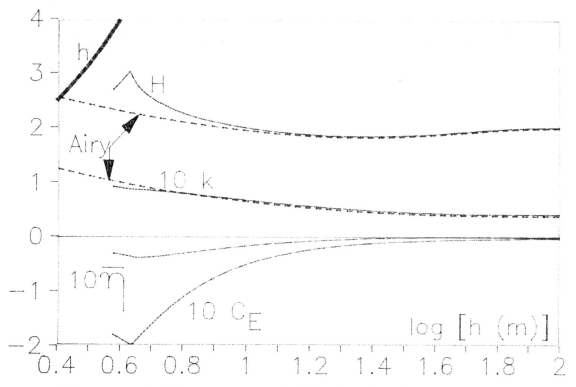

Figure 1: Dependent Variable Evolution

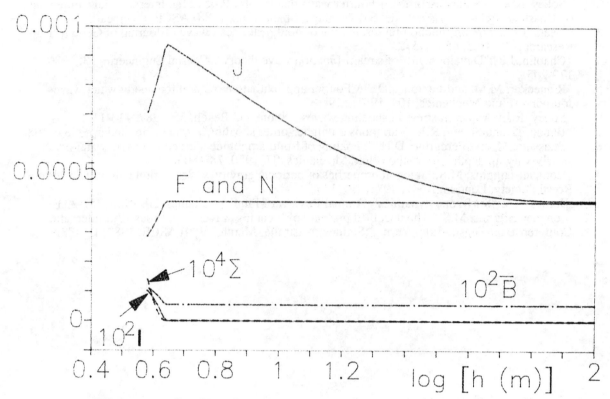

Figure 2: Evolution of Dimensionless Fluxes

Chapter 21
NUMERICAL MODELLING OF REFLECTIVE WAVES ON SLOWLY-VARYING CURRENTS

D Yoo, T S Hedges and B A O'Connor
University of Liverpool, UK

Summary

The hyperbolic-type refraction-diffraction-reflection model[3,4,5] is refined to take into account the effects of dissipation and wave-current interaction. The refined model is used to describe wave propagation around breakwaters, and the computed results are compared both with laboratory data and with other model results.

1. Introduction

In recent years, substantial advances have been made in the development of mathematical models for the description of wave climate in a coastal region, particularly since the hyperbolic and elliptic wave models were established by Ito & Tanimoto[1] and Berkhoff[2], respectively. Both models can take into account all essential features of wave behaviour such as refraction, diffraction and reflection. Berkhoff's elliptic scheme seems to have attracted greater attention than the hyperbolic scheme of Ito and Tanimoto, because the former is more accurate than the latter in describing waves in shallow water. However, major drawbacks of the elliptic model are its sensitivity to the boundary conditions and the requirement of a fully implicit technique for its solution.

The hyperbolic model has been refined to better represent shallow water effects on wave propagation by Nishimura et al.[3], by Watanabe and Maruyama[4], and by Copeland[5]. The three sets of equations proposed by these authors are virtually the same to a second order of accuracy, and the models have been applied to various engineering situations with considerable success. However, the effects of wave-current interaction on wave transformation have yet to be included. The significance of wave-current interaction on wave transformation and, in turn, on wave-induced currents was demonstrated by Yoo and O'Connor[6] using a 'period-average' wave model. This model describes well all important processes except reflection.

The refraction-diffraction-reflection model applicable to a non-uniform current field has been developed by Booij[7]. Booij's model is represented by one second-order elliptic differential equation which is of transient form, but it is more difficult to solve than the first-order differential equations used in the multi-equation hyperbolic model due to the elliptic type and higher order of the equation. In the present investigation, introducing a moving Lagrangian frame of reference, the first-order hyperbolic equations are modified to take into account the effects of wave-current interaction. In addition, the effect of bed frictional dissipation is also included in the description of wave transformation.

2. Wave motion on slowly-varying currents

The hyperbolic model developed by Ito and Tanimoto[1] derives from the combination of the conservation equations of continuity and momentum through depth-integration. Due to difficulties of manipulation, the second-order derivatives are ignored as usual, which then results in less accuracy in the description of waves in intermediate water depths. Copeland[5] solved this problem by employing the energy conservation equation instead of the continuity and momentum equations, giving:

$$\frac{\partial^2 \zeta}{\partial t^2} = \frac{1}{n} \frac{\partial}{\partial x_i} \ nC^2 \ \frac{\partial \zeta}{\partial x_i} \tag{1}$$

where $\zeta = \eta/\sigma$, η is the surface elevation, σ is the wave angular frequency, C is the wave celerity, n is the ratio of the group velocity to the wave celerity, i.e.

$$n = \frac{1}{2} \left[1 + \frac{2Kd}{\sinh (2Kd)} \right]$$

K is the wave number, d is the water depth, and $i = 1, 2$.

Equation (1), which is of elliptic type, is then decomposed into a pair of first-order hyperbolic equations for easier compution as follows:

$$\frac{\partial \zeta}{\partial t} + \frac{1}{n} \frac{\partial}{\partial x_i} (nR_i) = 0 \tag{2}$$

$$\frac{\partial R_i}{\partial t} + C^2 \frac{\partial \zeta}{\partial x_i} = 0 \tag{3}$$

where

$$R_i = \frac{1}{\sigma} \int_{-h}^{\eta} \tilde{u}_i \ dz \cong \frac{K_i}{K} \ \frac{\eta}{K}$$

h is the mean water depth and $\tilde{u}_i$ is the wave particle velocity. Equation (2) effectively represents conservation of mass and Eq. (3) represents conservation of momentum.

It is now assumed that waves propagate on a slowly-varying current, i.e. $|\partial U/\partial x| \ll KU$ and $|\partial U/\partial t| \ll \sigma U$. However, if the waves are viewed in a frame of reference moving with the current, then they appear to be propagating on quiescent water. Using this device, Eqs. (2) and (3) can then still be regarded as valid provided that values previously observed in a fixed frame of reference are replaced by values in the moving frame of reference.

Assuming that the vertical variation of current is negligible, the relationship between the horizontal co-ordinate in a fixed Eulerian frame of reference, x_i, and that in the moving Lagrangian frame of reference, r_i, is:

$$x_i = r_i + U_i t \tag{4}$$

where U_i is the current velocity tensor. Thus

$$\frac{\partial x_i}{\partial r_i} = 1 + \frac{K_i U_i}{\sigma_o} \qquad (5)$$

since

$$\frac{\partial r_i}{\partial t} = \frac{\sigma_o}{K_i}$$

where K_i is the wave number tensor and σ_o is the wave angular frequency observed in the moving frame of reference.

Employing Eq. (5), Eqs. (2) and (3) may now be rewritten as follows:

$$\frac{\partial \zeta}{\partial t} + \left[1 + \frac{K_i U_i}{\sigma_o} \right] \frac{1}{n} \frac{\partial}{\partial x_i} (nR_i) = 0 \qquad (6)$$

$$\frac{\partial R_i}{\partial t} + \left[1 + \frac{K_i U_i}{\sigma_o} \right] C^2 \frac{\partial \zeta}{\partial x_i} = 0 \qquad (7)$$

Here, ζ is now equal to η/σ_o instead of η/σ, and the wave number K is determined by the dispersion relationship:

$$\sigma_o^2 = gK \tanh Kd \qquad (8)$$

where g is the acceleration due to gravity. σ_o is related to the frequency, σ, observed in the fixed frame of reference by the Doppler equation:

$$\sigma = \sigma_o + K_i U_i \qquad (9)$$

The dispersion relation, Eq. (8), is derived for a flat bed. Therefore, the model represented by Eqs. (6), (7) and (8) may be valid only when $|\partial h/\partial x| \ll Kd$, i.e. for mild slope conditions. It is shown in the Appendix that the combination of Eqs. (6) and (7) reproduces Booij's second-order elliptic equation with minor differences. In the following section, it is also shown that the combination of the equations ensures conservation of wave action.

3. Dissipation due to bed friction

Dissipation terms have to be included only in the momentum equation. As usual, using a quadratic form for the dissipatation term, Eq. (7) is re-written:

$$\frac{\partial R_i}{\partial t} + \left[1 + \frac{K_i U_i}{\sigma_o} \right] C^2 \frac{\partial \zeta}{\partial x_i} + F|\hat{R}|R_i = 0 \qquad (10)$$

where F is the friction factor associated with R_i and $\hat{R}$ is the maximum value of R over a wave period, i.e. $\hat{R} \cong a/K$, where a is the wave amplitude.

Differentiating Eq. (6) with respect to time, employing Eq. (10) instead of Eq. (7), and introducing r_i gives the equation:

$$\frac{\partial^2 \zeta}{\partial t^2} - \frac{1}{n} \frac{\partial}{\partial r_i} \left\{ nC^2 \frac{\partial \zeta}{\partial r_i} + nF \frac{a}{K} C\zeta \right\} = 0 \qquad (11)$$

For sinusoidal linear waves, ζ can be replaced by

$$\zeta = \frac{a}{\sigma_o} \cos D \qquad \text{or} \qquad \zeta = \text{Real} \left[\frac{a}{\sigma_o} e^{iD} \right] \tag{12}$$

where D is the phase function. Substituting Eq. (12) into Eq. (11), taking the imaginary part and introducing wave action $A = \rho g a^2/(2\sigma_o)$ gives:

$$\frac{\partial A}{\partial t} + \frac{1}{n} \frac{\partial}{\partial r_i} (n\, C_i\, A) + F \frac{a}{K} A = 0 \tag{13}$$

Introducing Eq. (5) into Eq. (13) then yields:

$$\frac{\partial A}{\partial t} + \frac{1}{n} \frac{\partial}{\partial x_i} (n\, C_i\, A) + \frac{\partial}{\partial x_i} U_i A + F \frac{a}{K} A = 0 \tag{14}$$

Equation (14) is the well-known conservation equation of wave action[8] for dissipative shallow water waves, except that the second term is in small error; it should be $\partial/\partial x_i$ $(n\, C_i A)$. The effect of this error is expected to be very small, since n is of $O(1)$.

By considering the detailed flow mechanism near the bed in combined waves and currents, a method of estimating the friction factor associated with the wave amplitude has been given[9,10]. Using these lattermost results, the friction factor associated with R_i is estimated by:

$$F = 2K \frac{\delta}{g} \left[\frac{\sigma_o}{\sinh Kd} \right]^3 \tilde{C}_f \tag{15}$$

where δ is the averaging factor over a wave period and $\tilde{C}_f$ is the wave friction factor which is largely determined by the ratio of the roughness height of the bed to the maximum water-particle excursion at the bed. When waves are acting alone, the averaging factor δ is equal to $4/(3\pi)$ for linear waves; but when waves interact with currents, the factor δ increases from this value depending on the degree of interaction and the angle of intersection. Details of the computation of δ and $\tilde{C}_f$ are given elsewhere[9,10].

4. Model tests

Equations (6) and (10) form the basis of the present hyperbolic model for the description of wave refraction, diffraction, reflection and dissipation on slowly-varying currents. To solve the equations, the leap-frog explicit finite difference scheme is employed with appropriate boundary conditions. Wave breaking is accounted for as in previous work[1,5].

In the present study, consideration is given to the effects on waves of wave-induced currents. Therefore, it has also been necessary to incorporate in the model system a means of describing these currents, based upon calculations of the radiation stresses. Details of the current model are given elsewhere[11,12] and are not repeated here.

Figure 1 shows the layout of a physical model of a semi-detached breakwater[13] together with the finite difference grid used in the computational scheme. Waves of 91 mm height and with a period of 1.5 s were generated from offshore, normal to the breakwater axis. A training wall isolated most of the physical model area from the effects of wave reflection off the breakwater face. The finite difference grid had $\Delta x = 0.185$ m and $\Delta y = 0.2075$ m. The time step chosen was 0.005 s and a value of 1 mm was used for the effective roughness height of the concrete surface of the physical model.

300

A previous study using a 'period-average' wave model[6] showed that the computed wave and current patterns around the breakwater reached a steady state after about 40 seconds. The computation results at 50 seconds were chosen as the final solution. Since in the present 'inter-period' modelling, only discrete interaction between the wave and current fields was allowed at each wave period, rather than allowing simultaneous and continuous interaction, the time variations of the flow pattern were found to be somewhat different for the period-average and inter-period models. Nevertheless, the results were similar and values at 50 seconds were again chosen to represent the final solution. The computed current pattern was shown to be very different from the experimental results when wave-current interaction is ignored[6,14]. When this effect is incorporated, both inter-period and period-average schemes produce very good predictions of the current pattern as shown in Fig. 2.

Figure 3 shows the wave distribution measured in the experiment[13] and determined by the various schemes of wave modelling. The period-average wave model produces fairly good answers for the wave distribution[6,15] but there is some discrepancy in the shallow water at the corner behind the breakwater (see Fig. 3(b)). This is believed to be largely due to the neglect of wave reflection in the period-average modelling.

The inter-period wave model provides better answers for the wave distribution near the coastline, where the effect of reflection may not be negligible. But there is still some discrepancy in the wave distribution in the lee of the breakwater when the effect of wave-current interaction is ignored (see Fig. 3(c)). When wave-current interaction is taken into account (see Fig. 3(d)), better computational results are obtained.

A typical field situation of a detached breakwater on a sloping beach has also been investigated using the present inter-period scheme. The structure has plan dimensions of 30 m x 240 m and is located in about 8 m water depth on a uniform slope of 1:30. Solid walls are located on both lateral boundaries. The wave conditions at the inflow boundary are a = 1.0 m and T = 10.0 s with normal incidence, and the reflection coefficient assumed for the breakwater is 0.1. A finite difference mesh with Δx = 10 m and Δy = 20 m was used to model the field situation, and 0.2 s was chosen as the computational time step. A value of 20 mm was used for the effective roughness height of the bed which was assumed to be immobile.

Figure 4 shows the computational results for wave-height distributions and vector plots of wave-induced currents. The result shown on Fig. 4(a) is obtained at 100 second computation up to which time no wave-current interaction is allowed, while the result shown in Fig. 4(b) is obtained at 400 second computation when sufficient interaction between waves and currents has been allowed. Again, the results are found to be very different, which clearly indicates that the influence of currents on wave transformation is so significant that any computational results which ignore wave-current interaction must be carefully interpreted.

5. Concluding remarks

It has been shown that the first-order multi-equation hyperbolic model can easily be modified to account for the effects of wave-current interaction and bed friction. Comparisons with the results from a physical model study suggest that predictions are much improved by allowing for the effects of currents on wave propagation. The model has been applied also to a field situation for which it is shown that the influence of wave-current interaction is important to both the wave-height distribution and wave-induced currents.

6. Acknowledgement

The authors are grateful for the support of the Science and Engineering Research Council through its Marine Technology Directorate and Marinetech North West, U.K.

7. Appendix

In recent years Booij's second-order elliptic equation was corrected by employing a proper dynamic free surface boundary condition (Kirby[16]). The resulting equation is as follows:

$$\left[\frac{\partial}{\partial t} + U_i \frac{\partial}{\partial x_i} \right] \left[\frac{\partial \zeta}{\partial t} + U_i \frac{\partial \zeta}{\partial x_i} \right] + \frac{\partial U_i}{\partial x_i} \left[\frac{\partial \zeta}{\partial t} + U_i \frac{\partial \zeta}{\partial x_i} \right]$$

$$- \frac{\partial}{\partial x_i} nC^2 \frac{\partial \zeta}{\partial x_i} + (1 - n)\, \sigma_o \zeta = 0 \tag{A.1}$$

If waves are sinusoidal and nearly plane, i.e. $|\partial a/\partial x| \ll aK$ and $|\partial a/\partial t| \ll a\sigma$ or $a\sigma_o$, we find:

$$\frac{\partial^2 \zeta}{\partial t^2} = -\sigma_o^2 \zeta \tag{A.2}$$

Introducing Eq. (A.2) into Eq. (A.1) yields:

$$n \frac{\partial^2 \zeta}{\partial t^2} - \frac{\partial}{\partial x_i} nC^2 \frac{\partial \zeta}{\partial x_i} + f(\zeta, U_i) = 0 \tag{A.3}$$

where

$$f(\zeta, U_i) = U_i^2 \frac{\partial^2 \zeta}{\partial x_i^2} + U_i \frac{\partial U_i}{\partial x_i} \frac{\partial \zeta}{\partial x_i} + 2U_i \frac{\partial}{\partial x_i} \left(\frac{\partial \zeta}{\partial t} \right) + \frac{\partial U_i}{\partial x_i} \frac{\partial \zeta}{\partial t} \tag{A.4}$$

The additional terms $f(\zeta, U_i)$ represent the effects of wave-current interaction on wave transformation.

We now consider the first-order hyperbolic equations of the present model. Differentiating Eq. (6) with respect to time and substituting Eq. (7) yields:

$$n \frac{\partial^2 \zeta}{\partial t^2} - \frac{\partial}{\partial x_i} nC^2 \frac{\partial \zeta}{\partial x_i} + g(\zeta, U_i) = 0 \tag{A.5}$$

where

$$g(\zeta, U_i) = \frac{\partial}{\partial x_i} nU_i C_i \frac{\partial \zeta}{\partial x_i} + \frac{K_i U_i}{\sigma_o} \frac{\partial}{\partial x_i} nC^2 \frac{\partial \zeta}{\partial x_i} + \frac{K_i U_i}{\sigma_o} \frac{\partial}{\partial x_i} nU_i C_i \frac{\partial \zeta}{\partial x_i} \tag{A.6}$$

The effects of wave-current interaction are now represented by $g(\zeta, U_i)$ in Eq. (A.5). It is expected that $|\zeta(\partial nC/\partial x)| \ll |nC(\partial \zeta/\partial x)|$ for mild slope conditions. Therefore, Eq. (A.6) reduces to:

$$g(\zeta, U_i) = \frac{\partial nU_i}{\partial x_i} \frac{\partial \zeta C_i}{\partial x_i} + 2nU_i \frac{\partial^2 \zeta C_i}{\partial x_i^2} + nU_i^2 \frac{\partial^2 \zeta}{\partial x_i^2} \tag{A.7}$$

Again for mild slopes, $\zeta C_i \cong R_i$. Introducing Eq. (6) into Eq. (A.7), with some manipulation, we find:

$$g(\zeta, U_i) = nf(\zeta, U_i) \tag{A.8}$$

(It is, in fact, easy to evolve Eq. (A.4) to the form of Eq. (A.7) with the introduction of the continuity equation, Eq. (6)).

In shallow water, where the effects of wave-current interaction are expected to be strong, the ratio n is very close to unity. Thus $g(\zeta,U_i) \cong f(\zeta,U_i)$, and the present hyperbolic multi-equation model is almost equivalent to the second order elliptic single-equation model with regards to their accuracy. For a steady-state solution the latter is likely to be slightly more accurate than the former in intermediate water depths, i.e. when $n \neq 1$. However, for unsteady flow, both models contain errors. To solve the complex second order elliptic equation, Eq. (A.1), it has been necessary to ignore various second-order terms (Booij[7]). Consequently, the inherent better accuracy of the second-order single-equation model may not be achieved due to the difficulty in solving the complicated equation.

8. References

1. Ito, Y. and Tanimoto, K.: "A method of numerical analysis of wave propagation - application to wave refraction and diffraction". In: Proc. 13th International Conference on Coastal Engineering, ASCE, (Vancouver, Canada: July 10-14, 1972), pp. 503 -522.
2. Berkhoff, J. C. W.: "Computation of combined refraction diffraction". In: Proc. 13th International Conference on Coastal Engineering, ASCE, (Vancouver, Canada: July 10-14, 1972), pp. 471 - 490.
3. Nishimura, H., Maruyama, K. and Hiraguchi, H.: "Wave field analysis by finite difference method". In: Proc. 30th Japanese Conf. on Coastal Eng., 1983, pp. 123-127 (in Japanese).
4. Watanabe, A. and Maruyama, K.: Numerical model of wave transformation due to refraction, diffraction and dissipation". In: Proc. 31st Coastal Engineering Conference in Japan, 1984, pp. 406 - 410 (in Japanese).
5. Copeland, G. J. M.: "A practical alternative to the 'mild-slope' wave equation". Coastal Engineering, Vol. 9, 1985, pp. 125 - 149.
6. Yoo, D. and O'Connor, B. A.: "Mathematical modelling of wave-induced nearshore circulations". In: Proc. 20th International Conference on Coastal Engineering, ASCE, (Taipei, Taiwan: Nov. 9-14, 1986), pp. 1667 - 1681.
7. Booij, N.: "Gravity waves on water with non-uniform depth and current". 81 -1, Delft University of Technology, Dept. of Civil Eng., 1981.
8. Bretherton, F. P. and Garrett, C. J. R., "Wavetrains in inhomogeneous moving media", Proc. Roy. Soc. London A, Vol. 302, 1968, pp. 529 - 554.
9. Yoo, D. and O'Connor, B. A.: "Bed friction model of wave-current interacted flow". In: Proc. International Conference on Coastal Hydrodynamics, ASCE, (Delaware, U.S.A.: June 28 - July 1, 1987), pp. 93 - 106.
10. O'Connor, B. A. and Yoo, D.: "Mean bed friction of combined wave and current flow". Coastal Engineering, Vol. 12, 1988, pp. 1 - 21.
11. Copeland, G. J. M.: "Practical radiation stress calculations connected with equations of wave propagation". Coastal Engineering, Vol. 9, 1985, pp. 195 -219.
12. Yoo, D. and O'Connor, B. A.: "Numerical modelling of waves and wave-induced currents on groyned beach". In: Proc. IAHR Int. Symposium on Mathematical Modelling of Sediment Transport in the Coastal Zone (Copenhagen, Denmark: May 30 - June 1, 1988), pp. 127 - 136.
13. Gourlay, M. R.: "Wave set-up and wave generated currents in the lee of a breakwater or headland". In: Proc. 14th International Conference on Coastal Engineering, ASCE, (Copenhagen, Denmark: June 24-28, 1974), pp. 1976 - 1995.
14. Copeland, G.J.M.: "A numerical model for the propagation of short gravity waves and the resulting circulation around nearshore structures", Ph.D. Thesis, University of Liverpool, 1985.
15. Yoo, D. and O'Connor, B. A.: "Turbulence transport modelling of wave-induced currents". In: Proc. IAHR Int. Conf. on Computer Modelling in Ocean Eng. (Venice, Italy: September 19 - 23, 1988), (in press).
16. Kirby, J. T.: "A note on linear surface wave-current interaction over slowly varying topography", J. Geophy. Res., 1984, Vol. 89, pp. 745 - 747.

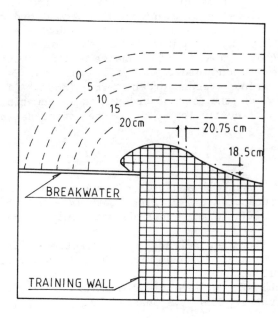

Fig. 1 Finite difference mesh system
for Gourlay's wave tank.

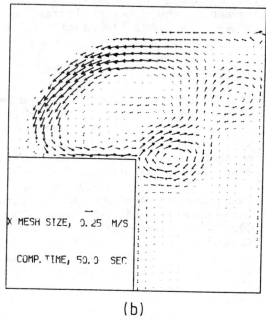

(b)

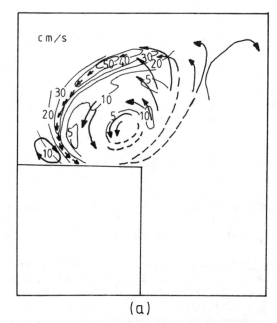

(a)

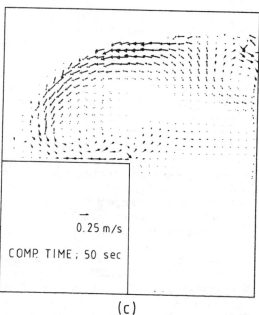

(c)

Fig. 2 Velocity distributions of wave-induced currents resulting from various wave
schemes compared with Gourlay's laboratory data
(a) experiment, after Gourlay [13]
(b) period-average model, after Yoo and O'Connor [15]
(c) present model with full wave-current interaction

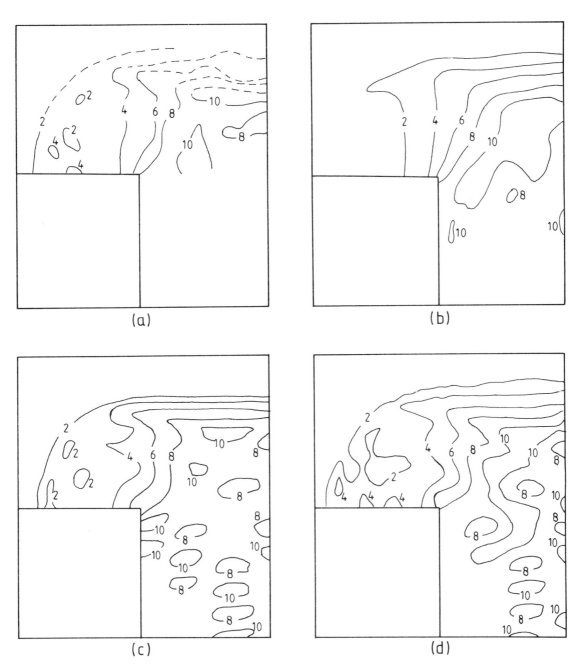

Fig. 3 Comparison of Gourlay's laboratory data with results from various computational schemes. Contours indicate wave heights in centimetres.
- (a) experiment, after Gourlay [13]
- (b) period-average model, after Yoo & O'Connor [6]
- (c) present model without full wave-current interaction
- (d) present model with full wave-current interaction

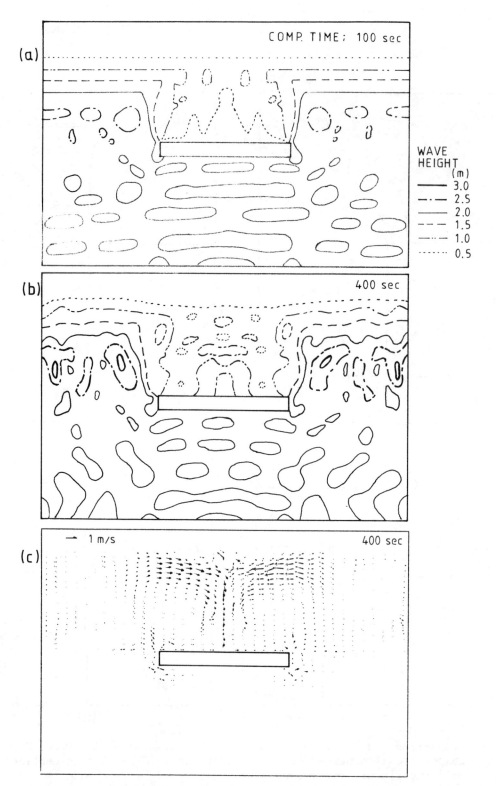

Fig. 4 Wave-height distribution and vector plot of wave-induced currents for a detached breakwater situation
(a) wave-height distribution without current-interaction
(b) wave-height distribution with current-interaction
(c) vector plot of wave-induced currents

Chapter 22
MILD-SLOPE MODEL FOR THE WAVE BEHAVIOUR IN AND AROUND HARBOURS AND COASTAL STRUCTURES IN AREAS OF VARIABLE DEPTH AND FLOW CONDITIONS

D P Hurdle, J K Kostense and P van den Bosch
Delft Hydraulics, The Netherlands

Summary

The wave behaviour in and around harbours and coastal structures can be
modelled in the frequency domain by means of the elliptic mild slope equation.
A finite element model has been developed using an extended version of the
equation, thus enabling the inclusion of the effects of refraction by ambient
currents and energy dissipation due to wave breaking and bottom friction. For
the numerical solution either a direct method, using gaussian elimination or
an indirect conjugate gradients method can be applied. For problems with large
numbers of nodal points the latter method is shown to be more efficient in
memory use and computer time. The results of several types of application are
presented, including an example of harbour seiching.

1. Introduction

For unidirectional monochromatic linear wave propagation in regions of variable depth, the mild slope equation, as originally derived by Berkhoff (1),
gives the most general description since both diffraction and refraction
effects are fully accounted for. For wave propagation in coastal areas an
efficient method is to apply a parabolic approximation of the mild-slope
equation, see e.g. Radder (2). The nature of the parabolic approximation is
such that diffraction and reflection are neglected in the main wave propagation direction (the computational direction) while diffraction effects are
accounted for in the lateral direction. Therefore, for regions with reflecting

and/or diffracting boundaries, such as harbours, it is essential to model the elliptic mild-slope equation. Kostense et al. (3) presented a finite-element model to solve the mild-slope equation which includes boundary conditions for partial reflection and transmission.

In harbour applications there is a tendency to include a part of the coastal region in the computational area if this has a distinct influence on the wave penetration characteristics, for example when an offshore breakwater or an entrance channel is present. In such coastal areas wave dissipation or refraction by currents are commonly important and thus these effects should also be modelled. Furthermore, wave energy dissipation due to bottom friction can have a significant effect on the resonant response of harbours, see e.g. Kostense et al. (3).

Booij (4) suggested how to extend the mild-slope equation with dissipation, and Kirby (5) gave the correct formulation for the inclusion of current refraction. These extensions have been implemented in the finite element model: for bottom dissipation and current refraction see Kostense et al. (3,6) and for wave breaking see De Giralomo et al. (7). The formulations for these extensions are summarized in section 2.

Boundary conditions are available to model both open boundaries (those through which incident waves pass) and fixed boundaries, where waves are partially reflected or transmitted. These are briefly described in section 3.

For the numerical solution, a finite element method is applied, based on linear triangular surface elements, which involves the solution of a large matrix equation. Until recently this equation has been solved using gaussian elimination to invert the matrix. The program now also includes the possibility of using a conjugate gradients method which can save considerable processor time and memory use for large problems. This method and conditions for improvement are described in section 4.

In section 5, the results of some applications are presented. These examples show the importance of modelling breaking, bottom friction, current refraction and directional spreading in engineering practice. These include an example of seiching response to low frequency waves.

2. Governing equations

The mild-slope equation was originally derived by Berkhoff (1) using the usual assumptions of linear potential theory. This models the wave potential field accounting for the effects of diffraction and refraction over a mildly sloping bottom as:

$$\frac{\partial}{\partial x_i} \left[cc_g \frac{\partial \phi}{\partial x_i} \right] + (cc_g k^2)\phi = 0 \qquad\qquad i = 1,2 \qquad\qquad (1)$$

Where c, c_g and k denote the phase velocity, the group velocity and the wave number, respectively. The two dimensional velocity potential $\phi(x,y)$ is related to the total potential $\Phi(x,y,z,t)$ according to:

$$\Phi(x,y,z,t) = f(z,h)\ \phi(\underline{x})\ e^{-i\omega t} \qquad\qquad\qquad (2)$$

where ω denotes the radian wave frequency, h is the local water depth and $\underline{x}$ is a vector representing the horizontal position. The function, f, is given by:

$$f(z,h) = \frac{\cosh[k(h+z)]}{\cosh[kh]} \qquad\qquad\qquad (3)$$

Equation (1) describes the behaviour of a monochromatic wave field in which the wave height is related to the potential by:

$$H = \frac{2k}{\omega}\ |\phi| \qquad\qquad\qquad (4)$$

In nature the incident wave field is random and has directional spreading. The behaviour of such wave fields can be described by solving equation (1) for a number of frequencies and directions and using linear superposition. In general it is found that for short wave penetration of harbours, frequency spreading has a limited effect. The importance of directional spreading depends on the modelled geometry. For many applications the wave behaviour in the computational area can be adequately represented using unidirectional monochromatic incident waves. In such a case the wave period is taken to be the period of peak spectral energy ($T_p = 1/f_m$) and the wave height to be the significant wave height.

Booij (4) suggested an additional term in the mild-slope equation to represent wave dissipation and Kirby (5) used a variational principle to derive an equation which also accounts for refraction effects due to an ambient current field. The equation used in the numerical model is obtained from the latter equation by assuming that the current is small - neglecting terms of second order in $\underline{U}$ - and by incorporating Booij's energy dissipation term, yielding:

$$\frac{\partial}{\partial x_i} \left[-cc_g \frac{\partial \phi}{\partial x_i} \right] - 2i\omega U_i \frac{\partial \phi}{\partial x_i} + \left[\omega_r^2 - \omega^2 - k^2 cc_g - i\omega \frac{\partial U_i}{\partial x_i} \right] \phi = i\omega_r W\phi \qquad (5)$$

where the relative frequency, ω_r, is given by:

$$\omega_r = \omega - \underline{k}.\underline{U} \qquad\qquad\qquad (6)$$

309

Advances in water modelling and measurement

A full description of the implementation of current effects in the numerical model is given by Kostense et al. (6).

The coefficient W in equation (5) represents the ratio of the rate of energy dissipation per unit area, D, to the local energy intensity, E:

$$W = D/E \qquad (7)$$

It is assumed that dissipation due to wave breaking and due to bottom friction can be superposed:

$$W = W_b + W_f \qquad (8)$$

The implementation of wave breaking in the numerical model is described by De Girolamo et al. (7), who also show that the simulations with the numerical model are in very good agreement with model experiments. The formulation used to quantify energy dissipation by wave breaking is that given by Battjes and Janssen (8):

$$D_b = \frac{\alpha}{8\pi} \rho g \omega_r Q_b H_{max}^2 \qquad (9)$$

where H_{max} is an estimate of the maximum height of a wave that does not break, Q_b is an estimate of the proportion of breaking waves, α is a constant usually taken to be 1, ρ is the water density and g is the acceleration due to gravity.

Wave energy dissipation by bottom friction is caused by bottom shear stress:

$$\underline{\tau} = - \frac{1}{2} f_w \rho \, |\underline{u}_b| \, \underline{u}_b \qquad (10)$$

where f_w is the wave friction coefficient and $\underline{u}_b$ is the water particle velocity just outside the bottom boundary layer. The appropriate value for f_w depends on the local sea bed roughness and $\underline{u}_b$ as described by Jonsson (9). Putnam and Johnson (10) showed that for regular waves the mean dissipated power per unit area is given by:

$$D_f = \frac{2}{3} \pi^2 \rho f_w \left(\frac{H}{T \sinh(kh)} \right)^3 \qquad (11)$$

Using equation (7) and expressing H in terms of the amplitude of the horizontal velocity at z = 0, the following expression for W_f is obtained:

$$W_f = \frac{4}{3\pi} f_w \frac{k}{\sinh(kh)} \frac{u_e}{\cosh^2(kh)} \qquad (12)$$

where u_e denotes the effective amplitude of the horizontal velocity, as given by Kostense et al. (3). The effective amplitude is equal to the amplitude for unidirectional waves where the horizontal velocity components are in phase. As shown by Kostense et al. (3), equation (12) can be directly derived for shallow water conditions from the linearised continuity and momentum equations. This equation is applicable -for computations of the seiching response of harbours to long wave exitation where consideration of monochromatic response is appropriate. Dingemans (11) adapted equation (11) for Rayleigh distributed waves. The corresponding expression for W is:

$$W_f = \frac{1}{\sqrt{2\pi}} f_w \frac{k}{\sinh(kh)} \frac{u_e}{\cosh^2(kh)} \tag{13}$$

This expression is used to simulate the effect of bottom friction on short period random waves, represented by the significant wave height. In the numerical model f_w is determined at each nodal point from the local velocity and sea bed roughness. This is particularly important for seiching response where the bottom velocities can vary substantially throughout the harbour. Further, both laminar and turbulent boundary layers can be accounted for as described by Kostense et al. (3). This allows comparison with experimental research as well as simulation of full scale effects for a range of conditions.

Thus in equation (5), the magnitude of W_f is dependent on the local velocity potential and further the value of ω_r is dependent on the local wave direction. As the solution method is linear, solution by iteration is necessary for problems involving dissipation or current effects. Note that the formulations should be treated with caution when both effects are relevant, because of the weakness in the physical understanding of dissipation with combined waves and current.

3. Boundary conditions

There are two basic types of boundary: open boundaries, through which incident waves pass, and fixed boundaries, which partially reflect or transmit waves. The implemented boundary conditions are as follows:

- a radiation condition;
- a wave maker condition;
- a condition for partial reflection;
- a condition for combined reflection and transmission.

The first two conditions are applicable for open boundaries and the last two are applicable for fixed boundaries such as beaches, rubble mounds, and

311

harbour walls. The wave maker condition is included to allow comparison with physical model tests and represents wave generation by a fully reflecting paddle.

The radiation condition matches the wave potential in the computational area to the potential of the incident and scattered waves in the outer area. The potential of scattered waves in the outer area can be found in terms of solutions which satisfy the Sommerfeld condition. This implicitly demands that the water depth and the current field should be uniform in the outer area. The requirement that the potential and its derivative normal to the boundary must be continuous on the boundary is then used to solve the system. Methods implemented to achieve this are the distribution of potential sources along the open boundary, and a series expansion in Hankel functions of the first kind. In the latter case the use of a circular shaped open boundary is required.

For fixed boundaries the first order conditions given by Berkhoff (1) for the partial reflection of normally incident waves are used:

$$\frac{\partial \phi}{\partial n} - ik(1-r)\phi = 0 \tag{14}$$

where n denotes the direction normal to the boundary, and r is a complex coefficient which can be related to the ratio of reflected wave height to incident wave height, R, and the phase shift at the boundary, ρ. For waves normally incident on a boundary this yields:

$$Re^{i\rho} = \frac{r}{2 - r} \tag{15}$$

A full discussion of reflection coefficients for oblique wave incidence is given by Behrendt (12) and Kostense et al. (3).

For transmitting boundaries Kostense et al. (3) show that the first order boundary condition can be expressed in terms of the potential on either side of the boundary (ϕ_1 and ϕ_2):

$$(1-s^2)\frac{\partial \phi_1}{\partial n} - ik(1-r)((1+st)\phi_1 - (s+t)\phi_2) = 0 \tag{16}$$

$$(1-s^2)\frac{\partial \phi_2}{\partial n} - ik(1-r)((1+st)\phi_2 - (s+t)\phi_1) = 0 \tag{17}$$

with s = (1-r)t, where r,s and t are complex coefficients which can be related to the physical reflection and transmission coeficients, R and T, and the related phase shifts, ρ and τ. For normal wave incidence this results in:

$$\text{Re}^{i\rho} = \frac{r}{2 - r} \qquad\qquad (18)$$

$$\text{Te}^{i\tau} = \frac{2t(1-r)}{2 - r} \qquad\qquad (19)$$

A full dicussion of the reflection and transmission coefficients for oblique wave incidence is given in Kostense et al. (3).

4. Numerical solution

The solution is based on a finite element method using a grid made up of triangular surface elements. The potential is assumed to vary linearly over each element and standard techniques are used to build a system of equations for the unknown potential at each nodal point of the form:

$$\mathbf{A} . \underline{\phi} = \underline{\psi} \qquad\qquad (20)$$

where $\mathbf{A}$ is an N by N matrix, $\underline{\phi}$ is a vector of length N containing the potential at each nodal point, $\underline{\psi}$ is a known vector of length N, and N is the number of nodal points. The matrix $\mathbf{A}$ is complex, non-symmetric and non-positive definite. In a grid with triangular elements each point has typically 6 neighbours, so each row in $\mathbf{A}$ has about 7 non-zero elements. For rows corresponding to the nodal points on an open boundary this number of non-zero elements is much larger because the values of the potential at all such points are related by the radiation condition.

Two different methods have been implemented to solve the matrix equation:
- a direct method using gaussian elimination;
- an indirect method using a conjugate gradients method.

The implemented gaussian elimination algorithm is based on a profile method. Gaussian elimination has a large storage requirement, which increases rapidly with the number of nodal points, N. Furthermore, the number of nodes along the open boundary has a considerable effect on the average bandwidth of the matrix, $\mathbf{A}$, and therefore on memory requirement. Thus, for large problems, fast secondary memory facilities or even disk storage must be used. This leads to I/O times which may exceed the processor time.

The indirect method used is the conjugate gradients squared algorithm described by Sonneveld (13) and Kaasschieter (14), which can be applied to non-positive definite matrices and, with some adaption, to complex matrices. Although this method is not guaranteed to converge, no problems have yet been encountered in this application. Each iteration involves two matrix multiplications of the form:

313

$$\underline{y} = A.\underline{x} \tag{21}$$

and also two of the form:

$$\underline{y} = K.\underline{x} \tag{22}$$

where K is an approximation to the inverse of A and $\underline{x}$ is a known vector of length N. The form of this approximation is known as the pre-conditioning. As would be expected, the accuracy of the pre-conditioning determines the number of iterations required. The most efficient pre-conditioning is that which gives the best balance between the increased number of iterations and the reduced complexity. However, for use on supercomputers with vector processors the ability to vectorise the computation also appears to be of major importance.

The operation in equation (21) does not vectorise well because each row of the matrix A has only about 7 elements, whereas long vectors are required for efficient vectorisation. To improve this, six of the non-diagonal elements in each row of **A** are stored in six vectors of length N and the remaining non-zero elements of **A** are stored in a number of variable length vectors. Each vector contains no more than one element from each row of **A** and if necessary zeros are supplied to elements of the vectors of length N. Index vectors are used to store column numbers and, for the variable length vectors, also the row numbers of the elements of **A**. The matrix multiplication can then be carried out as a combination of long-vector operations.

Several methods of pre-conditioning were tried and it was found that when using vector processors the most efficient of these was the most simple one. For example, a pre-conditioning of the following form gives a quite accurate estimate of A^{-1}:

$$K = (I + U)^{-1}.D^{-1}.(L + I)^{-1} \qquad (23)$$

where I denotes the unit matrix, U is the upper triangle of the matrix **A**, L is the lower triangle, and D is a diagonal matrix chosen such that elements in the leading diagonal of K.A are unity. Now, equation (22) can be computed from:

$$(L + I).\underline{z} = \underline{v} \qquad (24)$$

$$(U + I).\underline{y} = D.\underline{z} \qquad (25)$$

$\underline{y}$ and $\underline{z}$ can be determined using forward substitution in equation (24) and backward substitution in equation (25). However, these substitutions do not vectorise well as each step involves matrix multiplications in which the number of elements in each row is typically only four. To avoid vectorisation problems, the following pre-conditioning can be used:

$$K = E^{-1} \qquad (26)$$

where E is the leading diagonal of **A**. Because the system of equations (20) is scaled with E, this type of pre-conditioning requires no extra operations and gives a very efficient solution. For example, for a mesh with 70165 nodal points this pre-conditioning required twice the number of iterations required when using the pre-conditioning given in equation (23). However, because of increased simplicity of operations and the ability to vectorize the computation, it is considerably more efficient, requiring less than 25% of the processor time.

The efficiency of the conjugate gradients method, using the simple pre-conditioning defined above was compared with gaussian elimination for a number of problems with different mesh sizes on a Cray XMP. The results, given in table

315

1, show that gaussian elimination is more efficient for small problems but that for large mesh sizes the indirect method uses less processor time and considerably less I/O time offering a four-fold reduction in the combined processor and I/O time for the mesh with 70165 nodal points. However, the savings are even greater as the indirect method uses less memory and avoids the use of external fast storage facilities. The advantage of the indirect method is even greater when long open boundaries are applied, because of the increase in the average bandwidth of the matrix, A. On the other hand, if gaussian elimination is used the wave field for extra incident wave directions can be computed with little extra overhead, whereas with the method of conjugate gradients a new computation must be performed for each direction.

Table 2 shows the effect of increasing the number of nodal points on the computer time and efficiency for the indirect method. These computations were performed on a NEC SX/2 supercomputer.

5. Applications

The effect of directional spreading

The effect of modelling directional spreading in the incident wave field can be seen in a computation for the harbour of IJmuiden in the Netherlands. The modelled configuration and the bottom topography are shown in figure 1. Results with and without directional spreading are shown in figures 2 and 3 for a main wave direction from 270°N. The distribution of energy over direction was taken to be of the form $\cos^{4}(\theta-\theta_{o})$ (where θ is direction and and θ_{o} is the main wave direction ($|\theta-\theta_{o}| < 90^{\circ}$)). The qualitative pattern of wave height seen in both cases is similar but the predicted wave height is considerably lower at the eastern side of the harbour when the wave field is modelled with directional spreading.

The effect of bottom friction on seiching response

The effect of bottom friction on seiching response was established for the Rotterdam Europoort complex in the Netherlands. The modelled configuration and the bottom topagraphy are shown in figure 4. The computations were commissioned to examine the effect of a storm surge defence system on the low frequency response in the area. Such a response can be excited by incidence of waves of very long period or by non-linear interactions in random incident wave trains. For long waves, schematisations with a relatively small number of nodal points can be used but computations must be made at small frequency intervals to ensure that all the important resonant effects are found. The computations are then repeated at the resonant frequencies but including the effects of bottom friction. The amplitude ratio is shown plotted against frequency with and without bottom friction in figure 5 for one specific position. In the cases with bottom friction the initial wave height was taken to be 0.3 m and the Nikuradse roughness parameter was taken to be $k_N = 0.01$ m. Figure 5 shows that the effect of friction differs, depending on the oscillation mode. An instantaneous view of the water surface elevation for one of the resonant frequencies is shown in figure 6. Transmission through a boundary can also have considerable effect on the resonant response (see Kostense et al (3)).

The effect of an ambient current field

An example from engineering practice of the effect of an ambient current is the wave penetration of the Malamocco inlet to the Venice Lagoon, which was computed as part of the project to develop a flood defence system for Venice. The schematisation of the inlet and the bottom topography are shown in figure 7 and the ebb current field is illustrated in figure 8. The wave field was computed for an incident wave from a bearing of 137°N, with height 1.8 m and period 8 s. The results without current are shown in figure 9. The expected focussing of energy along the south side of the entrance channel is clearly shown. This results in a reduction of the wave height in the channel. Figure 10 shows the effect of including current in the computation. There is a significant increase in penetration of the inlet resulting from the effects of current refraction, which offset the effects of bottom refraction on the south slope of the channel.

The effect of wave breaking

An example of the effect of wave breaking was computed for the Chioggia inlet to the Venice Lagoon, also computed as part of the Venice project. The schematisation of the inlet and the bottom topography is shown in figure 11. The wave field was computed for an incident wave coming from a bearing of 104°N with a significant height of 5.0 m and a period T = 14 s. The results of the computation including breaking are shown in figure 12. The effects of breaking can clearly be recognized on the area of shallow water to the South of the inlet where because of its shape it would normally be expected to see the concentration of energy by bottom refraction. However, the wave height is instead considerably reduced because of breaking, see also De Girolamo et al. (7).

6. References

1. Berkhoff, J.C.W.: "Computation of combined refraction-diffraction". In Proc. 13th Int. Conf. Coastal Engineering (Vancouver: 1972), pp 471-490.

2. Radder, A.C.: "On the parabolic equation method for water-wave propagation". Journal Fluid Mechanics 95, 1, Nov. 1979, pp. 159-176.

3. Kostense, J.K.; Meijer K.L.; Dingemans M.W.; Mynett A.E. and van den Bosch P.: "Wave energy dissipation in arbitrarily shaped harbours of variable depth". In Proc. 20th Int. Conf. Coastal Engineering (Taipei: 1986), pp. 2002-2016.

4. Booij, N.: "Gravity waves on water with non-uniform depth and current". Thesis, Delft Univ. of Technology, May 1981.

5. Kirby, J.T.: "A note on linear surface wave current interaction over slowly varying topography". J. Geophysical Res., 89:C1, 1984, pp. 745-747.

6. Kostense, J.K.; Dingemans, M.W. and van den Bosch, P.: "Wave-current interaction in harbours". Proc. 21st Int.Conf. Coastal Eng.(Malaga: 1988).

7. De Girolamo, P., Kostense, J.K. and Dingemans M.W.: "Inclusion of wave breaking in mild-slope wave propagation model". In Proc. Int. Conf. Computer Modelling in Ocean Engineering (Venice, Italy: 1988).

8. Battjes, J.A. and Janssen, J.P.F.M.: "Energy loss and set-up due to breaking of random waves". In Proc. 16th Int. Conf. Coastal Engineering (Hamburg, Germany: 1978), pp 569-587.

9. Jonsson, I.G.: "A new approach to oscillatory rough turbulent boundary layers". Inst. of Hydrodyn. and Hydraulic Engng., Techn. Univ. Denmark, Series Paper No. 17, 1978.

10. Putnam, J.A. and Johnson, J.W.: "The dissipation of wave energy by bottom friction". Trans. Am. Geophysics Union, 30, 1, 1949, pp. 67-74.

11. Dingemans, M.W.: "Verification of numerical wave propagation models with field measurements - CREDIZ verification Haringvliet". Delft Hydraulics, report W 488 part 1, Dec 1983.

12. Behrendt, L.: "A finite element model for water wave diffraction including boundary absorption and bottom friction". Inst. of Hydrodynamics and Hydraulic Engng., Techn. Univ. Denmark, Series Paper No. 37, June 1985.

13. Sonneveld, P.: "CGS, a fast Lanczos-type solver for nonsymmetric linear systems". Delft Univ. of Technology, Report 84-16, 1984.

14. Kaasschieter, E.F.: "The solution of non-symmetric linear systems by bi-conjugate gradients squared". Delft Univ. of Techn., Report 86-21, 1986.

Number of nodal points	gaussian elimination			conjugate gradients		
	C P (sec)	I/O (sec)	operatn. rate (Mflops)	C P (sec)	I/O (sec)	operatn. rate (Mflops)
2787	1	2	30	3	1	35
24965	50	35	45	48	28	87
70165	396	598	71	210	22	95

Mflop = million floating point operations per second
C P   = processor time
I/O   = input and output time

Table 1. Comparison between efficieny of gaussian elimination
and conjugate gradients method on the Cray XMP.

Number of nodal points	Number of iter- ations	Processor time (s)	Operation rate (Mflops)[*]
70165	1429	88	241
109120	2546	194	255
148793	1368	166	265
249088	3468	657	249

Table 2. Performance of the conjugate gradients method for various
mesh sizes on a NEC SX/2

319

Water depth (m)

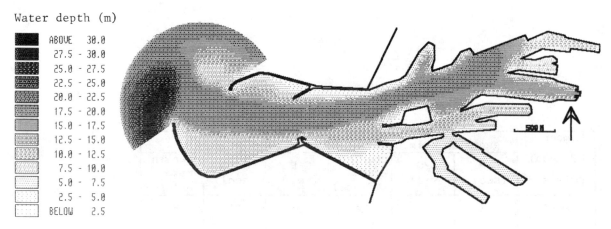

Figure 1   Configuration and bottom topography of the harbour of IJmuiden,
            The Netherlands

Wave height (%)

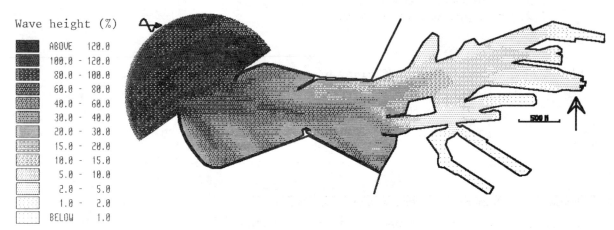

Figure 2   Percentage ratio of wave height to incident height,
            unidirectional incident waves, T = 10 s.

Wave height (%)

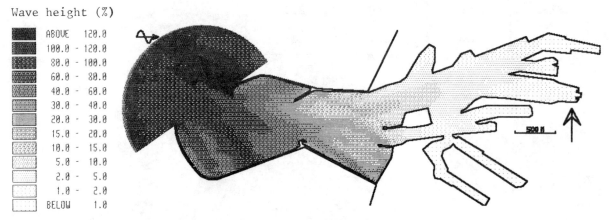

Figure 3   Percentage ratio of wave height to incident height,
            directional spreading $\cos^4(\theta - \theta_o)$, T = 10 s

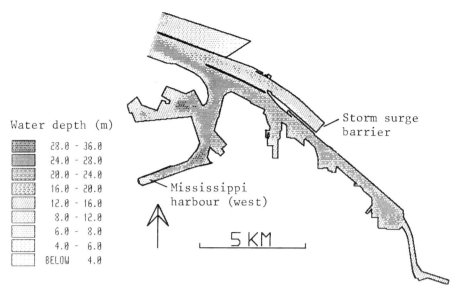

Water depth (m)

	28.0 - 36.0
	24.0 - 28.0
	20.0 - 24.0
	16.0 - 20.0
	12.0 - 16.0
	8.0 - 12.0
	6.0 - 8.0
	4.0 - 6.0
	BELOW 4.0

Storm surge barrier

Mississippi harbour (west)

5 KM

gure 4 Configuration and bottom topography of the Rotterdam Europoort area

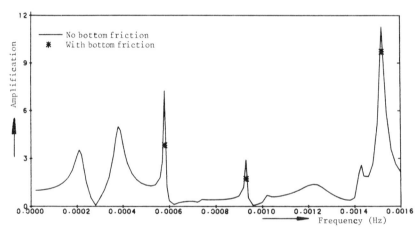

gure 5 Computed response amplitude ratio at Mississippi harbour
with and without wave friction, H = 0.3 m, k_N=0.01 m

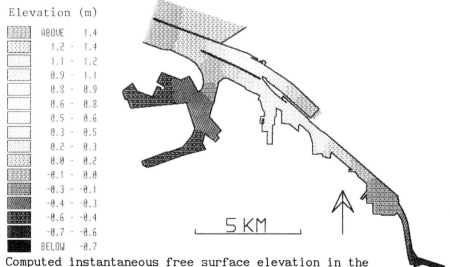

Elevation (m)

	ABOVE 1.4
	1.2 - 1.4
	1.1 - 1.2
	0.9 - 1.1
	0.8 - 0.9
	0.6 - 0.8
	0.5 - 0.6
	0.3 - 0.5
	0.2 - 0.3
	0.0 - 0.2
	-0.1 - 0.0
	-0.3 - -0.1
	-0.4 - -0.3
	-0.6 - -0.4
	-0.7 - -0.6
	BELOW -0.7

5 KM

gure 6 Computed instantaneous free surface elevation in the
Rotterdam Europoort area, f = 0.00058 Hz, H = 0.3 m, k_N = 0.01 m

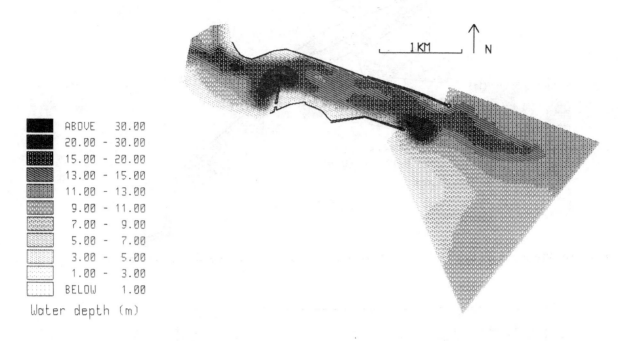

Water depth (m)

Figure 7   Configuration and bottom topography for the Malamocco
inlet to the Venice lagoon

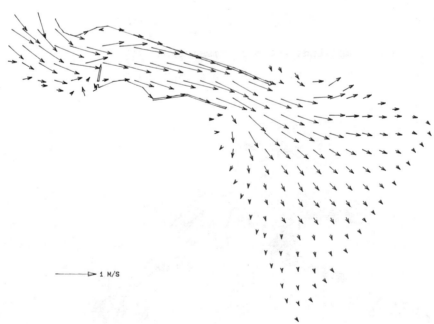

Figure 8   Applied current field

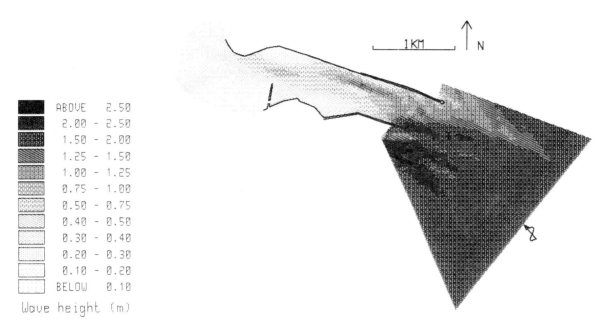

Figure 9   Computed waveheight for the Malamocco inlet without
current, T = 8 s

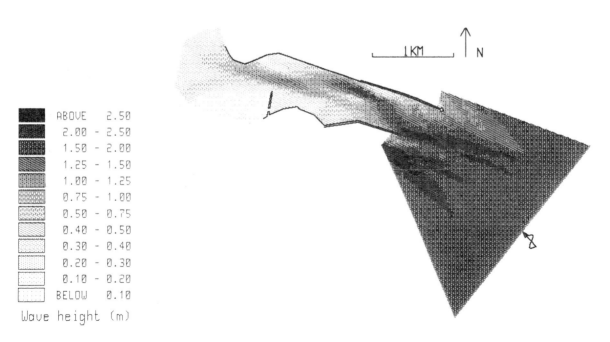

Figure 10   Computed waveheight for the Malamocco inlet with current, T = 8 s

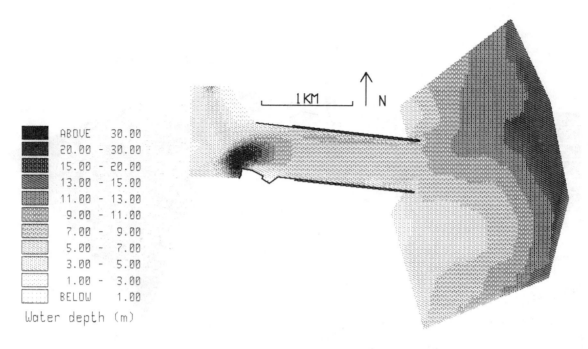

Figure 11   Configuration and bottom topography for Chioggia inlet to
            the Venice Lagoon

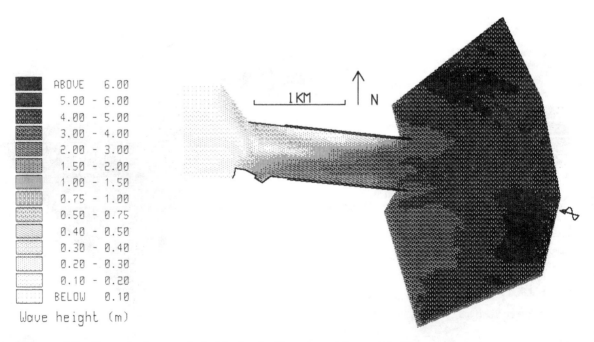

Figure 12   Computed wave height including breaking effects, T = 14 s

Chapter 23
NUMERICAL MODELLING OF WATER WAVES USING BOUNDARY ELEMENT METHOD

T T Al-Shemmeri
North Staffordshire Polytechnic, UK

SUMMARY

This paper presents a Numerical Technique to model water waves of an infinitesimal amplitude. The Method utilises the direct approach of the Boundary Element Method. The accuracy and stability of this technique to time-varying water waves is demonstrated by comparison with the linear wave theory.

KEYWORDS

    Mathematical modelling
    Water waves
    Computational Fluid Dynamics
    Boundary Element Method
    Offshore structures and design

NOMENCLATURE

Symbol	Definition
a	wave amplitude
g	acceleration due to gravity
h	water depth
k	wave number
K	time step iteration counter
Pa	Pressure term
t	time
T	wave period
x	horizontal coordinates
y_c	crest elevation
y_t	trough elevation
λ	wave length
α	wave reflection coefficient
β	angle of free surface with horizontal
Γ	boundary surface
ς	surface elevation
$\theta_1 \ \theta_2$	weighting factors
ρ	density
σ	wave property
Φ	potential function
Ω	domain

1. INTRODUCTION

Theoretical analyses of wave motion are based on the equations of motion, continuity and the boundary conditions. The last are determined by the existance of a "free surface", i.e. an interface between air and water, and by the rigid walls within which the fluid is bounded. Numerical modelling provides a fast, cheap and efficient method of studying water waves. In order to represent accurately the flow domain in such application, a large number of nodes are normally required. This is the case with both Finite Difference and Finite Element Methods. Another setback for such methods is created by the nature of the free surface of the flow. The Boundary Element Method (BEM) is a fairly recent technique based on a weighted residual formulation [2]. This method as the name implies requires the outer surface of the boundary, only, to be discretised and hence a drastic savings in data handling, computer store and CPU time is achieved.

This paper will present a description of the implementation of BEM to water waves. It also discusses computed results of modelling linear waves and compares results with theory.

LAWS OF MOTION OF WATER WAVES

Free surface flows are governed by the energy equation of a non-steady incompressible fluid given by:

$$\frac{\partial \Phi}{\partial t} + \frac{1}{2} |\nabla \Phi^2| + g \varsigma = - \frac{Pa}{\rho} \qquad (1)$$

Coupled with :

$$\frac{\partial \varsigma}{\partial t} = \frac{\partial \Phi}{\partial n} \Big/ \cos \beta \qquad (2)$$

which represent the instantaneous free surface elevation ς in terms of the velocity potential Φ.

Rewriting equations (1) and (2) into finite time difference form using weighting factors θ_1 and θ_2 to specify solution scheme. In the present paper Crank-Nicholson scheme was chosen (ie $\theta_1 = \theta_2 = 0.5$)

Hence

$$\frac{\Phi^{K+1} - \Phi^K}{\Delta t} = \left[\tfrac{1}{2} |\nabla \Phi^2|^K + \theta_1 (g\varsigma + \frac{P_a}{\rho})^{K+1} \right.$$
$$\left. + (1 - \theta_1)(g\varsigma + \frac{P_a}{\rho})^K \right] \tag{3}$$

and

$$\frac{\varsigma^{K+1} - \varsigma^K}{\Delta t} = \left[\theta_2 (\frac{\partial \Phi}{\partial n})^{K+1} + (1 - \theta_2)(\frac{\partial \Phi}{\partial n})^K \right] / \cos \beta^K \tag{4}$$

Elimination of ς^{k+1} between equations (3) & (4) yields :

$$\Phi^{k+1} = \Phi^k - g(\Delta t)^2 \theta_1 \theta_2 \frac{[\partial \Phi/\partial n]^{k+1}}{[\cos \beta]^k}$$

$$- \Delta t \left\{ \frac{1}{2} \left[\left[\frac{\partial \Phi}{\partial x}\right]^2 + \left[\frac{\partial \Phi}{\partial x}\right]^2 \right] \right\}^k$$

$$+ g\varsigma^k + \Delta t (1-\theta_2) \theta_1 g [\partial \Phi/\partial n]^k/[\cos \beta]^k + \theta_1 [P_a/\rho]^{k+1} +$$
$$(1-\theta_1) [P_a/\rho]^k) \tag{5}$$

2. BEM FORMULATION

Using Green's second identity, Φ_i the potential (5) at selected points is given by:

$$-\alpha \Phi_i = \int_\Omega (\Phi \frac{\partial \Phi^*}{\partial n} - \Phi^* \frac{\partial \Phi}{\partial n}) d\Omega \tag{6}$$

where $\Phi^* = \frac{1}{2\pi} \ell_n (\frac{1}{r})$ \hfill (7)

is the fundamental solution of a 2-D isotropic case.

In order to solve equation (6) for Φ_i, the values of Φ and $\partial \Phi/\partial n$ are required on the entire boundary.

The procedure is thus to discretize the surface Ω into boundary elements. The variation of Φ and $\partial \Phi/\partial n$ is assumed to be linear.

By choosing nodal points in succession and numerically integrating equation (6), then applying the appropriate

327

boundary conditions, a system of N equations is formed such that :

$$G \; \Phi^{K+1} \; = \; H(\frac{\partial \Phi}{\partial n})^{K+1} \tag{8}$$

in which (K + 1) belongs to time step (t + Δ t).

Rearranging (8) to set all known arrays in the RHS vector and all unknowns on the LHS vector thus, obtain a system matrix of the following form:

$$Ax = b \tag{9}$$

which is solved by Gaussian elimination procedure.

Once the solution $(\frac{\partial \Phi}{\partial n})^{K+1}$ is obtained on the free

surface, Φ^{K+1} can be calculated using equation (5) and also the new surface elevation ζ^{K+1} using equation (4). The solution then proceeds to the next time step and so on.

2.1 Standing Waves of Infinitesimal Amplitude

Analytically, the simplest case of water waves is the linear wave theory. This theory is valid only when

$\frac{a}{\lambda} \left[\frac{\lambda}{h}\right]^3$ is small, i.e. for waves of small amplitude.

A standing wave (Fig. 1) travelling from left to right in a tank can be defined in terms of its spatial location (x) and time (t) as

$$\zeta = a \sin (kx) \sin (\sigma t) \tag{10}$$

where ζ is the local free surface height at distance x from the source of the wave origin. k is the wave number and σ is the reciprocal of the wave length.

A standing wave of the form given in equation (10) can be generated by three possible methods.

a. Assuming a starting elevation ζ on free surface,

b. Assuming known velocity on the free surface by differentiating (10) :

$$\frac{\partial \emptyset}{\partial n} = a \; \sigma \sin (kx) \cos (\sigma t) \tag{11}$$

c. Assuming a pressure distribution on the free surface given in reference [3] as :

$$P = - \frac{ag}{\sigma} \sin (kx) \tag{12}$$

Therefore, three alternative test cases are available to check the applicability of the present approach to model water waves.

2.2 Description of the Test Problem

The problem of particular interest seen in Figure 1 consists of a tank 1m long, 0.5 m deep. The amplitude chosen as 1 cm,

$$\left[U_R = \frac{0.01}{2.0} \left[\frac{2.0}{0.5}\right]^3 < 1 \right]$$

in order to ensure that infinitesimal amplitude linear wave theory is valid. The circular frequency for this wave is $\sigma = \sqrt{gk \tanh(kh)} = 5.3166$, and the time period is

$$T_p = \frac{2\pi}{\sigma} = 1.1818 \text{ seconds.}$$

The bottom corners suffer no discontinuity and therefore only free surface corners were treated for singularity [1]. The total number of nodes used are 27, 12 of which one of the free surface.

In solving the problem, after each time step the elements on the sides of tank were updated in accordance with the current location of the free surface. In particular the end nodes (i.e. the crest and trough) were adjusted so that the nodes on the sides were maintained with equi-distant elements. This procedure was applied to all free surface problems in the present work.

3. RESULTS AND DISCUSSION

The three initial conditions described earlier were applied to obtain solutions for the standing wave problem of infinitesimal amplitude (0.01 m amplitude in a tank of 0.5 metre depth and wave length of 2 metres) with N = 25. (cf. Fig. 1).

Figs. 2a, 3a and 4a indicate that apart from the error in the free surface height, the wave period was also shifted. The error in computed wave period was 10.96% for elevation start, 7.84% for velocity start, and 9.79% for pressure start. The average error of computed crests was 12.21%, 5.22% and 4.28% respectively and of troughs was 11.05%, 4.05% and 4.36% respectively. Velocity start, with an initial horizontal free surface, appeared to be the best of the three approaches for this particular test case. Results obtained with singularity at corners being correctly treated (N = 27) are depicted in Figures 2b, 3b and 4b. A glance at the graphical representation of these results immediately demonstrates the drastic improvements in these solutions. Errors in elevation of the wave at the crests were only 0.0725%, 0.4425% and 0.1675% for the three initial conditions of elevation start, velocity start and pressure start respectively and 0.41%, 0.07% and 0.3425% at the troughs. Errors in wave periods computed were 0.335%, 0.363% and 0.448% respectively. Thus the results were dramatically improved compared with linear wave theory.

Mesh refinement (N = 52) appeared to have very little influence on the results when the time step unaltered. For this case the error in elevation was 0.195%, 0.31% and 0.08% at crest; and 0.33%, 0.19 and 0.45% at troughs for the three starts respectively. This result can be accounted for by the fact that the solutions were compared for accuracy with linear wave theory, which is in itself subject to error. All the above tests were conducted with a fixed time step of 0.02 seconds.

A time step of 0.02 sec. (in 1.18 wave time period) implies that 60 computed values would represent the wave over one wave period. This compares with 20 free surface elements per wave length. Thus, apart from the inaccuracy encountered in using an inappropriate time step, the choice of a very small time step implies unnecessary waste of computer time. The time step used must be reasonably small, such that a fairly considerable number of steps would represent the wave period. Hence for a given problem an optimum time step exists for a fixed number of boundary elements. This depends on the accuracy required and the permissible computer time allocated. For the present problem no specific optimum time step has been deduced, since the errors indicated for the various solutions (based on different time steps) are already based on the approximations of linear wave theory. However a conclusion can be drawn that the optimum time step lies in the range

$$TS < \frac{TP}{2 \times IS}$$ where IS is the number of free surface nodes per

wave length provided that the spatial discretisation is sufficient for reasonable accuracy. The choice of a time step within the range 0.02–0.06 seconds did not appear to affect the accuracy of solution of the problem under consideration. Time steps larger than 0.06 seconds were not checked for obvious reasons.

4. REFERENCES

[1] AL SHEMMERI, T.T.
     "A Time-varying Boundary Element Technique for
     Computing Unlinearized Water Waves"
     PhD Thesis, UMIST, 1983

[2] BREBBIA, C.A.
     "The Boundary Element Method for Engineers"
     Pentech Press, London, 1978.

[3] Sir Horace Lamb
     "Hydrodynamics"
     6th Edition, Cambridge University Press 1932

[4] LIU, P.L-F.
     "Integral Equation Solutions to Non-Linear Free
     Surface Flows"
     Finite Elements in Water Resources, Vol. 2/Eds.
     Brebbia, C.A., et al) Pentech Press, 1978, pp.
     4.87–4.98

[5] BETTS, P.L., and AL-SHEMMERI, T.T.
     "Water Waves: A Time-varying Unlinearized Boundary
     Element Approach", 4th International Symposium on
     Finite Element Methods in Flow Problems, Tokyo,
     Japan, July 1982.

[6] AL-SHEMMERI, T.T.
"Unlinearized   Time-Varying   Boundary   Element
Approach"
2nd International Conference on Numerical Methods
for Non-Linear Problems.   Barcelona, Spain 9-13,
April 1984.

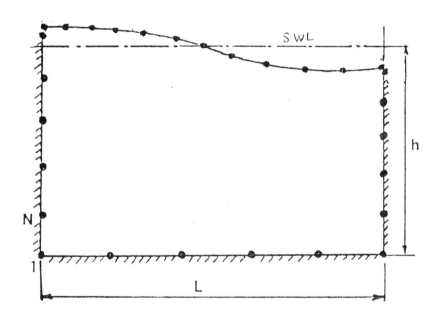

Fig. 1   Definition of   free surface geometry

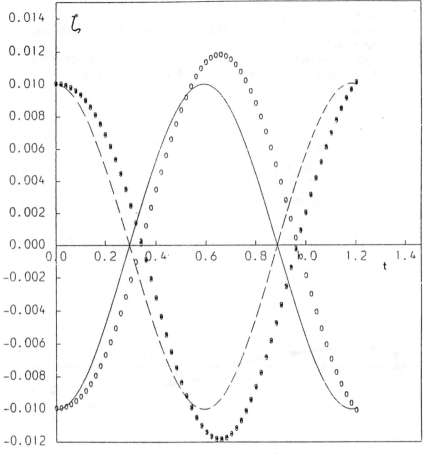

Fig.2a Elevation start, untreated

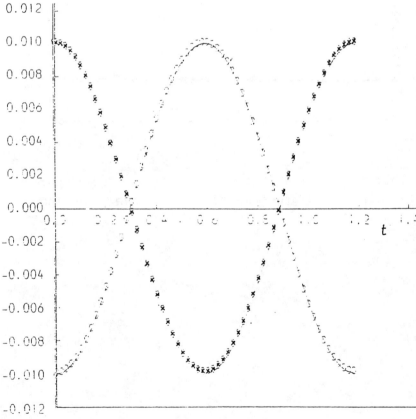

Fig.2b Elevation start, treated

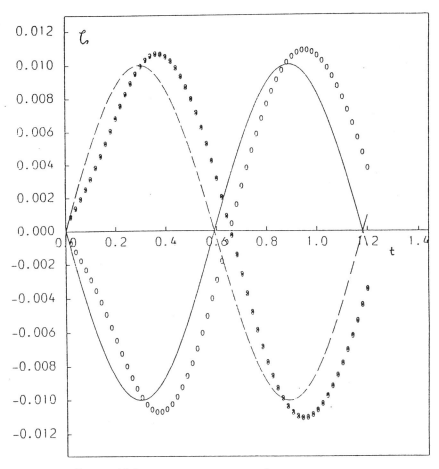

Fig.3a Velocity start, untreated

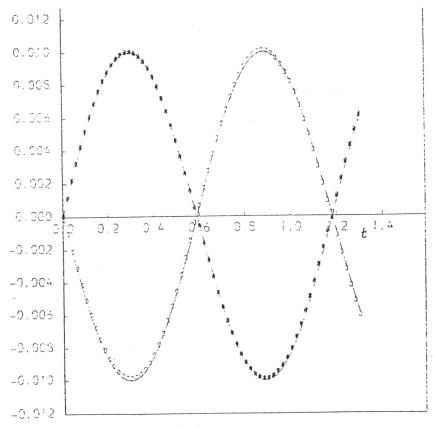

Fig.3b Velocity start, treated

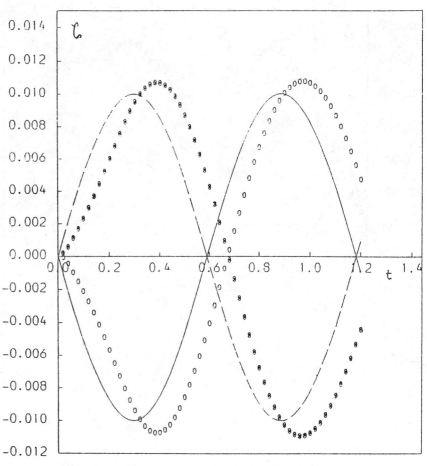

Fig.4a   Pressure start,untreated

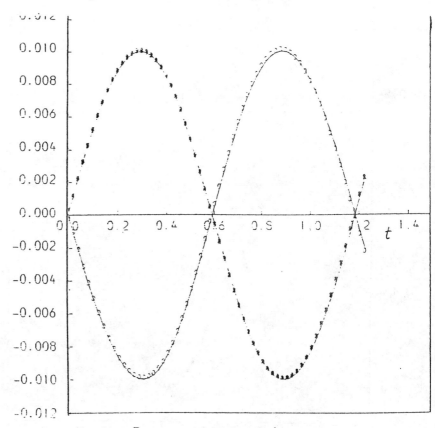

Fig.4b   Pressure start,treated

Chapter 24
COMPARISON OF SIMPLIFIED WAVE PREDICTION METHODS AS RECOMMENDED IN THE SHORE PROTECTION MANUAL

P Parle
Posford Duvivier, UK
R Burrows
University of Liverpool, UK

Abstract

Wave conditions hindcast using the simplified method of wave prediction outlined in the latest edition (1984) of the US Army, Coastal Engineering Research Centre Shore Protection Manual (SPM '84) differ substantially from wave conditions hindcast using the simplified method of wave prediction outlined in the 1977 edition of the Shore Protection Manual. (SPM '77).

A comparison of the two methods was carried out during a re-appraisal of the wave climate categorisation model using storm events (Waccas) which had been developed at the University of Liverpool during 1985. This Waccas package contained a wind/wave hindcasting model which was based on the simplified wave prediction method of SPM '77 enhanced to allow for energy accumulation in the sea state and for changes in wind speed and direction.

This paper outlines the reasons for the comparison, sets out the results of the comparison, discusses these results and explains why it was decided to retain the use of the SPM '77 simplified wave prediction method over that of SPM '84.

1.1 Introduction

The paper compares the simplified method of wave prediction outlined in the 1977 edition of the US Army Coastal and Engineering Research Centre Shore Protection Manual (SPM '77) (Ref. 1), with the simplified method outlined in the 1984 edition (SPM '84) (Ref. 2).

A wave climate categorisation model using storm events (Waccas) had been developed at the University of Liverpool, during 1985, for the low cost long term monitoring of coastal defence works (Ref. 3). This Waccas package contained a wind/wave model based on the simplified method of wave prediction outlined in SPM '77. This wind/wave model was used to hindcast wave data at the Mersey Bar Light Vessel in Liverpool Bay for a wave climate study. One years wave height predictions were compared with recorded values. The correlation was good. This paper discusses SPM '84 and SPM '77 simplified wave prediction methods in the context of this Liverpool Bay Study.

The simplified method of wave prediction recommended in SPM '77 is known as the SMB method, being based on the work of Sverdrup, Munk and Bretschneider. The equations recommended in SPM '84 are based on work carried out by Hasselman et al during the Jonswap experiment.

SPM '84 also differs from SPM '77 in the method recommended for calculating fetch values and in the wind speed values used in the equations. SPM '84 fetch calculations are based on the arithmetic mean of a substantially narrower spread of directions than is used in the effective fetch method of SPM '77, 24° as opposed to 90° respectively. SPM '84 gives adjustment factors to be applied to the wind speed U before it is used in the equations as a wind stress factor U_a.

The adjustment factors are for:

(1)	observational bias,
(2)	anemometer elevation,
(3)	reading duration,
(4)	air-sea temperature differences,
(5)	gauge location
(6)	conversion to wind stress factor.

SPM '77 wind speed adjustment factors are for gauge location and air-sea temperature differences only. During the Liverpool Bay study, adjustment factors for gauge location and air-sea temperature differences were not used.

Preliminary calculations indicated considerable differences in wave conditions predicted by the two methods, and, though wave height predictions due to the wind/wave model within the '85 Waccas model correlate well with wave height measurements, it was considered useful to compare the results of the two methods.

The paper compares wave height, wave period and minimum duration values calculated using SPM '84 and SPM '77 for the range of wind speeds, water depths and fetch lengths likely to be used in the wind/wave model.

SPM '84 equations are examined for their sensitivity to the various adjustment factors, and, in the context of the Liverpool Bay Study, for their sensitivity to the different methods of calculating the fetch lengths.

The paper also explains how the algorithm within the Waccas wind/wave model which accounts for energy accumulation as wind speed and direction varies, exaggerates the differences between the two methods.

2. Simplified Wave Prediction Methods

2.1 Simplified Wave Prediction Method SPM '77

SPM '77 used the SMB method of predicting wave conditions from wind data. The significant wave height H_s and the significant wave period T_s are determined directly in terms of wind speed U, fetch F, water depth d and duration t over which the wind acts. Transitional and shallow water versions of the SMB equations are used in the 1985 wind/wave model and are as follows:

$$\frac{gH_s}{U^2} = 0.283 \tanh(0.530(gd/U^2)^{0.75}) \tanh\left[\frac{0.0125(gF/U^2)^{0.42}}{\tanh(0.530(gd/U^2)^{0.75})}\right] \qquad 1$$

$$\frac{gT_s}{2\pi U} = 1.20 \tanh(0.833(gd/U^2)^{0.375}) \tanh\left[\frac{0.077(gF/U^2)^{0.25}}{\tanh(0.833(gd/U^2)^{0.375})}\right] \qquad 2$$

$$\frac{gt_{min}}{U} = K \exp\left[(A(\ln(gF/U^2))^2 - B\ln(gF/U^2) + C)^{0.5} + D\ln(gF/U^2)\right] \qquad 3$$

Where

$K = 65882$, $A = 0.0161$, $B = 0.3629$, $C = 2.2024$ and $D = 0.8798$

and d = water depth(m), F = fetch length(m) , U = wind speed (m/s)

H_s = Significant Wave Height (m)

T_s = Significant Wave Period (s)

t_{min} = minimum duration necessary to reach fetch limited conditions (s)

Equation 3 is a deep water equation and is used to check if the sea is duration or fetch limited.   For given U and F values, t_{min} is calculated.   If t the duration for which the wind acts is less than t_{min} the sea is said to be duration limited and an equivalent fetch (F_{eq}) must be found using t instead of t_{min} in equation 3. F_{eq}, U and d are then inserted into equations 1 and 2 to give the duration limited values of H_s and T_s respectively.   For comparison purposes it is assumed that fetch limited conditions exist for all U, F and d values used.   Wind data used in the '85 Waccas model is assigned to one of 16 directions before being input to the wind/wave model.   Direction 1 is North, direction 16 is NNW with the directions increasing clockwise and 22.5° between consecutive directions.   For each direction a representative depth and fetch value is estimated.

Equations 1 to 3 are incorporated in an algorithm enabling wave heights, periods and wave directions to be calculated for wind speed and direction values given at 3 hourly intervals, fetch and depth values for each direction having been input to the model initially.   The algorithm allows for:

1.   changes in wind speed and direction
2.   accumulation of energy in a sea state
3.   the fact that the direction of the largest wave does not necessarily coincide with the predominant wind direction.

2.2   Simplified Wave Prediction Method '84

2.2.1   SPM '84 Equations

The SPM '84 simplified wave prediction equations are:

$$\frac{g H_{m0}}{U_a^2} = 0.283 \tanh(0.530 (gd/U_a^2)^{0.75}) \tanh \left[\frac{0.00565 (g F/U_a^2)^{0.5}}{\tanh(0.530(gd/U_a^2)^{0.75}} \right] \qquad 4$$

$$\frac{g T_m}{2\pi U_a} = 1.20 \tanh(0.833 (gd/U_a^2)^{0.375}) \tanh \left[\frac{0.0379 (gF/U_a^2)^{1/3}}{\tanh(0.833(gd/U_a^2)^{0.375}} \right] \qquad 5$$

where $T_m = T_s/0.95$

$$\frac{g t_{min}}{U_a} = 5.37 \times 10^2 (g T_m/U_a)^{7/3} \qquad 6$$

d = water depth, F = fetch length,

H_{m0} = Spectrally based wave height,

T_m = Period of the peak of the wave spectrum

t_{min} = Minimum duration necessary for fetch limited conditions

U_a is the wind stress factor

H_{m0} the spectrally based significant wave height is four times the square root of the variance of the sea surface elevation and for deep water is equal to H_s. For comparison purposes it is assumed that $H_s = H_{m0}$ in shallow water.

2.2.2   Wind stress factor

Before a measured wind speed is used in the SPM '84 simplified wave prediction method, various correction factors should be applied, and the adjusted wind speed should then be converted into a wind stress factor U_a. These correction factors are briefly outlined below along with the factors used for the Liverpool Bay Case. A fuller description is given in Chapter 3 of SPM '84 (Ref. 2).

1.   Observational bias

According to SPM'84, Cardone states that ship observed wind speeds are biased and should be corrected using the formula $W = 1.864 \times U^{0.7778}$, where U is the ship reported wind speed in m/s and W is the actual recorded wind speed in m/s. As the Liverpool Bay Wind Data is ship reported, this correction has been applied. This correction factor has the effect of reducing high wind speeds (greater than 16.5m/s) and increasing low wind speeds (lower than 16.5m/s).

2.   Elevation

Having been corrected for bias, the wind speed should now be corrected for elevation of the measuring instrument. The wind speeds used in the SPM '84 formulae are those that act at 10m above sea level. The correction factor RE is $RE = (10/Z)^{0.1429}$, where Z is the elevation of the wind speed measuring instrument.

The elevation of the instrument for the Liverpool Bay Data was estimated to be 6.7m above sea level and so RE = 1.06.

3.   Temperature differences between the air and water cause stability effects and should be corrected for. From SPM '84 (fig. 3-14) if the air temperature is 10° below the sea temperature, the wind speeds should be increased by a factor of 18%. When the air temperature is above that of the sea the wave generating capability of the wind is lowered.

As no temperature readings were available for the Liverpool Bay Case, a stability correction factor RT of 1.1 has been used as recommended by SPM'84.

4.   Duration averaged wind speed.

Each wind speed measurement aboard the Mersey Bar Light Vessel (MBLV) took 1-2 minutes to carry out. These values were considered to be the mean for the 3 hour interval between measurements. That is to say, the duration average correction RT = 1.0.

If the wind speeds were the fastest mile values for the 3 hour period, figure 3.12 and 3.13 of SPM '84 would have to be used to convert these values into average values for the 3 hour period. The Met. Office generally supplies mean hourly wind speeds at amenograph stations.

5.   Location

Offshore winds are influenced by the land over which they have been blowing and need to travel some distance over water before stabilising. A graph by Resio and Vincent shown in figure 3.15 SPM '84 gives the ratio of U_w (wind speed over water) to U_L (wind speed over land). RL = U_w/U_L. RL varies from greater than 1.73 for U_L <2.0m/s to 0.9 for U_L >18.5m/s.

For the Liverpool Bay case the MBLV is at sea and the winds of interest have been blowing across the sea for anything from 50 to over 200 km and therefore no correction for location is used, ie RL = 1.0.

Having corrected the wind speed for items 1-5 the wind stress factor is calculated using the formula

$$U_a = 0.71 \; W_s^{1.23} \qquad\qquad\qquad 7$$

U_a is in m/s and W_s is the corrected wind speed in m/s and for the Liverpool Bay case

$$W_s = RL \times RD \times RT \times RE \times 1.864 \; U^{0.7778}$$

Letting K = RL x RD x RT x RE

$$U_a = 1.527 \times K^{1.23} \times U^{0.9567}$$

For the Liverpool Bay Case

$$K = 1.0 \times 1.0 \times 1.1 \times 1.6 = 1.166$$

therefore
$$U_a = 1.844 \; U^{0.9567}$$

For comparison purposes a value of K = 1.166 is used in the SPM '84 equations. The sensitivity of wave conditions to K is examined in Section 3.2.

3. Comparison of Prediction Methods

Using equations 1 to 3 (the SMB method given in SPM '77) values of gH_s/U^2, $gT_s/2 \; U$ and gt_{min}/U^2 depend on the values of gd/U^2 and gF/U^2. From equations 4 to 6 and 7 (the SPM '84 method) it can be seen that gH_s/U^2, $gT_s/2 \; U$ and gt_{min}/U^2 depend not only on gd/U^2 and gF/U^2, but, also on the values of U and K. A K value of 1.166 is used in all the comparison calculations. To enable meaningful comparisons to be made it is necessary to hold U constant, calculate wave height, period and minimum duration values using SPM '84 and SPM '77 for a range of water depth, and fetch lengths, compare these values and repeat for different values of U. For the Liverpool Bay Case, SPM '77 and SPM '84 methods of calculating fetch lengths result in different values, but, for comparison purposes the fetch values used in both sets of equations are the same. Section 3.3 examines the sensitivity of the SPM '84 equations to the different methods of calculating the fetch values as outlined in SPM '84 and SPM '77. As explained in Section 2.1 fetch limited conditions are assumed to exist for all U, F and d values used for comparison purposes. Table 1 gives an explanation of the symbols used in this paper.

3.1 Differences due to Equations

Wave height, wave period and minimum duration values are calculated using SPM '77 and SPM '84 equations. These values are estimated for the following range of water depth, wind speed and fetch values.

Water Depth	10 to 100m	
Wind Speed	10 to 30 m/s	K = 1.166
Fetch Length	10 to 200 km	

From these estimates the ratios of H84, T84 and TM84 values to H77, T77 and TM77 values respectively are calculated. These ratios RH(H84/H77), RT(T84/T77) and RTM(TM84/TM77) are plotted against fetch for water depths of 10, 25 and 100m and for wind speeds of 10, 20 and 30 m/s. These plots are shown on figures 1 to 3. These ratios do not vary with water depth, wind speed and fetch in the same way as H, T and TM. They vary in the following manner

Wave height, RH:

as U decreases	RH increases
as F increases	RH increases
as d increases	RH decreases DF $<$DF$_{ch}$
	RH increases DF $>$DF$_{ch}$

where DF$_{ch}$ = 0.1 to 0.2 x 10^4 for the range of F and U values used.

Therefore the lower the wind speed, the greater the fetch and the deeper the water (provided U and F are such that DF is greater than DF$_{ch}$), the greater is the increase in wave height due to using SPM '84 instead of SPM '77.

Wave period, RT:

as U increases	RT increases
as F increases	RT increases
as d increases	RT decreases DF $<$DF$_{ct}$
	RT increases DF $>$DF$_{ct}$

DF$_{ct}$ = 0.2 to 0.25 x 10^4 for the range of F and U values used.

RT ranged from less than 1 to greater than 1 for the values of U, F and d examined.

Minimum Duration RTM:

as F decreases	RTM increases
as d increases	RTM increases

for d > 25m	as U increases	RTM increases
at d = 20m	as U 10 -> 20m/s	RTM increases for all F
	as U 20 -> 30m/s	RTM increases for F > 75km
		RTM decreases for F < 75km

for d < 20m	as U 10 -> 20m/s	RTM increases
	as U 20 -> 30m/s	RTM decreases

SPM '84 minimum duration predictions are always lower than the SPM '77 value and the greatest underestimate occurs at the lowest d and U values and the highest F value. Therefore for the same wind data and bathymetry, fetch limited sea states are predicted more frequently using the SPM '84 equations than when using the SPM '77 equations.

Table 2 gives maximum and minimum values of RH, RT and RTM, and, the values of U, d and F at which they occur, for the range of U, d and F values considered.

3.2   Sensitivity to K

Sensitivity to K of wave height, wave period and minimum duration is measured by calculating each parameter (H, T, TM), using the SPM '84 equations, with K values of 0.9 and 1.2 and obtaining their ratio, eg H84(K = 1.2)/H84(K = 0.9). These calculations are carried out for the same range of d, U and F values as in section 3.1. A summary of the results of these calculations is given in Table 3. Both H84 and T84 are most sensitive to the value of K at those values of U, F and d that give maximum RH and RT ratios.

3.3   Effect of SPM '84 Fetch Calculation Method

The recommended method of calculating fetch lengths for use with the SPM '84 equations consists of constructing nine radials, (at 3 degree intervals) from the point of interest, the central radial being in the direction of interest. These radials are extended until they first intersect the shoreline. The length of each radial is measured and the arithmetic average is used as the fetch for the direction given by the central radial. Section 3-5 part 4 of SPM '84 also states, during its discussion of narrow fetch conditions, that 'the effective fetch should not be used with the growth curves presented therein'.

In its discussion of the origin of the effective fetch concept, SPM '84 states that early users of the SMB method overpredicted H values for small values of F (this occurred when the SMB method was applied to inland lakes and reservoirs). The effective fetch concept was used to give a fetch adjustment to compensate for this overestimate. If the SPM '84 wave prediction curves are used with effective fetch values in this context, wave heights will be underestimated states SPM'84.

Fetch values using the SPM '84 method are calculated for directions 13 to 1 of the Liverpool Bay Case (North West Sector). These SPM '84 fetch values are greater than the corresponding values used in the '85 Waccas model, though this does not always have to be the case. Wave height, wave period and minimum duration values are estimated using the SPM '84 equations with both the '85 Waccas fetch values (Present Model, P.M.) and the SPM '84 fetch values. These estimates are carried out for U = 10m/s and 30m/s. Ratios are calculated of H, T and TM estimated using SPM '84 fetch values to those estimated using the '85 model fetches.

The maximum wave height ratio occurs when U = 10m/s, and, the maximum T and TM ratios occur at U = 30m/s. Table 4 gives the maximum and minimum ratios for directions 13 to 1 of the Liverpool Bay Case. SPM '77 and SPM '84 fetch values for directions 13 to 1 of the Liverpool Bay Case are shown on a map of the Irish Sea in Figure 4.

4.   Discussion

This paper has examined the effects of using the simplified method of wave prediction as outlined in SPM '84, on wave height, wave period and minimum duration values for various fetch lengths, wind speed and minimum duration values, instead of the simplified method as outlined in SPM '77. Effects due to 3 causes were examined :

1.   the effects due to the equations themselves, using K = 1.166 in the SPM '84 equations and using the same fetch value in both.
2.   the effect on H84, T84 and TM84 of different K values.
3.   the effect of the different methods of calculating fetch values on H84, T84 and TM84 using the Liverpool Bay Case as an example.

These effects are 'multiplicative' and their total effect for the Liverpool Bay Case (North West Sector) can be seen in table 5.

The above 'effects' were examined assuming that fetch limited conditions existed for every U, F and d value considered. That is to say, in all the comparison calculations (regarding H and T) the wind is assumed to act at a constant speed over a given depth and fetch for a time period greater than the tmin for that wind speed, fetch and depth. In the wind/wave program of the '85 Waccas model the wind speed, fetch and water depth is assumed constant for a 3 hour period (after 3 hours U, F or d may change). This 3 hour period may not be sufficient for fetch limited conditions to be attained and the sea state is then said to be duration limited. If the sea is duration limited an equivalent fetch is calculated and this equivalent fetch is used to calculated the duration limited wave height and period. These duration limited waves are smaller than the fetch limited ones. Using SPM '84 equations the time necessary to reach fetch limited conditions is a factor ranging between 0.28 to 0.81 of the time necessary to reach fetch limited conditions using SPM '77 equations, for the range of U, F and d considered. For the same wind data therefore, fetch limited conditions are reached more often if the SPM '84 equations (same fetches, K = 1.166) are used in the wind/wave model. The effect of these lower values of minimum duration (due to SPM '84) in the wind/wave model would be to exaggerate further the increase in wave height and period over the increases found for fetch limited conditions. The actual increase, due to the algorithm within the wind/wave model is quantifiable only by inserting the SPM '84 equations in the model.

For the Liverpool Bay Case (North West Sector), assuming fetch limited conditions, the use of SPM '84 can result in an increase in wave height of 30 to 95% using K = 1.166. Using the same K value wave period can increase by 4 to 37% and minimum duration can decrease by 23 to 42%. A lower value of K dampens the increases and decreases. For example, for K = 0.9 : wave height increases are 2 to 40%; wave period can decrease by up to 7% or increase by up to 19% and minimum duration decreases by 9 to 35%. Conversely, a higher value of K exaggerates the increases and decreases. With K = 1.2 : wave height increases are 34 to 102%; wave period increases are 5 to 39% and minimum duration decreases by 24 to 43%. From Table 5 it is evident that the effect of using the simplified wave prediction method outlined in SPM '84 over the method presently in use in the 1985 Waccas wind/wave model would be to increase the wave heights and period for a given set of wind data.

A paper by Bishop, (Ref. 4), compares the SMB method with two other wave prediction methods, one based on a method developed by Donelan at Canada's National Water Research Institute and a method based on the results of the Jonswap experiment. The three wave prediction methods are compared for fetch limited pseudo steady state conditions using meteorological and wave data from Lake Ontario in 1972. Bishop found that each of the three methods predict wave conditions with similar accuracy.

The wave prediction method based on Jonswap, used by Bishop in his comparison, uses the following equations for significant wave height and peak spectral period.

$$H_s = 1.6 \times 10^{-3} \ g^{-0.5} \ U \ X^{0.5} \qquad\qquad 8$$
$$g = 9.81 \ m/s$$
$$T_m = 0.286 \ g^{-0.67} \ U^{1/3} \ X^{1/3} \qquad\qquad 9$$

Where U (m/s) is the measured wind speed W_m factored by 1.1 to correct for elevation.

X is the fetch length (m)

Equations 8 and 9 are the same as the SPM '84 deep water equations for H_s and T_m save that U is used instead of U_a.

Using K = 1.1 the SPM '84 U_a values for Bishop's measured wind speed (W_m) values are given by the following formula.

$$U_a = 0.71 \ (1.1W_m)^{1.23} = 0.789 \ W_m^{1.23}$$

SPM '84 U_a values for Liverpool Bay measured wind speed (W_m) values are given by the following formula.

$$U_a = 0.71 \ (1.166 \times 1.864 \times W_m^{0.7778})^{1.23} = 1.845 W_m^{0.957}$$

Table 6 gives, for various measured wind speed values; the U values used in Bishop's equations based on Jonswap; the U_a values that SPM '84 would use for Bishop's data; the U_a values that SPM '84 would use for Liverpool Bay data and the ratios U_a/U and $(U_a/U)^{0.333}$ for the Bishop's and Liverpool Bay data.

U_a/U and $(U_a/U)^{0.333}$ values given the ratios of deep water H_s and T_m values estimated using the SPM '84 method and the Jonswap methods used by Bishop. For the range of wind speeds examined use of SPM '84 results in increases of 5 to 58% and 2 to 16% in estimates of significant wave height and peak spectral period respectively over estimates based on Bishop's Jonswap method. Ratios of U_a/U and $(U_a/U)^{0.333}$ for Liverpool Bay U_a values are also shown in Table 6. These ratios show the effect of the correction factor for bias recommended by SPM '84. The bias correction reduces the height and period estimates for high values of wind speed and increases these estimated for low values of wind speed.

Other studies have shown that once the overwater wind speed can be accurately predicted, wave hindcasting techniques based on the SMB method give good comparisons with recorded data. During the Canadian Coastal Sediment Study an extensive review of existing procedures for estimating sand transport was undertaken (Ref.5). Part of this review involved examining different methods of estimation of deepwater wave climate using single point parametric wave hindcast models. The parametric equations considered for use in the hindcast models were:

1.   the SMB equations
2.   the Jonswap equations
3.   the Darbyshire and Draper equations
4.   the simplified equations of SPM '84

It was found that the SMB equations (as set out in SPM '73, which are the same as those in SPM '77) gave the best overall results.

A similar wind wave hindcast technique based on the SMB equations has been used by Baird and Glodowski (Ref. 6) to hindcast wave conditions for use in estimations of wave power off the Canadian coast. This technique also gave good results when compared with recorded data, provided the overwater wind conditions could be accurately estimated.

Wave direction has a significant influence on many coastal engineering structures. In wave hindcasting wind and wave directions generally are assumed coincident. However, according to Donelan (Ref. 7), wind and wave direction differences of 50 degrees are not uncommon. The findings of Donelan have been incorporated in the wind/wave model of the University of Liverpool. The use of the findings of Donelan are the subject of ongoing research and are not discussed in this paper.

5.   Conclusion

The 1984 edition of the Shore Protection Manual (SPM '84) recommends the use of a wind stress factor calculated from an adjusted wind speed in their simplified method of wave prediction which is based on the results of the Jonswap experiment.

Using the wave prediction equations outlined in the simplified method of wave prediction of the 1977 edition of the Shore Protection Manual (SPM '77) a wind/wave model developed at Liverpool University during 1985 was used to predict wave heights, periods and directions at the Mersey Bar Light Vessel for the period September '65 to September '66. Measured wave heights and periods at the Mersey Bar Light Vessel were available for the same period. Good correlation was found to exist between measured and predicted wave heights and this can be seen in Figure 5 which is a plot against time of some of the predicted and measured wave heights at the Mersey Bar Light Vessel (Ref.3). Figure 6 gives a plot of the corresponding measured and predicted wave period against time at the Mersey Bar Light Vessel.

During a study of the wave climate at Sheerness, carried out by Posford Duvivier using the Liverpool University wind/wave model, a correlation similar to that found at the Mersey Bar Vessel was found to exist between measured and predicted wave heights at the Tongue Light Vessel. Other studies (Ref. 5 & 6) also show that wave hindcasting techniques based on the simplified equations of SPM '77 give good estimates of wave conditions.

It is clear from Table 5 and the discussion in Section 4, that the use of SPM '84 simplified method of wave prediction within the wind/wave program of the Waccas model could result in considerable over estimates of wave heights and periods. Over estimates of the order of 25 to 100% for wave heights and of the order of 0 to 30% for wave periods could result from the use of SPM '84 in the Liverpool Bay

343

Study.   Further evidence of over estimation of wave conditions by SPM '84 is provided in a paper by Bishop.   Bishop found that a simple method of wave prediction based on Jonswap and similar to the SPM '84 method in all respects, save for use of wind speed instead of wind stress, gave wind predictions of similar accuracy to those of SMB.   Bishop used meteorological and wave data collected on Lake Ontario during 1972.   Use of wind stress (for Lake Ontario Case) instead of wind speed in the Jonswap equations resulted in increases in the estimates of significant wave height of 5 to 60% and in peak spectral period of 2 to 16% as the measured wind speed ranged from 5 m/s to 30 m/s.

The findings of this paper and other studies indicate that, of the equations examined, the best equations to use in a simple wind/wave hindcast model are the SMB equations of SPM '77.   This casts doubts on the simplified method of wave prediction outlined in the 1984 edition of SPM.   As many Coastal Engineers use the SPM as a reference work, it is hoped that this paper will cause a reappraisal of the method by the Coastal and Engineering Research Centre.

References

1.   US Army Coastal Engineering Research Centre
- "Shore Protection Manual, Vol. 1, 3rd Edition, 1977", US Government Printing Office.

2.   US Army Coastal Engineering Research Centre,
- "Shore Protection Manual, Vol. 1, 4th Edition, 1984", US Government Printing Office.

3.   Burrows, R., Hedges, T.S., Barber, P.C. and Dickinson, P.J., (1985), "Wind/Wave Analysis and Storm Simulation : Liverpool Bay", Research Report, Dept. of Civil Engineering, University of Liverpool.

4.   Bishop, C.T., 1983, "Comparison of Manual Wave Prediction Models", J. of Waterway, Port, Coastal and Ocean Engineering, Vol. 109, No.1, Feb. 1983.

5.   Readshaw, J.S., Glodowski, C.W., Chartrand, D.M., Willis, D.H., Bowen, A.J., Piper, D. and Thibault, J., 1987, "A Review of Procedures to Predict Alongshore Sand Transport", Coastal Sediments '87, New Orleans, Vol. 1, p738-755.

6.   Baird, W.F. and Glodowski, C.W., 1978, "Estimation of Wave Energy Using a Wind Wave Hindcast Technique", International Symposium on Wave and Tidal Energy, Canterbury.

7.   Donelan, M.A., 1980, "Similarly Theory Applied to the Forecasting of Wave Heights Periods and Directions", Proceedings of the Canadian Coastal Conference.

Table 1

Symbols used in the comparison of SPM '84 and SPM '77 simplified wave prediction methods.

H77	Significant Wave Height	SPM '77
H84	Significant Wave Height	SPM '84
T77	Significant Wave Period	SPM '77
T84	Significant Wave Period	SPM '84
TM77	Minimum Duration	SPM '77
TM84	Minimum Duration	SPM '84
RH	Ratio of Wave Height	H84/H77
RT	Ratio of Wave Period	T84/T77
RTM	Ratio of Minimum Duration	TM84/TM77

$$DF = gF/U^2 \qquad \text{F fetch length}$$
$$\text{U wind speed}$$

Table 2

Maximum and minimum values of Height, Period and Minimum Duration ratios for the range of wind speeds, fetch lengths and depths examined.

Wave Height

$$RH_{max} = 1.83 \quad \text{at} \quad U = 10m/s, F = 200km, d = 100m$$

$$RH_{min} = 1.05 \qquad U = 30m/s, F = 10km, d = 100m$$

Wave Period

$$RT_{max} = 1.32 \qquad U = 10m/s, F = 200km, d = 100m$$

$$RT_{min} = 0.82 \qquad U = 30m/s, F = 10km, d = 100m$$

Minimum Duration

$$RTM_{max} = 0.81 \qquad U = 30m/s, F = 10km, d = 100m$$

$$RTM_{min} = 0.28 \qquad U = 10m/s, F = 200km, d = 10m$$

Range of input variables

Wind Speed (U)	10 to 30m/s	
Fetch Length (F)	10 to 200km	K = 1.166
Water Depth (d)	10 to 100m	

345

Table 3

Sensitivity to K

K = RL x RD x RT x RE

Wave Height

Max H84(K = 1.2)/H84(K = 0.9) = 1.49 at U = 10m/s. F = 200km, d = 100m
Min H84(K = 1.2)/H84(K = 0.9) = 1.20 at U = 30m/s, F = 200km, d = 10m

Wave Period

Max T84(K = 1.2)/T84(K = 0.9) = 1.19 at U = 10m/s, F = 200km, d = 100m
Min T84(K = 1.2)/T84(K = 0.9) = 1.11 at U = 30m/s, F = 200km, d = 10m

Minimum Duration

Max TM84(K = 1.2)/TM84(K = 09) = 0.94 at U = 10m/s, F = 200km, d = 100m
Min TM84(K = 1.2)/TM84(K = 09) = 0.80 at U = 30m/s, F = 200km, d = 10m

Liverpool Bay
K = 1.2

	U = 10m/s, F = 200km, d = 100m	U = 30m/s, F = 300km, d = 10m
H84(K = 1.2)/H84(K = 1.166)	1.04	1.02
T84(K = 1.2)/T84(K = 1.166)	1.02	1.01
TM84(K = 1.2)/TM84(K = 1.166)	0.99	0.98

K.09

H84(K = 0.9)/H84(K = 1.166)	0.70	0.85
T84(K = 0.9)/T84(K = 1.166)	0.85	1.00
TM84(K = 0.9/TM84(K = 1.166)	1.05	1.22

RL   factor for Location

RD   factor for duration

RT   factor for air-sea temperature difference

RE   factor for elevation

TABLE 4

Ratios of H84, T84 and TM84 due to different fetch estimates

Liverpool Bay Case

Direction Number	depth (m)	Fetch (km) Present Model	SPM'84	Fetch (SPM84) Fetch (PM)	H84 (SPM'84) H84 (PM)		T84 (SPM'84) T84 (PM)		TM(SPM'84) TM (PM)	
					Max	Min	Max	Min	Max	Min
13	50	120.3	177	1.47	1.17	1.13	1.10	1.09	1.24	1.22
14	48	131	176	1.34	1.12	1.10	1.07	1.06	1.17	1.16
15	35	101	112	1.12	1.04	1.03	1.02	1.02	1.06	1.06
16	31	111	149	1.34	1.11	1.07	1.07	1.06	1.16	1.15
1	20	65	78	1.20	1.07	1.04	1.04	1.04	1.10	1.10

Direction 1 is North

Direction 13 is West

TABLE 5

Some RH, RT and RTM ratios for Liverpool Bay Study

Direction	Depth	Fetch (km) SPM'84	SPM'77	Parameter	Wind Speed	R	ΔK 1.166-1.2	1.166-0.9	ΔF	SPM'84/SPM'77 K=1.2	K=1.166	K=0.
13	50	177	120	H	10	1.67	1.036	0.719	1.17	2.02	1.95	1.40
				T	10	1.26	1.015	0.868	1.09	1.39	1.37	1.19
				TM	10	0.55	0.987	1.098	1.22	0.66	0.67	0.74
				H	30	1.28	1.030	0.757	1.13	1.49	1.45	1.10
				T	30	1.03	1.012	0.896	1.10	1.14	1.13	1.01
				TM	30	0.62	0.981	1.184	1.24	0.76	0.77	0.91
14	48	176	131	H	10	1.68	1.036	0.719	1.12	1.95	1.88	1.35
				T	10	1.26	1.015	0.867	1.06	1.36	1.34	1.16
				TM	10	0.54	0.988	1.096	1.16	0.62	0.63	0.73
				H	30	1.29	1.030	0.761	1.10	1.42	1.42	1.08
				T	30	1.04	1.012	0.896	1.07	1.13	1.11	1.00
				TM	30	0.61	0.981	1.184	1.17	0.70	0.71	0.85
15	35	112	101	H	10	1.62	1.034	0.728	1.04	1.74	1.69	1.23
				T	10	1.23	1.014	0.873	1.02	1.27	1.26	1.10
				TM	10	0.55	0.986	1.115	1.06	0.57	0.58	0.65
				H	30	1.27	1.029	0.767	1.03	1.35	1.31	1.00
				T	30	1.02	1.012	0.898	1.02	1.05	1.04	0.93
				TM	30	0.60	0.980	1.189	1.06	0.62	0.64	0.76
16	31	149	111	H	10	1.62	1.034	0.732	1.11	1.86	1.80	1.32
				T	10	1.24	1.014	0.873	1.06	1.33	1.31	1.15
				TM	10	0.52	0.986	1.115	1.15	0.59	0.60	0.67
				H	30	1.27	1.027	0.777	1.07	1.40	1.36	1.06
				T	30	1.03	1.012	0.899	1.07	1.12	1.10	0.99
				TM	30	0.57	0.980	1.191	1.16	0.65	0.66	0.79
1	20	78	65	H	10	1.52	1.033	0.742	1.07	1.68	1.63	1.21
				T	10	1.18	1.013	0.882	1.04	1.24	1.23	1.08
				TM	10	0.55	0.984	1.142	1.10	0.60	0.61	0.69
				H	30	1.25	1.027	0.783	1.04	1.34	1.30	1.02
				T	30	1.00	1.012	0.901	1.04	1.05	1.04	0.94
				TM	30	0.58	0.980	1.198	1.10	0.63	0.64	0.76

for explanation of Table 5 see next page

Table 5 Cont'd

Parameters

	H	Wave Height
	T	Wave Period
	TM	Minimum Duration

Direction 1 North, 13 West. The North West Quadrant is divided into segments of 22.5°.

d waters depth, F (SPM'77) fetch corresponding to direction calculated using '85 Waccas model method.

$$R = \frac{\text{parameter (SPM'84 equations, SPM'77 fetch)}}{\text{parameters (SPM'77 equation, SPM'77 fetch)}}$$

$$\Delta K = \frac{\text{parameters (SPM'84 equation, SPM'77 fetch, K = 1.2 or 0.9)}}{\text{parameters (SPM'84 equation, SPM'77 fetch, K = 1.166)}}$$

$$\Delta F = \frac{\text{parameters (SPM'84 equation, SPM'84 fetch)}}{\text{parameters (SPM'84 equation, SPM'77 fetch)}}$$

$$\text{SPM'84/SPM'77} = \frac{\text{parameters (SPM'84 equation, SPM'84 fetch, K)}}{\text{parameters (SPM'77 equation, SPM'77 fetch)}}$$

Table 6

Comparison of Wind Speed values used in Jonswap Method and Wind Stress values used on Bishops and Liverpool Bay Data.

W_m	U	U_a		(U_a/U)		$(U_a/U)^{0.333}$	
		Bishop	LBC	Bishop	LBC	Bishop	LBC
m/s							
30	33	52.3	47.8	1.58	1.45	1.16	1.13
25	27.5	41.8	40.2	1.52	1.46	1.15	1.13
20	22	31.8	32.4	1.45	1.47	1.13	1.14
15	16.5	22.3	24.6	1.35	1.49	1.11	1.14
10	11	13.6	16.7	1.24	1.52	1.07	1.15
5	5.5	5.8	8.6	1.05	1.56	1.02	1.16

LBC Liverpool Bay Case

$U = 1.1 W_m$

$U_a \text{ (Bishop)} = 0.798 W_m^{1.23}$

$U_a \text{ (LBC)} = 1.845 W_m^{0.957}$

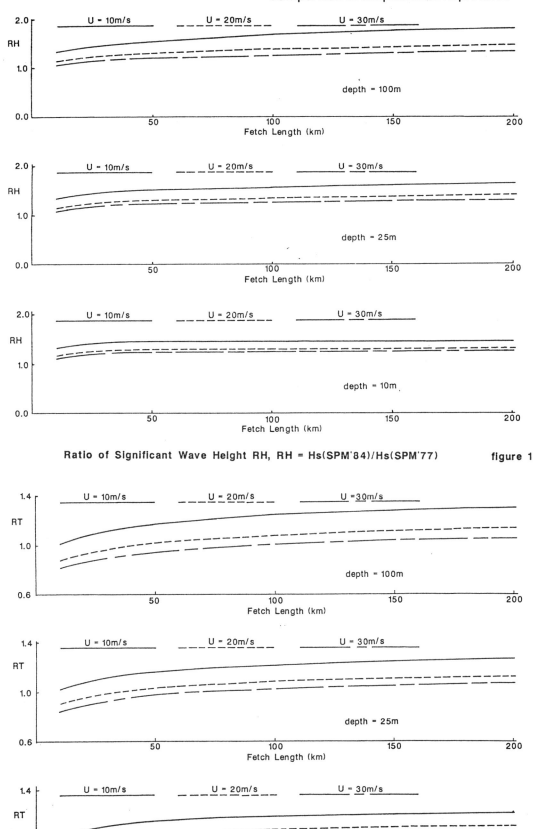

Ratio of Significant Wave Height RH, RH = Hs(SPM'84)/Hs(SPM'77)   **figure 1**

Ratio of Wave Period RT,   RT = Ts(SPM'84)/Ts(SPM'77)   **figure 2**

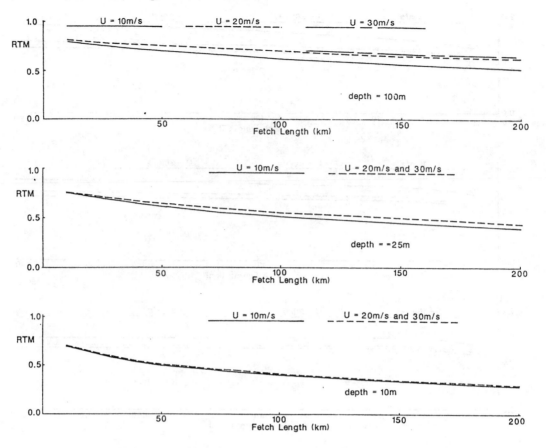

Ratio of Minimum Duration RTM, RTM = TM(SPM'84)/TM(SPM'77)   figure 3

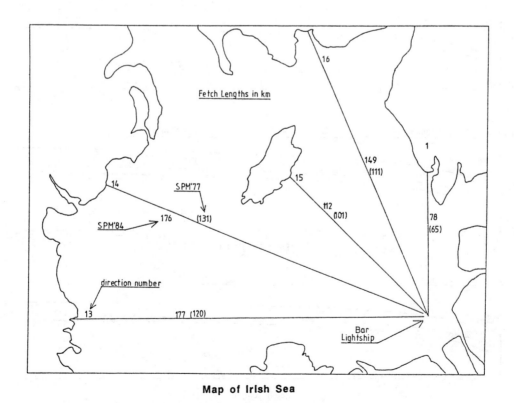

Map of Irish Sea

Showing SPM'77 and SPM'84 fetch lengths for the Liverpool Bay Case   figure 4

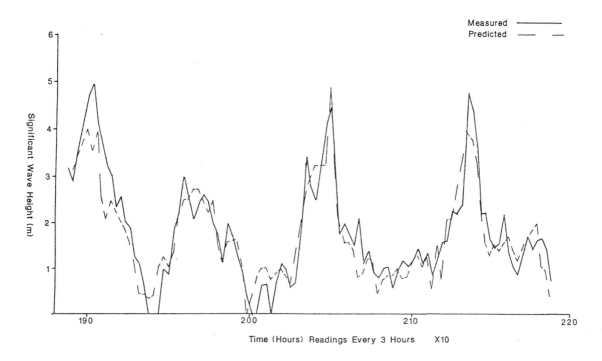

Measured ———
Predicted — —

MEASURED AND PREDICTED WAVE HEIGHTS AGAINST TIME AT THE MERSEY BAR LIGHT VESSEL   Figure 5

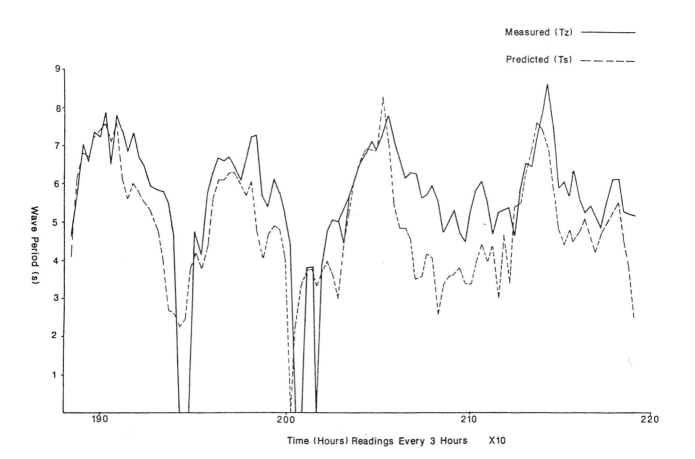

Measured (Tz) ———
Predicted (Ts) — — — —

MEASURED AND PREDICTED WAVE PERIOD AGAINST TIME AT THE MERSEY BAR LIGHT VESSEL   Figure

351

Chapter 25
WAVE TESTING OF REFLECTIVE COASTAL STRUCTURES

G N Bullock
Plymouth Polytechnic, UK
G J Murton
British Maritime Technology, UK

SUMMARY

The special problems associated with hydraulic model tests of highly reflective coastal structures are described. Particular attention is drawn to the way in which reflection and re-reflection between the test structure and a conventional wavemaker can produce systematic changes in the incident wave spectrum. Measurements are presented which demonstrate how this problem may be overcome by use of an absorbing wavemaker.

Laboratory measurements of the wave-overtopping rate over low coastal structures can be highly sensitive to re-reflection effects and special consideration is given to this problem. A system for continuously monitoring the overtopping of a low-crested breakwater is described and data presented which extends the range of those previously available.

Predetermined test conditions are shown to be easier to achieve when an absorbing wavemaker is used. This also reduces the risk of obtaining erroneous results. Consequently, it is recommended that an absorbing wavemaker should always be used when testing highly reflective structures.

NOMENCLATURE

a_i, A_i	=	variation in water surface elevation due to incident waves in the time (a_i) and frequency (A_i) domains
C	=	real part of cross-spectrum S_{xy}
$C_r = H_r/H_i$	=	reflection coefficient
F	=	transfer function of filter
f	=	frequency (Hz)
g	=	acceleration due to gravity (m^2/s)
H_i	=	incident wave height (m)
H_r	=	reflected wave height (m)
H_s	=	significant wave height (m^2/s)
h_{max}	=	range of water surface motion at a partial anti-node
h_{min}	=	range of water surface motion at a partial node
$k=2\pi/L$	=	wave number
L	=	wave length (m)
m_0	=	variance of incident wave motion (m^2)
N%	=	percentage of waves overtopping a structure
Q	=	imaginary part of cross-spectrum S_{xy}
$\bar{Q}$	=	mean overtopping discharge ($m^3/s/m$)
Q*	=	dimensionless discharge
R_c	=	elevation of breakwater crest above still water level (m)
R*	=	dimensionless freeboard
S_{ii}	=	incident wave spectrum (m^2s)
S_{rr}	=	reflected wave spectrum (m^2s)
S_{xx}	=	spectrum from measurements taken at location x (m^2s)
S_{xy}	=	cross-spectrum from measurements taken simultaneously at locations x and y (m^2s)
S_{yy}	=	spectrum from measurements taken at location y (m^2s)
s	=	distance between locations x and y (m)
T_z	=	mean zero-crossing period (s)
v_c, V_c	=	variation in paddle control signal in the time (v_c) and frequency (V_c) domains
W	=	transfer function of a wavemaker

1.   INTRODUCTION

Hydraulic models are often used to investigate the likely performance of coastal structures that will be subject to significant wave action.   As with all model studies, the reliability of the results is directly related to the accuracy with which the prototype conditions can be reproduced.   Many of the fundamental problems associated with modelling water waves in a laboratory channel have been described by Svendsen (1).   The present paper is concerned with some of the additional problems that arise when testing highly reflective structures.

The amount of wave energy reflected from a structure depends both on the character of the waves and on factors such as the slope, roughness and permeability of the structure.   Estimates of the reflection coefficient (C_r) for smooth slopes, sand beaches and rubble-mound breakwaters may be found in the Shore Protection Manual (2) where:-

$$C_r = H_r/H_i \hspace{6cm} 1$$

in which H_i is the height of the incident waves and H_r is the height of the reflected waves.   Typically, C_r tends to 1.0 as the seaward face of the structure steepens and both its permeability and roughness decrease.

In the prototype, waves reflected by a structure generally propagate out to sea and are eventually lost from the coastal system.   However, in a conventional laboratory channel, waves reflected by a model structure are re-reflected back towards the model by the paddle of the wavemaker.   A system of positive feedback is thus established in which the waves incident on the test structure become the sum of the new input from the paddle and the re-reflected waves.   The energy of the incident waves increases in an uncontrolled way until losses within the channel dissipate energy at a rate equal to the primary input from the paddle. At best this makes the results of the test more difficult to interpret. At worst the results are incorrect or misleading and this may not be recognised.

2.   SOLUTIONS TO THE PROBLEMS OF MEASUREMENT AND CONTROL

Two prerequisites to having confidence in the results of laboratory tests on highly reflective structures are:-

a.   The ability to distinguish between incident and reflected waves.

b.   The ability to maintain the required incident wave climate throughout the test period.

Recent developments in analysis and control have made these requirements easier to meet.

2.1   Incident and Reflected Waves

When regular waves are reflected back on themselves in a straightforward manner, the interaction of incident and reflected waves form a standing wave. Within a laboratory channel the envelope of crests and troughs may easily be detected by traversing a wave gauge along the channel. Figure 1 shows a typical trace, in this case obtained from tests on a 1:15 scale model of Plymouth Breakwater. Without a significant amount of other knowledge, the incident and reflected wave heights cannot be determined by measuring the variation in water surface elevation at any one point. However H_i and H_r can be estimated by measuring the difference in elevation between the envelope of crests and the envelope of troughs at both a partial node (h_{min}) and a partial anti-node (h_{max}). Then, on the basis of linear theory:-

$$H_i = 0.5 \ (h_{max} + h_{min}) \hspace{4cm} 2$$

and $$H_r = 0.5 \ (h_{max} - h_{min}) \hspace{3.5cm} 3$$

As a first approximation, reflection of each frequency component of an irregular wave train may be considered to form a pattern of nodes and anti-nodes similar to that described above. However, because the distance between a node and an anti-node is frequency dependent, the combination of frequencies in irregular waves does not produce simple envelopes of crests and troughs. Thus, the incident and reflected spectra cannot be estimated from relationships like equations 2 and 3. Figure 2 illustrates the way in which the 'hidden' standing waves cause spectra calculated from the temporal variation in water surface elevation at a point to vary with the location of the point. As may be seen, these so called 'point' spectra can differ significantly from the incident wave spectrum.

Various methods have been devised for calculating directional wave spectra of which the Modified Maximum Likelihood Method (3) has been proposed for the general case of an incident and reflected wave field. For waves in a laboratory channel where both incident and reflected waves propagate in known directions, the Frequency Response Function Method (4) can be used. This enables both incident (S_{ii}) and reflected (S_{rr}) wave spectra to be calculated from the variation in water surface elevation recorded simultaneously at two locations x and y separated by a distance s measured along the length of the channel. The relationships are:-

$$S_{ii} = (S_{xx} + S_{yy} - 2C \cos ks - 2Q \sin ks)/4 \sin^2 ks \hspace{2cm} 3$$

$$S_{rr} = (S_{xx} + S_{yy} - 2C \cos ks + 2Q \sin ks)/4 \sin^2 ks \hspace{2cm} 4$$

where S_{xx} and S_{yy} are the 'point' spectra for x and y respectively; C is the real part and Q the imaginary part of the cross-spectrum S_{xy} and k is the wave number. Unfortunately, both relationships become unrealiable in the vicinity of the singularities that occur when sin ks = 0, i.e. when:-

$$ks = n\pi, \quad n=0,1,2, \ldots\ldots\ldots \qquad\qquad 5$$

The authors overcome this problem by using three wave gauges rather than the theoretical minimum of two. Two gauges (gauges 1 and 2) are spaced to give maximum resolution at frequencies near that at which the peak of the incident wave spectrum is anticipated to occur, ie:-

$$s_{12} \approx \pi/(2k_p) = L_p/4 \qquad\qquad 6$$

where k_p and L_p are the values of k and L corresponding to the spectral peak. Gauge 3 is located such that:-

$$s_{23} \approx (\sin^{-1} 0.5)/k_p \qquad\qquad 7$$

$$s_{13} = s_{12} + s_{23} \qquad\qquad 8$$

The success of this arrangement in defining an incident wave spectrum is illustrated by Figure 3. Similar results are achieved for reflected spectra.

2.2 Absorbing Wavemakers

Figure 4 illustrates the characteristic effect of re-reflection from a conventional wavemaker when regular waves are in use. The particular results shown were obtained by supplying a wedge-type paddle with a sinusoidal control signal of constant amplitude at frequencies in the range 0.1Hz to 1.2Hz (5). The incident waves were reflected by a test structure with a C_r of between 0.6 and 0.7. Once steady conditions had been achieved, the amplitude of the resultant incident waves was determined by traversing a wave gauge as previously described.

The high peaks in the (wave amplitude)/(input voltage) curve occur where the distance between the test structure and the paddle was an integer multiple of half wave lengths for the prevailing wave frequency. In these circumstances the partial standing wave has anti-nodes at both the test structure and the paddle. The low points in the curve correspond to situations where there was an anti-node at the test structure (as must always be the case) but a node at the paddle.

The general form of the curve in Figure 4 can be explained by recalling the characteristics of the water particle motion under a standing wave (2). At a node the particle motion is primarily vertical. Thus, in circumstances where an anti-node should form at the paddle, the paddle should have little or no horizontal motion once the required waves have been established. If the paddle continues to have a significant horizontal motion, an additional unwanted motion will be imposed on the water and the incident waves will continue to grow until sufficient energy is dissipated by breaking or by some other means. In circumstances where a node should form at the paddle, the horizontal motion of the paddle should increase to match the increased excursion of the water particles once the standing wave has formed. If the paddle's motion does not increase the re-reflection will reduce the height of the incident waves. Hence the low points in Figure 4.

Situations similar to that described above were frequently encountered when it was conventional to use regular waves in hydraulic model investigations. Typically, iterative adjustments would be made to the paddle motion until a satisfactory approximation to the required incident wave height was achieved. Now that irregular waves are used, it is not always realised that the same underlying problems still exist. That they do is illustrated by curve C in Figure 5. This incident wave spectrum was measured under similar conditions to the data in Figure 4 and the frequencies at which maxima and minima occur in the curve correspond to the formation of partial anti-nodes and nodes at the paddle.

Curve A in Figure 5 was calculated for waves generated using the same paddle control signal as was used for curve C, but with a reasonably effective spending beach in the channel. In these circumstances, the performance of a conventional servo-controlled wavemaker can be idealised by the relationship:-

$$A_i = W \, V_c \qquad\qquad 9$$

which links the Laplace transforms of the incident wave motion (A_i) and the paddle control signal (V_c) by means of the paddle transfer function (W). A relationship of this type is used to predict the performance of a conventional wavemaker when setting up the irregular-wave climate required for a particular test. Indeed, such a relationship was used to produce the 'required conditions' represented by curve A.

The lack of similarity between curves A and C serves to illustrate the way in which the 'required conditions' can be corrupted when a highly reflective structure is under test. Changes to the magnitude and shape of the 'required spectrum' are only two of the more obvious aspects of the problem. Other re-reflection effects, such as changes in the characteristics of wave groups, are less easy to quantify.

The problems caused by re-reflection can be ameliorated by changing the paddle control signal so that the primary input from the paddle and the re-reflected waves combine to form the required incident waves. With a conventional paddle, determination of the 'correct' control signal for a particular test is likely to be a long and not altogether successful process. Furthermore, it is a process that should be repeated each time the reflection characteristics of the structure are changed. An alternative, and in many cases a more cost effective approach, is to convert the conventional wavemaker into an 'absorbing' wavemaker; so called because an additional loop in its control circuit enables the paddle to move in a way that absorbs rather than reflects approaching waves.

The theory behind absorbing wavemakers has been outlined by Gilbert (6) and is set out in greater detail in Reference 5. In terms of equipment, the essential requirements beyond those of a conventional wavemaker are:-

a.   A means of detecting reflected waves as they approach the paddle.

b.   A means of making the paddle move in a way that cancels out the reflected waves as they reach the paddle.

Such items were first incorporated in the new designs of wavemaker developed for offshore engineering research (7). Typically, the paddles are of the flap type and thus best suited to the generation of 'deep-water' waves.

Most of the wavemakers used for coastal engineering research have piston or wedge type paddles as these forms lend themselves to the generation of 'shallow-water' waves. It has recently been shown (5) that many of these existing installations could be made capable of absorption by adding an extra feedback loop to their control circuits. Furthermore, the input to the additional loop can be derived from a conventional wave gauge.

The main features of a possible conversion are shown schematically in Figure 6. Here, the variation in water surface elevation is measured immediately in front of the paddle by a wave gauge. The conventional paddle control signal $v_c(t)$ is then compared with the filtered output from the wave gauge $v_f(t)$ and the difference signal $v_c'(t)$ supplied to the wavemaker's servo-control system in place of $v_c(t)$. The resulting incident waves will be uncorrupted by reflection effects provided the transfer function of the filter satisfies the relationship:

$$F = 1/W \qquad\qquad 10$$

In practice, good results can be achieved even when this idealised relationship is not perfectly satisfied (5). The success of the authors' conversion of a conventional wedge-type wavemaker may be judged from the agreement between curves B and D in Figure 5. Both of these spectra were obtained with the same primary control signal. Like curve A, curve B was obtained with the spending beach in the channel and, like curve C, curve D was obtained when the highly reflective structure was present.

It should be emphasised that no iterative adjustments were required to achieve the level of agreement exhibited by curves B and D. Nor was the filter characteristic in exceptionally close agreement with Eqn. 10. The filter could have been improved but the level of absorption already achieved was deemed to be adequate. To test this hypothesis and to gain valuable additional data, a series of overtopping tests was undertaken.

3.   WAVE OVERTOPPING

Very little data are available concerning wave overtopping over low-crested structures. A previous attempt to quantify wave pumping over a reef, using regular waves in a channel equipped with a conventional wavemaker, gave results that were very sensitive to reflection effects (7). The tests described below were conducted to see whether reflection effects could be detected in overtopping measurements made with irregular waves. A variety of test procedures were used with the wavemaker working in both absorption and conventional modes.

3.1   Experimental Arrangement

All the tests were conducted in a 20.7m long, 0.9m wide by 1.2m deep channel equipped with the wedge-type wavemaker that was used to obtain the results described in Section 2. The depth of water was always within 30mm of 800mm.

Test structures in the form of smooth impermeable breakwaters were fabricated from varnished marine plywood The seaward face of the basic structure sloped up from the channel bed at a gradient of 1:1 to a horizontal crest set 800mm above the bed. The crest spanned the width of the channel and extended for 300mm in the direction of wave propagation. A seaward slope of 1:2 was achieved by attaching an additional sheet of plywood to the basic structure. Sheets of plywood were also used to increase the crest elevation in increments of 50mm to 950mm. Finer adjustments in crest elevation relative to still water level (SWL) were made by changing the depth of water.

A large plastic sheet was used to form a watertight 'pond' behind the model breakwater and overtopping waves flowed into this area. Water was discharged from the pond over two lm long sharp crested weirs set perpendicular to the breakwater crest. The height of the weirs was adjustable so that the water level in the pond could be held close to SWL. This, together with keeping wave reflection within the pond to a minimum, ensured that there was no seaward flow across the breakwater crest.

The flow from the pond passed out of the wave channel via a short pipe and was ultimately discharged into a sump after passing over another weir. The variable head over this weir was monitored by means of a water level gauge linked to a computer. This enabled the fluctuating discharge over the breakwater to be calculated with reasonable accuracy. Only a relatively small volume of water was passing through the overtopping apparatus at any one time and a sensibly constant SWL was maintained by pumping water from the sump back into the wave channel.

The irregular waves were generated using pseudo-random control signals based on feedback through a shift register (8). The signals were both repeatable and periodic, with a period that could be set anywhere from a few seconds up to several hundred years. Furthermore, the magnitude of V_c could be closely defined for signals taken over one or more complete periods.

Most of the present tests were conducted with irregular wave sequences of between 200 and 300 seconds period. Overtopping rates were calculated from measurements taken over a full period which typically contained around 160 waves. A range of wave energy levels were considered, generally with energy spectra of the Pierson-Moskowitz type. To determine whether any wave-run dependent effects could be detected, up to four different wave sequences were generated for each combination of sequence period and spectral characteristics. A few measurements were made using samples from wave sequences with a very long period.

An array of three wave gauges was used to determine the incident and reflected wave spectra as previously described. The gauge closest to the breakwater was positioned directly over the toe of the structure. An additional gauge was mounted on the crest of the breakwater. This enabled overtopping waves to be detected and hence the percentage of waves overtopping the structure to be calculated. All these gauges were linked to the same computer as was used to determine the discharge over the breakwater.

3.2   Analysis and Results

In all over 100 wave overtopping tests were carried out on the two breakwater profiles for a range of crest elevations between 40mm and 100mm above SWL.

Overtopping is an extremely complex phenomenon which depends on at least as many factors as wave reflection. Consequently, the interpretation of data is not easy and the various methods proposed for estimating overtopping can yield results which vary by more than an order of magnitude (9). So that some comparison could be made with other published data, the results from the present tests were analysed in terms of the dimensionless discharge Q* and the dimensionless freeboard R* used by Owen (10,11). The dimensionless discharge is defined as:-

$$Q* = \overline{Q} / (T_z \, g \, H_s) \qquad\qquad 11$$

where $\overline{Q}$ is the mean overtopping discharge in terms of the volume/unit time/unit length of breakwater; T_z is the mean zero-crossing wave period; g is the acceleration due to gravity and H_s is the significant wave height. The dimensionless freeboard is defined as:-

$$R* = R_c/(T_z \, g^{\frac{1}{2}} \, H_s^{\frac{1}{2}}) \qquad\qquad 12$$

where R_c is the elevation of the breakwater's crest above SWL. Owen has also published information concerning the percentage of waves overtopping a structure (N%) as a function of R*.

Unfortunately, it is not easy to evalute Q*, R* and N% accurately when the incident and reflected wave trains are combined. The problem, similar to that outlined in Section 2.1, is the determination of H_s and T_z for the incident waves rather than for the combined motion. It is not known how Owen overcame this problem. Here H_s was estimated from the incident wave spectrum on the assumption that:-

$$H_s = 4(m_0)^{\frac{1}{2}} \qquad\qquad 13$$

where m_0 is the variance of the incident wave motion. T_z could also have been estimated from spectral moments but for present purposes it was deemed adequate to use the value of T_z calculated from measurements made above the toe of the breakwater. This was a less than ideal solution and in future work it would be appropriate to address the whole problem of choosing a characteristic wave period for use in parameters such as Q* and R*.

The curves obtained for the variation if Q* with R* are shown in Figure 7. As is usual in tests of this nature, the raw data exhibited a fair amount of scatter with some values of Q* up to a factor of 1.5 away from the appropriate curve. However, this degree of scatter would seem to be somewhat less than that obtained by Owen who placed the 95% confidence bands at about 3Q* and Q*/3. Curves for the variation of N% against (R*)^2 are shown in Figure 8.

Most of the results used to define the curves shown in Figures 7 and 8 were obtained with the wavemaker working in its absorption mode. In these circumstances the incident wave spectrum, calculated from measurements taken over the full period of an irregular wave sequence, was always close to the spectral form targeted when the paddle control signal was set up. When different wave sequences were generated with the same period and spectral charcteristics, very similar values of Q* were obtained for effectively the same values of R*. This suggests that the mean overtopping discharge is not highly sensitive to variations in the wave pattern.

When tests were conducted using samples from very long wave sequences, there was generally a significant difference between the incident spectrum calculated from measurements taken during the sample interval and the target spectrum. Thus, it was not surprising to find that when measurements were taken for completely different portions of the same long sequence, the mean overtopping discharges for the samples were quite variable. Had the values of Q* and R* been calculated on the basis of the targeted wave characteristics, the general scatter of results would have been increased. No such increase was detected when the Q* and R* relating to a particular sample were calculated on the basis of the wave characteristics measured during that sample interval.

With the wavemaker working in its conventional non-absorbing mode, all the incident wave spectra were subject to the re-reflection effects described in Section 2.2 These effects were worse with the 1:1 slope for which $C_r \approx 0.85$ than with the 1:2 slope for which $C_r \approx 0.60$. Typically, the incident spectrum was very spiked and in rather poor agreement with the target spectrum. Despite this, the values calculated for Figures 7 and 8 on the basis of the prevailing conditions all fall within the scatter band formed by the other results. The operators of conventional wavemakers may be tempted to derive some comfort from this, but it would be unwise to conclude that uncontrolled deviations from the intended test conditions are of little importance. Indeed, it may be more appropriate to question the methodolgy that leads to scatter bands wide enough to embrace this last set of results.

A range of lower freeboards was covered in the present tests than in the tests described by Owen (10, 11). Relevant curves are compared in Figures 9 and 10 where the inset boxes contain the results of the present study. Given that Owen was concerned with the earth embankment type of sea defences and conducted his tests in a large wave basin using JONSWAP spectra, the agreement between the curves is remarkably good.

4.   CONCLUSIONS

It has been shown that a fairly simple technique can be used to distinguish between the incident and reflected wave spectra within a channel. However, knowledge of the incident wave spectrum solves only part of the problem if traditional parameters like H_s and T_z are to be used to define the wave climate. In a reflective environment accurate values for these parameters cannot be obtained by measuring the variation in water surface elevation at a single point and further work is required to determine how best to characterise the incident waves in such circumstances. In the present context it would be interesting to know how accurately period parameters like T_z could be derived from spectral moments.

When highly reflective structures are tested using a conventional wavemaker, the re-reflection of reflected waves can significantly alter the intended incident wave conditions. In general the full effect of such alterations is unknown and the changes are difficult if not impossible to eliminate. Fortunately, this problem can now be ameliorated by incorporating an additional feedback loop in the servo-control system of a conventional wavemaker to convert it into an absorbing wavemaker.

Use of absorption enabled the required incident wave conditions to be established for the investigation of wave overtopping over low-crested breakwaters despite there sometimes being over 90% reflection from the structure. Close control over the irregular wave test conditions could be maintained by taking measurements throughout an interval equal to the period of the pseudo-random paddle control signal. The data collected in these circumstances appeared to exhibit less scatter than that reported by Owen. However, the resultant curves were very similar and the new data justifies the extrapolation of Owen's results to lower dimensionless crest elevations than previously considered.

The intended incident wave conditions were not established during overtopping tests conducted with the wavemaker working in its conventional non-absorbing mode. Despite the incident wave spectra being distorted by re-reflection effects, values of the dimensionless overtopping parameters calculated on the basis of actual rather than intended wave conditions were found to lie within the bounds of the other results. Further work is required to determine whether this was fortuitous, whether it stems from the insensitivity of either the analytical or experimental technique or whether wave overtopping is not particularly sensitive to changes in the shape of the incident wave spectrum.

Only some of the problems associated with wave testing of reflective structures have been considered in this paper. The very fact that tests on a particular structure are considered necessary indicates the existence of some problem or unknown and, as the overtopping tests have served to illustrate, the results are likely to be difficult both to obtain and to interpret. In such circumstances it seems foolish to complicate the situation in ways that can now be avoided. Thus, use of an absorbing wavemaker is strongly recommended.

5.   ACKNOWLEDGEMENTS

The assistance of Hydraulics Research Ltd during the development of the absorbing wavemaker is gratefully acknowledged.

6.   REFERENCES

1.   Svendsen, I.A.:   "Physical modelling of water waves".   In: "Physical Modelling in Coastal Engineering".   Rotterdam, Netherlands, A.A. Balkema, 1985, pp13-47.

2.   "Shore Protection Manual". Coastal Engineering Research Center, U.S. Army Corps of Engineers, Washington, DC, 1984.

3.   Isbe, M. and Kondo, K.: "Method for estimating directional wave spectrum in incident and reflected wave field".   In: Proc. 19th International Conference on Coastal Engineering (Houston, Texas, USA : Sept. 1984), ASCE, New York, U.S.A., 1985, Chapter 32, pp467-483.

4.   Gilbert, G. and Thompson, D.M.:   "Reflections in random waves: the frequency response function method".   Rept. No. IT 173, Hydraulics Research Station, Wallingford, UK, March 1978.

5.   Bullock, G.N. and Murton, G.J.:   "Performance of a wedge-type absorbing wavemaker", To be published by the Waterway, Port, Coastal and Ocean Engineering Division, ASCE.

6.   Gilbert, G.: "Absorbing wave generators", In: , Hydraulic Research Station Notes, No. 20, HRS, Wallingford, UK, 1978, pp3-4.

7.   Rayner, R.F.:   "Aspects of the oceanography of two mid-Indian Ocean coral atolls", Ph.D. Thesis, Department of Civil Engineering, University of Salford, UK, 1983.

8.   Fryer, D.K., Gilbert,G. and Wilkie, M.J.: "A wave spectrum synthesizer", Journal of Hydraulics Research, 11, 3, 1973, pp193-204.

9.   Douglass, S.L.:   "Irregular wave overtopping rates".   In: Proc. 19th International Conference on Coastal Engineerg (Houston, Texas, USA: Sept. 1984), ASCE, New York, USA, 1985, Chapter 22, pp316-327.

10.   Owen M.W.:   "Design of seawalls allowing for wave overtropping".   Rept. No. EX 924, Hydraulics Research Station, Wallingford, UK, June 1980.

11.   Owen, M.W.: "Overtopping of sea defences".   In: Int. Conf. on the Hydraulic Modelling of Civil Engineering Structures (Coventry, England: Sept. 1982) BHRA, Cranfield, England, 1982, Paper H3, pp469-479.

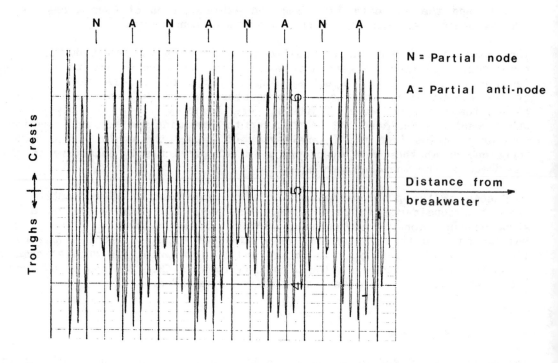

Fig. 1   The variation in wave height within a partial standing
          wave.

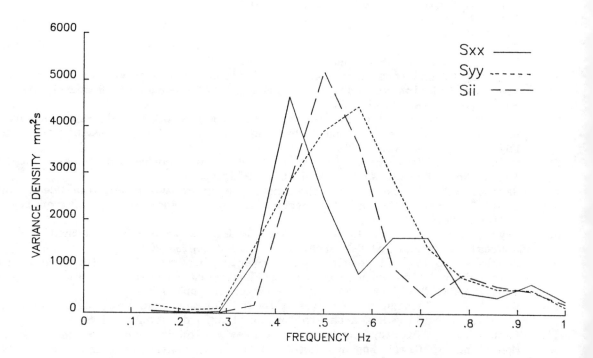

Fig. 2   Comparison of point and incident spectra for a
          reflective wave field.

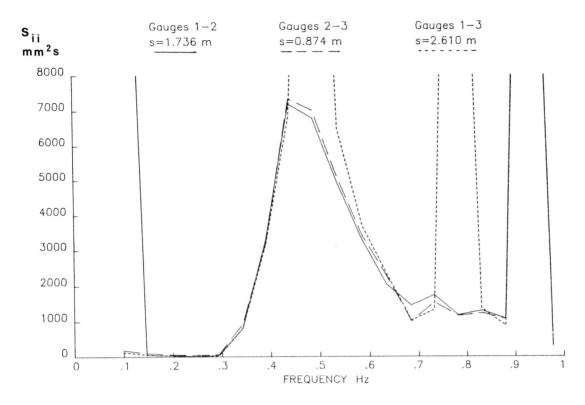

Fig. 3   Definition of an incident spectrum using data from three wave gauges.

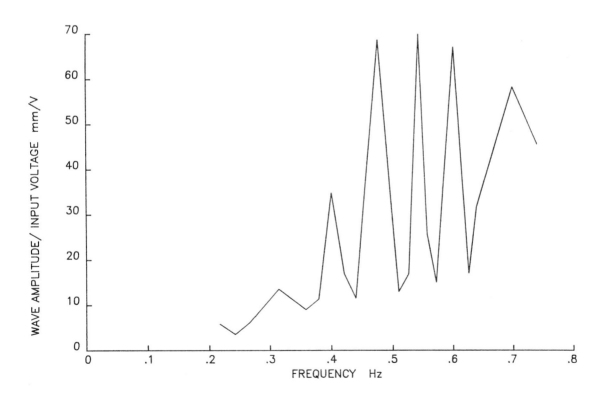

Fig. 4   The effect of reflection and re-reflection on the performance of a conventional wavemaker.

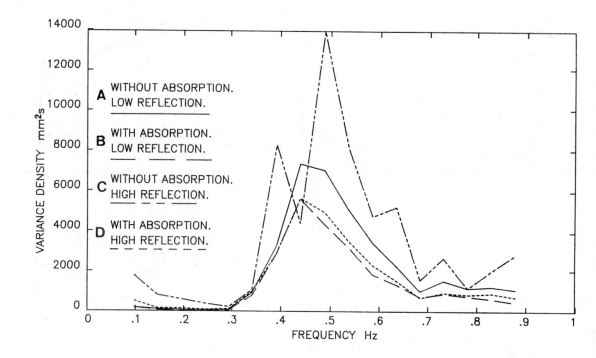

Fig. 5   Incident wave spectra both with and without absorption.

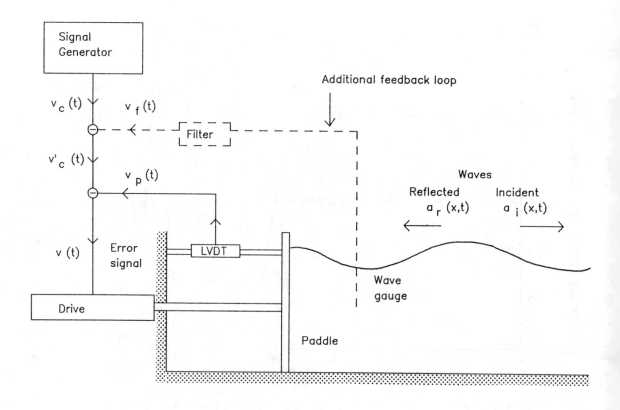

Fig. 6   Schematic representation of an absorbing wavemaker.

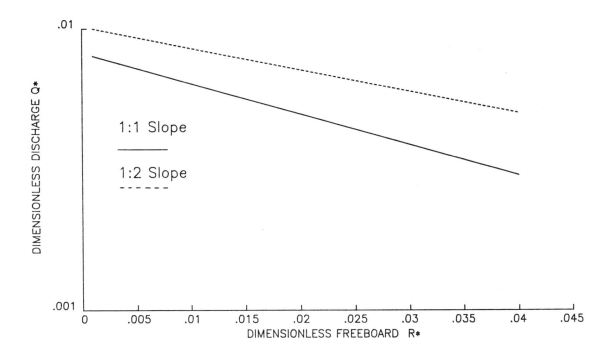

Fig. 7   Dimensionless discharge against dimensionless freeboard
          for a low-crested breakwater.

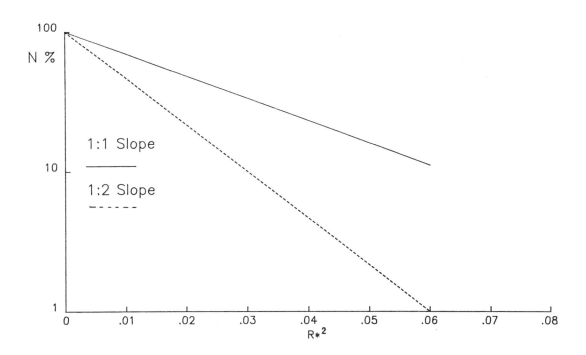

Fig. 8   Percentage of waves overtopping a low-crested breakwater.

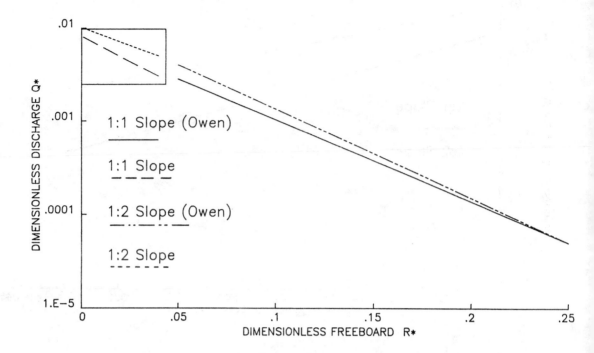

Fig. 9   Comparison with the results obtained by Owen.

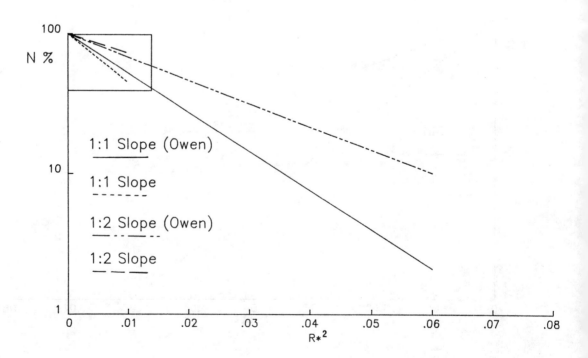

Fig. 10   Comparison with the results obtained by Owen.

Chapter 26
WAVE ACTION ON MULTI-RISER MARINE OUTFALLS

R Burrows and R B Mort
University of Liverpool, UK

Summary

Both experimental and numerical studies into the effects of wave
action on the operation of sewage outfalls discharging into shallow
water are reported. The inducement of internal circulation within the
multi-riser diffuser systems associated with the seaward discharge
manifold is of major concern to the operational performance of these
systems, evidence suggesting that resulting saline intrusion may
ultimately lead to partial blockage. Results presented demonstrate
that the pressure differentials between risers created by the motion
of waves in shallow receiving waters can significantly exacerbate
these problems when discharges fall below the 'design' capacity.

1. Introduction

The disposal of sewage to the sea via long-sea outfalls is generally
considered, within the U.K., to be a cost effective and
environmentally acceptable practice. Its introduction in the absence
of primary, or indeed secondary treatment, is, however, receiving
criticism internationally (1). With the point of discharge suitably
located perhaps several kilometres offshore, beaches will be protected
from pollution provided that the sewage has been suitably screened. A
further prerequisite is that the effluent should be subjected to a
high degree of dilution in the receiving water to ensure the rapid
depletion of pathogen concentrations and to avoid slick formation and
consequential environmental stress.

The dilution is achieved by staging the discharge from a series of
diffuser ports from risers spaced out at suitable distances along the
seaward end of the sub-sea pipeline. Typical installations may have
manifolds incorporating between, perhaps, 3 and 25 risers. For the
purposes of initial dilution the efflux jet is normally sized on the
basis of provision of a densimetric Froude number, F_D, of unity at the
diffuser exit port under maximum design flow, Q_D. This flow

characteristic is defined as follows, with input parameters for this application given in square brackets,

$$F_D = V/(\epsilon g L)^{1/2} \qquad (1)$$

where V = velocity of flow [$\equiv Q_D/(\pi D^2/4)$]
g = gravitational acceleration
L = relevant length dimension [$\equiv D$, port diameter]
$\epsilon = (\rho_2 - \rho_1)/\rho_2$
ρ_2 = density of heavy fluid [saltwater]
ρ_1 = density of light fluid [freshwater/sewage]

An approximate flow balance from the risers can be achieved under the required design flow rate Q_D using standard methods of pipe flow hydraulics, to ensure that the appropriate dilution is achieved for each diffusing plume of effluent. Unfortunately, as flows Q drop below Q_D there is an increasing tendency for the discharge from the seaward risers to reduce and eventually reverse, setting up internal circulations within sections of the manifold system. This behaviour has been demonstrated visually by Charlton (2,3) and Wilkinson (4) from small scale experimental studies, and also numerically by Larsen (5) from a computer model of the system. As a result of the density excess of the saline influx a stratification wedge may form in the outfall and suspended sediments will tend to settle in the pipe invert. These particulate may either be drawn in with the saline water or fall out of the sewage flow above the wedge. If not purged out at times of diurnal maximum flow these deposits may accumulate and eventually lead to partial blockage or modify the hydraulic characteristics of the system so as to affect flow balances.

Purging velocities required to expel intruded seawater can be computed in the manner put forward by Wilkinson (6). Unfortunately, it may often be the case when outfalls are designed on future flow forecasts, that daily peak flows are inadequate to accomplish this flushing action in the early years of operation unless substantial storage is provided in the headworks. In these circumstances, it may be advisable to seal off the most landward risers initially and to bring them into operation only when flows build up to a point where daily purging of intruded seawater can be achieved. Sadly, this is not a common practice and apparent malfunctions observed in numerous outfalls (7) may well be partly attributable to these effects.

Charlton (3,8) has suggested several means for restricting saline intrusion including the introduction of venturi-type constrictions within the diffuser ports or the main outfall itself. These do, however, carry with them additional head losses and consequential pumping requirements, since to be effective they should be sized to provide adequate flow velocities under the conditions of low flow. For purposes of preventing or arresting saline intrusion a densimetric Froude number exceeding unity is again sought using equation (1). Unfortunately, although this requirement ($F_D > 1.0$) can be demonstrated for stratified flows in open channels, its justification is less clear for enclosed flow in pipes of circular section, where the selection of the appropriate length dimension, L, appears to be open to intuitive judgement (9). Diffuser ports sized in this manner may produce velocities under peak flows in excess of those optimal for plume dilution and the small size may lead to increased risk of blockage. Incorporation of valves on the diffuser ports could eliminate the intrusion problems. Simple flaps have been suggested (10) and flexible rubber 'duck-bill' arrangements have been employed (11) in several cases but no system has found extensive application to date.

From the above discussion it has been established that internal circulations are likely to exist in the normal operation of multi-riser outfalls as presently designed, and that these potentially lead to operational problems. A logical extension is, therefore, to investigate whether the situation is exacerbated by wave action. This is only likely to be a factor in shallow receiving waters where pressure fluctuations resulting from surface wave activity extend down to bed level. However, in these circumstances it is possible that the wave induced agitation of the sea bed may also give rise to significant influx of sea bed sediments if intrusive conditions result. The potential influence of waves has been suggested previously by Charlton (12) but no systematic experimental study of these effects has been reported and this deficiency inspired the research programme reported here.

2. Experimentation

The experimental installation used for the study is illustrated schematically in figure 1. The seaward end of an outfall was represented by a 5m length of acrylic pipe of 105mm internal diameter on to which were attached four 50mm internal diameter vertical acrylic riser pipes 400mm long, set at 500mm spacings to form the discharge manifold. Small diameter diffuser ports normally installed on the top of the risers were not included in the tests reported here. Flows were supplied from a header tank and measured either by V-notch in an intermediate stilling tank or (for high flows) by venturi-meter installed at the head of the model outfall section.

The complete pipe system was located within a wave flume 12m long, 750mm wide, with operational water depths up to 900mm. A 'Keelavite' wave generator at one end of the flume was capable of producing regular or random waves, the latter being defined by a target energy spectrum. Wave motion was recorded in the vicinity of the manifold section by surface piercing capacitance gauges and reflective interference of the wave trains was suppressed by a slatted wooden spending beach at the 'landward' end of the flume.

The oscillatory flow velocities in the risers, which occur as a result of the wave action were measured with a 'sensordata' ultrasonic velocity probe. This had to be inserted at a central section in each riser sequentially during each test run and dummy transducer arms were retained in the other risers to eliminate any differential effects on head losses within the risers. Whilst this procedure had some drawbacks, not least the experimental inconvenience, no alternative system could be found, hot-wire anemometry being unsuited to the reversing flows, and financial constraints prevented the acquisition of multiple ultrasonic probes. Visualisation of the oscillations and internal circulations under steady flow could be achieved either by release of dye films in the risers or by the complete colouration of the freshwater flows. The latter technique was also used extensively in a parallel study into the intrusive saline wedges which form in the pipe invert. This will be reported later by Mort (13). Pressures have also been recorded at five sections along the outfall, as shown (PT/-) in figure 1, using Druck PDCR42 miniature transducers set in housings attached to the pipe section. This data has yet to be fully utilized in the final calibration of the numerical model but should also provide empirical measures of the head losses across the pipe riser junctions, an aspect for which guidance is deficient.

Data collection and analysis was conducted using an Eclipse Computer system capable of receiving instantaneously up to 32 channels of information at a sampling frequency of 100Hz. In the present experiment a maximum of 11 channels were used (1 - velocity probe; 6 - pressure transducers; 3 - wave gauges and 1 - wave generator) with sampling of oscillatory signals selected at 20Hz. Computer software was developed specifically for the graphical presentation of the results (sample time series and statistics) from runs of 100 second duration.

In the planning of the study no attempt was made to replicate a typical outfall configuration. The aim was simply to demonstrate characteristic flow phenomena in such systems. Relative to existing outfalls, from geometric scaling, the spacing between risers is rather short in the model but this was constrained by the limited extent of the working section in the wave flume and the need to include at least four risers to provide scope for various alternative internal circulation loops. It was, nevertheless, necessary to select a flow rate at which the riser system should be hydraulically balanced and this was set at 2 litres/sec. Based on scale modelling to the densimetric Froude number, and using a model saline density of approximately 1.015, it was found that this flow would represent about 60% of the design capacity of a specific prototype, in which minimum flows would fall to about 10% of this capacity. A range of model flows spanning 0.3 - 2.0 litres/sec would, therefore, be representative of practical situations. Since the flow balance was set below the equivalent ultimate capacity it was recognised that the inter-riser flow variations experienced in the model would consequentially be less than those in a corresponding prototype. Flow balance itself was achieved by the trial insertion of orifice rings of differing size into the lower sections of the riser pipes.

The full programme of tests conducted covered 8 flow rates in the range 0.19 to 0.94 litres/sec to represent situations where major flow imbalances might be expected and at 2.0 litres/sec, the balanced flow Q_D. At each flow rate, tests were run with quiescent receiving water and with five different wave conditions, ranging in height between 3.2 and 6.5cm and in period between 0.67 and 2.0 secs.

3. Theoretical Modelling

The basis of the mathematical model developed for the outfall system operating under the influence of wave action follows from the earlier work of Larsen (5). It results from the application of the continuity and momentum equations to elements of flow within the pipe system and employs finite difference methods for solution.

3.1 Basic equations

Following directly from derivations in Steeter and Wylie (14), the continuity and momentum equations may be written respectively and for pipes of circular section, as

$$\frac{a^2}{g}\frac{\partial V}{\partial x} + \frac{V\partial H}{\partial x} + \frac{\partial H}{\partial t} + V\sin\theta = 0 \qquad (2)$$

$$\frac{g\partial H}{\partial x} + \frac{V}{}\frac{\partial V}{\partial x} + \frac{\partial V}{\partial t} + \frac{fV|V|}{2D} = 0 \qquad (3)$$

where $a = (k/\rho)/[1+(k/E)(D/t')]$; k and ρ are bulk modulus and density of water respectively; E is Youngs modulus of the pipe material; D and t' are the pipe diameter and thickness respectively; and these terms account for potential expansion of the pipe and fluid compressibility brought about changes in pressure. With reference to figure 2, x is a distance along the outfall, V represents mean pipe flow velocity, H measures the elevation of the hydraulic grade line above the datum and can be expressed as $H = \{(p/\rho g) + z\}$ where p is the hydrostatic pressure and z the position head at that section of the pipe. The inclination of the outfall is given by θ, f is a friction factor taken from the Colebrook-White equation and t is time.

The equations can be expressed in finite difference form to represent flow in sub-elements of the pipe system of length Δx. This sub-division applied to the experimental configuration is indicated in figure 2. Solution to the problem can then be achieved, following specification of the relevant boundary conditions given below.

3.2 Boundary Conditions

- Upstream: this can be taken as a directly connected pump supply for which, at station i = 0, instantaneous velocity V_o and head H_o remain constant. Alternatively, supply to the outfall may be received from a dropshaft as shown in figure 2 which would act as a surge chamber and where the boundary conditions becomes the continuity requirement,

$$\frac{dH_D}{dt} = \frac{1}{A_D} (Q_p - AV_o) \qquad (4)$$

- Downstream: at the point of discharge from the riser port the pressure in the discharging fluid must be equal to that within the denser receiving water, which is subjected to attenuated oscillations as a result of the surface wave action. For regular waves at riser J in figure 2, the pressure can be expressed, from Ippen (15), as

$$P_J = \rho_2 g \left[y_J - \frac{H_W}{2} \frac{\cosh\{ 2\pi(d - y_J)/L\}}{\cosh\{ 2\pi d/L\}} \sin\{(2\pi x_J/L) - (2\pi t/T)\}\right] \qquad (5)$$

where H_W, t and L are the wave height, period and length respectively, the latter being obtained from $L = (gT^2/2\pi) \tanh (2\pi d/L)$; d is the water depth; and y_J is the depth of submergence of the riser ports.

3.3 Solution Method

Equations (2) and (3) written in finite difference form and applied to the discretised system of elements (of length Δx), with the above boundary conditions introduced, can be solved for V and H by various methods. Herein the method of characteristics has been used, with time steps Δt set at $(\Delta x/a)$ secs following from the recommendations of Streeter and Wylie (14).

Flow conditions in the risers have not been solved by an extension of the finite difference scheme, but instead, are dealt with by a lumped inertia method. This is appropriate since flow changes in these short narrow pipes will follow almost instantaneously as a result of wave induced pressure changes. Here, the net upward force exerted on the fluid contained in the riser must balance the rate of change of its momentum. Using the dimensions in figure 2 for riser J, this requirement becomes

$$(P_I^* - P_J) A_J - \frac{f L_J V_J |V_J|}{2g D_J} = \rho_1 L_J A_J \left[\frac{dV_J}{dt} + g\right] \qquad (6)$$

where A_J, D_J and L_J are the riser area, diameter and length respectively. The second term on the left hand side represents frictional resistance forces. Pressure P_I^* at the base of the riser must be established from the total head at station i = I in the outfall but accounting for the head losses through the pipe junction. This also creates a head loss for flows continuing down the outfall as indicated as ΔH_I in figure 2. Presently, these losses are accounted for using the empirical results of Miller (16) but pressure measurements from the experimental studies will later enable an improved calibration of the numerical model to the experimental configuration tested. As an alternative to the use of P_I^* the value of P_I computed from the upstream outfall pipe element can be substituted with an additional term of $(-\Delta H_J \rho_1 g A_J)$ introduced to the left hand side of equation (6). ΔH_J then represents the required head loss associated with the outfall/riser junction. In this form, numerical calibration can be used to effect hydraulic balances in the mathematical model by the trial choice of ΔH_J at each junction, thereby modelling the effect of the orifice plates introduced for the same purpose in the physical model.

The main limitation of the model in its present form is that it is not able to account for stratification in the outfall and the density changes in the discharging fluid that result from intrusive flow conditions. Empirical means for specification of both the saline wedge profiles in the outfall pipe and the scale of mixing are required to improve the performance of the model.

4. Results and Discussion

Sample output from the experimental model discharging a low flow of 0.355 litres/sec ($Q/Q_D = 0.18$) under regular waves of height 6.4cm and period 1.43 seconds is illustrated in figure 3. This shows velocity oscillations in each riser over a duration between 25 and 40 seconds of a 100 second test run. Inserted on the plots are also the mean of the oscillating velocities (computed from the complete 100 second sample and indicated by a broken line) and the steady state condition in the absence of waves (shown chain dotted). Under these conditions it is clear that intrusion through seaward risers 1 and 2 occurs under steady flow conditions and that this is enhanced by wave action. Compensating increases in discharge from risers 3 and 4 leads to an overall continuity flow balance.

It must be appreciated that the time series for each riser were obtained from a sequence of repeated runs of the same conditions, since the velocity meter had to be transferred from riser to riser. A consequence of this is that slight variation in the repetition of conditions may give rise to apparent imbalance between inflows and outflows from the system. A further restriction is that since the time origin in the plots is not unique, instantaneous comparison of relative flows in each riser has no physical justification.

It was found from observation of the complete data series collected that the landward riser, number 4, consistently shows the greatest range of oscillation in both the experimental and numerical models. Note that the maximum and minimum values of velocity quoted on figure 3 also represent the statistics for the entire sample. They therefore indicate, by interpolation against the time series plotted, the presence or otherwise of long period oscillations possibly caused by reflective resonance in the wave flume. This feature was most apparent for the shorter wave periods when the resulting velocity variations, in risers 3 and 4 in particular, lose the characteristic sinusoidal form and show oscillations of apparently random amplitude over a range of frequencies.

Under a flow of 0.944 litres/sec for the same wave conditions as those in figure 3 the net effect of wave action appeared to be concentrated in the two most seaward risers as seen in figure 4.. Here increased wave induced intrusion in riser 1 is compensated for by an increased discharge from riser 2. This condition, representing $Q/Q_D = 0.47$ together with a range of intermediate flows down to $Q/Q_D = 0.18$, are represented in figure 5. This plots the mean flow rates through each riser (+ve discharging; -ve intrusive) for the different wave conditions tested. The gross disparity in flow distribution even at a flow of $Q/Q_D = 0.47$ is worth emphasis bearing in mind that the riser system was set to an approximate balance for $Q/Q_D = 1.0$. Furthermore, for flows of $Q/Q_D \cong 0.25$, which may loosely represent typical minimum conditions of discharge in prototype systems, only the two landward risers may be expected to be in a discharging condition.

The effect of waves on the behaviour shown in figure 5 is not consistent in terms of changes from riser to riser and this may be in large part explained by the above-mentioned inadequacy of the velocity measuring system. Nevertheless, it is clear that the effects are greatest for conditions of low flow in the outfall and that flows in all risers are generally affected. The general trend is for waves to increase intrusion within the seaward risers with the landward risers being forced to increase discharging flows to satisfy the continuity requirement.

The scale of the wave induced changes can be better appreciated in percentage terms as shown in figure 6. Whilst too much credibility should not be placed on these values because of the potential experimental errors, it is quite clear that for this model at least, the degree of intrusion of saline water into the outfall has been greatly increased by the wave action. It would appear from the results that the larger wave heights with associated longer periods generally prove to be most detrimental in this respect. However, no simple rule for practical application could be contemplated from such a limited data base since, in addition to scaled equivalents of H_W and T, the water depth and riser spacing will also be primary factors in governing the behaviour. These latter parameters were not varied in this programme of tests.

Although not covered in figures 5 and 6, tests have also been conducted with the outfall in a shut-down condition ($Q = 0$) when subjected to wave action. Internal circulations are again induced but these are weak and were found to be highly unstable. The systematic velocity measurements taken successively in each riser in general failed to demonstrate the required continuity balance. This arose partly because of this instability and partly because the scale of the velocities often approached the 2mm/sec resolution of the ultrasonic velocity probe, thus yielding inadequate time series. More reliable, but inherently qualitative evidence of the internal circulations was obtained by dye injection into each of the risers. A log of the motion of the dye films then illustrated the modes of flow and a sample of these results is presented in Table 1, where D denotes a discharging riser and I an intrusive situation.

It is either under shutdown or near design flow conditions that the numerical model in its present form is best able to represent the physical situation as no density stratification will take place within the pipe system. Figure 7 shows a sample output of the model under shutdown conditions which demonstrates features of the observed behaviour, the landward riser again being subjected to the greatest oscillations. These traces also demonstrate a weak longer period oscillation of about 4.4 second period, which matches the oscillations computed in the dropshaft modelled as part of the headworks. Although a larger period oscillation was noticed in the experimental data, this was not nearly so strong and was possibly induced from the wave field itself. The most likely explanation for the absence of this effect in the experimental model is the suppression of landward motion in the outfall caused by the venturi and the reduced pipe diameter upstream, which was not built into the mathematical description. Earlier steady flow testing of the computer model had demonstrated a rapid transient decay of numerical instabilities arising from assumed initial conditions in the time simulation and similar behaviour would therefore be expected when the model is run with wave action present. Another unknown factor which might influence the performance of the numerical model under these circumstances is the precise form of minor losses created at the pipe/riser junctions at such low flow velocities (low Reynolds Number). Future analysis of the pressure transducer records should potentially shed some new light in this area.

Notwithstanding the limitations of the numerical model when intrusion leads to density differentials and stratification, figure 8 is included for conditions closely matching those of figure 3. Whilst similar intrusive behaviour is observed between the two sets of results, the numerical disparities place into perspective the further advances necessary in the theoretical description before it could be considered for reliable synthesis of prototype systems.

5. Conclusions

1. At flows significantly below the ultimate capacity of an outfall system, it has been demonstrated that intrusive conditions are likely to occur in certain risers forming the seaward discharge manifold. There is evidence to suggest that this saline influx may lead to operational problems and possible malfunction in the long-term under conditions where this is not purged during regular outfall operation.

2. Wave action over the discharge manifold, in conditions where water depths are relatively shallow, has been shown to increase the scale of this intrusion and also to initiate intrusive internal circulations when the outfall is in a shut-down condition.

3. The data acquired and the range of conditions investigated in the work reported are inadequate to enable any quantitative assessment of the likely effects in practical outfall systems. Improved experimental techniques enabling instantaneous velocity measurement in each model riser are essential to improve the quality of results.

4. No attempt has been made to account for the presence of diffuser heads, with multiple ports, as incorporated on most riser systems. This will be investigated in later physical model tests. The presence of a significant flow constriction in such diffuser systems would be expected to suppress to some degree the scale of wave induced variations.

5. A complementary computer model developed as part of the study demonstrates similar behaviour to that observed in the experiment but with deficiencies in calibration in its present form. However, substantial empirical developments are necessary if saline wedge formation in the outfall pipe and density mixing of discharging fluid is to be realistically represented.

6. References

1. Mandl, V., "European community activities towards the protection of the marine environment". Proc. ICE Conference on Marine treatment of sewage and sludge, Brighton, 1987, 1-10.

2. Charlton, J.A., "Salinity intrusion into multiport sea outfalls", Proc. 18th A.S.C.E. International Conference on Coastal Engineering, Capetown, 1982.

3. Charlton, J.A., "Sea Outfalls". Developments in Hydraulic Engineering" edited by P. Novak, Elsevier, Amsterdam, 1985.

4. Williamson, D.L., "Seawater circulation in sewage outfall tunnels". Journal of Hydraulic Engineering, A.S.C.E., Vol. 111 No. 5, May 1985.

5. Larsen, T., "The influence of waves on the hydraulics of sea outfalls". Proc. 20th A.S.C.E. International Conference on Coastal Engineering, Taiwan, November, 1986

6. Williamson, D.L., "Purging of saline wedges from ocean outfalls".
    Journal of Hydraulic Engineering, A.S.C.E., Vol. 110, No. 12,
    December, 1984.

7. Grace, R.A., "Sea outfalls - a review of failure, damage and
    impairment mechanisms". Proc. I.C.E. Part 1, Paper No. 8766,
    February, 1985.

8. Charlton, J.A., "The venturi as a saline intrusion control for sea
    outfalls". Proc. I.C.E. Part 2, Paper No. 8980, December 1985.

9. Davies et al, "A laboratory study of primary saline intrusion in a
    circular pipe". I.A.H.R. Journal of Hydraulic Research, Vol. 26,
    No. 1, 1988

10. Hansen et al, "San Francisco ocean outfall port valve
    development". Proc. A.S.C.E. Hydraulics Speciality Conference,
    Chicago, 1980

11. Roberts, D.G.M. et al, "Weymouth and Portland marine treatment
    scheme: tunnel outfall and marine treatment works". Proc. I.C.E.
    Part 1, Paper No. 8749, February, 1984.

12. Charlton, J.A., "Outfall design guide for environmental
    protection: Section 6", edited by Neville-Jones & Dorling, W.R.C.
    publication, November, 1986.

13. Mort, R.B., "Investigation into the effects of wave action on long
    sea outfalls", Ph.D. thesis, University of Liverpool, 1988 (in
    preparation)

14. Streeter, V.L., and Wylie, E.B., "Fluid Mechanics", pub.
    McGraw-Hill, New York, 1975

15. Ippen, A.T., "Estuary and Coast-line Hydrodynamics", pub.
    McGraw-Hill, New York, 1966

16. Miller, D.S., "Internal flow systems", B.H.R.A. Fluid Engineering,
    1978.

Table 1   **Motion in risers under shutdown conditions (Q = 0) from observation of dye movements.**

WAVE CONDITIONS		OBSERVED FLOWS*			
H_W (cm)	T (secs)	Riser 1	Riser 2	Riser 3	Riser 4
6.1	1.0	0	I	I	D
6.1	0.8	I	I	0	D
6.1	0.67	0	I	I	D
5.49	2.5	0	D	I	I
7.16	2.5	D	D	I	I
9.35	2.50	D	D	I	I
9.97	3.33	D	D	D	I
5.01	5.00	0	I	D	D

* D - discharging; I - intrusive   0 - zero.

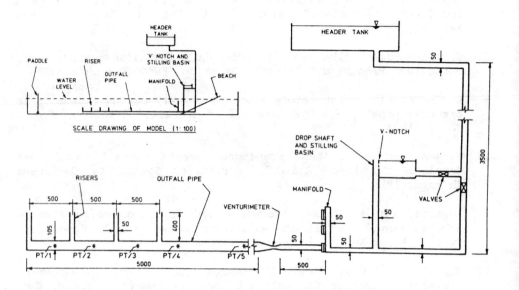

FIGURE 1: General arrangement of experimental apparatus

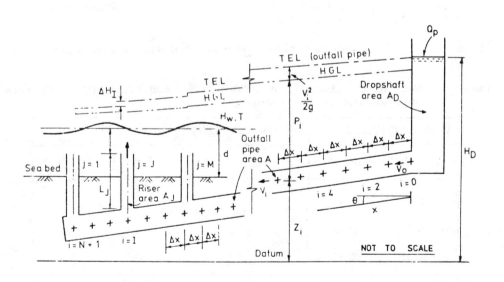

FIGURE 2: Definition sketch for numerical model

376

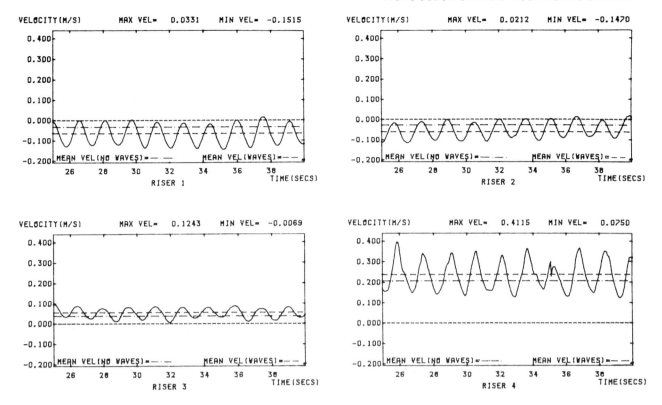

WAVEHEIGHT= 0.0640 M   WAVEPERIOD= 1.4290 S   FLOW RATE=0.35500 L/S

FIGURE 3: Experimental velocity fluctuations in risers for Q/Q_D = 0.18

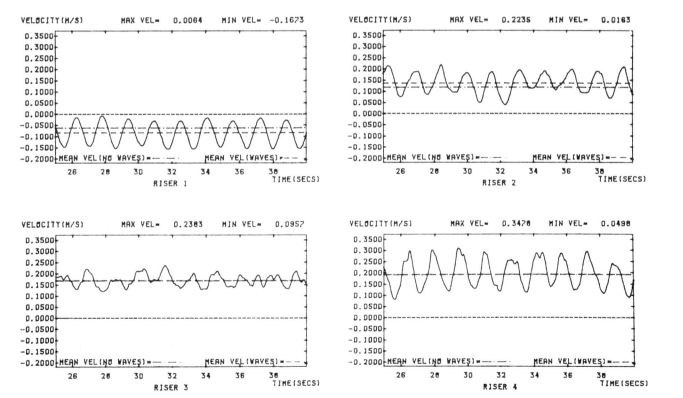

WAVEHEIGHT= 0.0640 M   WAVEPERIOD= 1.4290 S   FLOW RATE=0.94470 L/S

FIGURE 4: Experimental velocity fluctuations in risers for Q/Q_D = 0.47

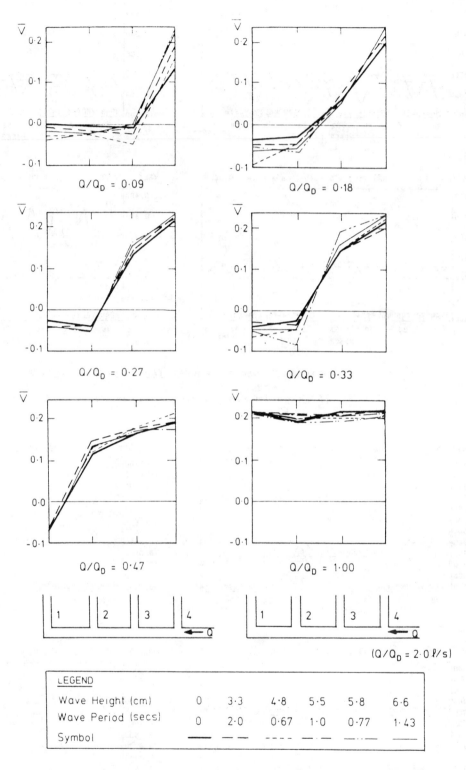

FIGURE 5: Experimental mean of riser velocities ($\overline{V}$) over full range of test conditions

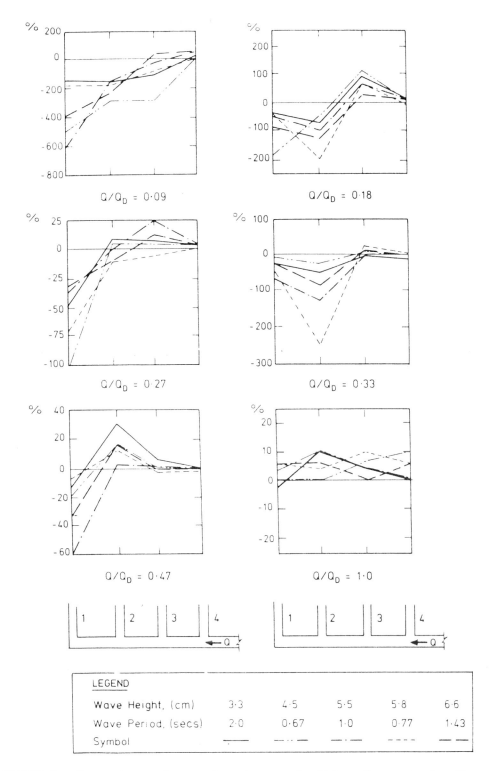

FIGURE 6: Percentage change in riser velocities from steady flow (quiescent receiving water) over full range of test conditions

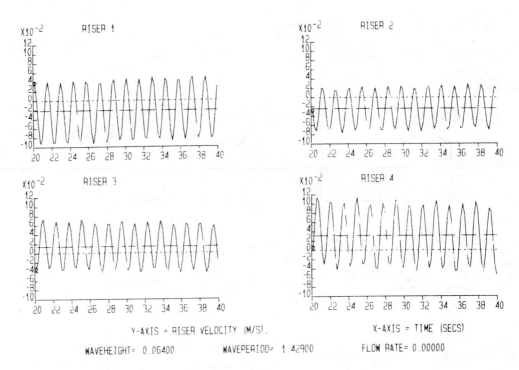

FIGURE 7: Theoretical velocity fluctuations in risers under shutdown conditions (Q = 0)

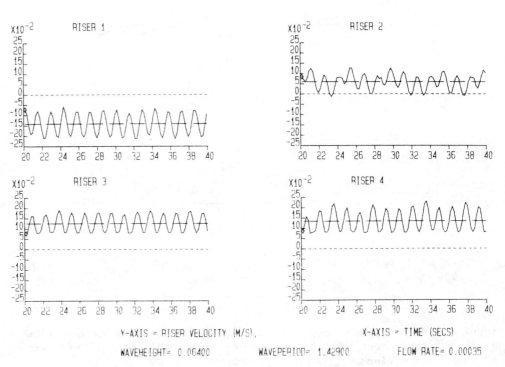

FIGURE 8: Theoretical velocity fluctuations in risers from existing numerical model for conditions matching figure 3

Chapter 27
FIELD MEASUREMENTS AND NUMERICAL MODEL VERIFICATION OF COASTAL SHINGLE TRANSPORT

A J Chadwick
Brighton Polytechnic, UK

Summary

Currently, the application of numerical models of beach evolution
to coastal engineering schemes is hindered by the imprecision of
the core sediment transport equations. This leads to the
requirement for calibration of the models to every new scheme. To
improve both the predictive performance of these models and their
universality, better longshore transport equations are required.
This paper describes the methods and results of an intensive field
measurement programme of coastal shingle transport. In addition, a
numerical model of beach evolution has been calibrated against a 12
year record of aerial survey data, using three different transport
equations. These results have been compared with those predicted
using a transport equation derived directly from the field
measurements and one derived from an hydraulic model study. The
implications for future numerical modelling and field measurements
are discussed.

1. Introduction

In recent years there have been many advances in our understanding
of and ability to model wave climate and wave/current interaction.
These models are being used in the design process for many coastal
engineering schemes. However, many of the models have not been
calibrated through lack of field data. In contrast, advances in
sediment transport modelling in the coastal zone have been sparse
compared with those in wave modelling (for a very recent review of
all of these models see McDowell(1). The urgent need for
comprehensive field measurements to validate existing numerical
models has been highlighted by McDowell and O'Connor(2) and by
ICE(3). In consequence, the hydraulic engineering research unit at
Brighton Polytechnic instigated a field measurement programme
concerned with sediment dynamics on open shingle beaches.

2. Fieldwork activities

2.1 Beach Site

The beach site chosen was Shoreham Beach, West Sussex , on the south coast of England (see Plate 1). This is an open beach of length 2.0 km between groynes to the west and Shoreham harbour to the east. Its composition is mainly shingle, with a sand foot at about the low water mark.

2.2 Measurement programme

This was carried out between 2nd. September 1986 and 28th. November 1986. During this period, the following sets of measurements were made.

- Water surface elevation traces sampled in 10 minute intervals at 5 Hertz approximately six times daily, giving 191 data sets in total.

- Alongshore and on/offshore shingle transport on 18 occasions, together with concurrent measurement of wave height,angle and speed.

- Beach profiles on six fixed lines from beach head to below the low water mark on 16 occasions.

- Hydrographic surveys on six fixed lines extending 600m to sea on two occasions.

The weather conditions were either excellent or abysmal for fieldwork, with very little in between. The sea state was similarly afflicted, being characterised by calms and small swells or storms. These extreme conditions caused numerous fieldwork measurement difficulties and the loss of much valuable equipment.

2.3 Measurement techniques

The water surface elevation measurements were taken by an automated system comprising sensor (SEM), data logger and computer system; the shingle transport by surface and submerged traps; the wave angles and speed by the use of an array of visual wave poles; the beach profiles by stadia tacheometry and the hydrographic surveys using a boat equipped with echo sounder, chart recorder and range finder gun.

Full details of the field measurement techniques are given in (4) and (5). Plate 2 shows the SEM and the visual wave pole array used initially. Plate 3 shows a surface trap in operation.

The SEM, data logging and computer system worked very reliably until the failure of the SEM on 21st. October, following a sequence of severe gales which began on 19th. October. The visual wave pole array was also lost in this storm. Subsequently, a new triangular wave pole array was designed and constructed (see Plate 4). This proved to be far more robust than the original array. Unfortunately, it was not possible to replace the SEM and hence wave parameters were visually estimated thereafter.

2.4 Problems associated with shingle trapping

The majority of the tests were those in which one or two surface mounted traps were set to measure longshore transport. In these tests, the traps were positioned approximately on the waterline two hours before high water and the volumes of shingle collected were measured as the tide receded (approximately five hours later). Simultaneous measurements of wave height, period, speed and angle were taken during these tests.

To obtain useful results from such tests, several environmental conditions needed to be simultaneously met. These are:

- a five hour range centred on high water to be within daylight hours,

- waves approaching the shoreline at a significant angle,

- sea state preferably not complex, ie. wave angle from one general direction not several,

- significant wave height to be between 0.2m to 1.0m,

- beach profile not to alter significantly in the vicinity of the traps.

During the fieldwork period, the sea state was far from ideal for shingle trapping. In consequence, of the 18 tests attempted only 7 were successful.

3. Derivation of the longshore transport rates from field data

3.1 Observed physical transport mechanisms

Coastal shingle transport is mainly effected by bed load transport in the swash zone. Seaward of the swash zone transport is generally negligible in comparison. This was confirmed by visual observation and by analysis of the measured beach profiles, which showed negligible movement below the low water mark.

In the swash zone, transport is primarily due to the translation of the water mass in the breaking wave. The translation takes place between the point of wave breaking and the limit of the uprush. The transport rate in the swash zone varies rapidly, from a maximum in the vicinity of the wave break point to zero at some point below the limit of uprush due to the threshold of movement criteria.

3.2 The distribution of longshore transport flux across the swash zone

Ideally, the distribution of transport flux across the swash zone should be measured in the field. Such measurements are extremely problematic. The swash zone position and length up the beach vary with every wave and in tidal waters the still water level varies continuously with time. When field transport rates are measured using trapping techniques, the volume of material collected in any given time cannot, therefore, be simply related to transport rate.

If, however a distribution of transport flux is assumed, a priori, then the instantaneous transport rates may be derived. These can then be compared, via a beach evolution model, with the recorded long term transport rates and associated wave climate. In this way the validity of the a priori assumptions can be tested.

A simple distribution of the flux rate $q(m^3/s/m)$ has been assumed in which q is a maximum at the break point (q_m) reducing linearly

Advances in water modelling and measurement

to zero over a distance l_u, the length over the beach of the swash zone (work is currently in hand to establish a better relationship, based on the velocity distribution across the swash zone).

The total transport rate Q (m³/s) is defined as:

$$Q = \int_0^{l_u} q_1 \, dl$$

where $q_1 = q_m (l_u - 1)/l_u$

Hence $Q = q_m l_u / 2$

The length l_u is a function of the breaking wave height (H_b) and the beach slope $(\tan \alpha)$. Assuming that the wave breaks in a depth $d_b = 1.28 H_b$ then:

$$l_u = (1.28 H_b + H_\omega)/\sin \alpha$$

where H_ω is the runup height.

Numerical values for l_u have been estimated using Fig. 15 in BS 6349(6) which relates runup height to beach slope and wavelength.

3.3 The influence of tidal rise and shingle trap size

Fig. 1. shows the sequence of events for a rising tide and fixed wave height and period. Theoretically, the total volume collected (V_{th}) is given by:

$$V_{th} = 2(V_1 + V_2 + V_3)$$

(assuming equal rates of tidal rise and fall (R m/s))

where

$$V_1 = \int_0^{T_1} dt \int_{l_b}^{l_u} q_1 \, dl$$

$$T_1 = l_t \sin \alpha / R$$

$$l_b = l_u - Rt$$

$$V_2 = \int_{T_1}^{T_2} dt \int_{l_b}^{l_b + l_t} q_1 \, dl$$

384

$$T_2 = T_1 + (l_u - l_t) \sin \alpha / R$$

$$l_b = l_u - l_t - R(t-T_1)$$

$$V_3 = \int_0^{T_3} dt \int_0^{l_t - l_b} q_1 dl$$

$$T_3 = T_2 + l_t \sin \alpha / R$$

$$l_b = R(t-T_2)$$

These definite integrals can be solved explicitly, given the incident wave height (H_b), wave period, trap length (l_t), rate of tidal rise (R) and beach slope (tan α). The solution is of the form:

$$V_{th} = \text{constant} \times q_m$$

Hence the ratio $Q_{th}/(V_{th}/(2T_3))$ may be evaluated as a pure number (η). This ratio can then be used to convert the trapped volumes (V_f) to the equivalent field transport rate (Q_f) as follows:

$$Q_f = \eta V_f/(2T_3)$$

3.4 The influence of wave height distribution

To calculate the field transport rate from the trapped volume requires a representative wave height as one of the input parameters. The true solution is one which accounts for the entire distribution of wave heights. If a particular wave height distribution function is assumed, then the true solution can be found.

In this instance, the Rayleigh distribution has been used, for which the probability of wave height exceedence is given by:

$$P(h>H) = \exp - (H/H_{rms})^2$$

and the probability density function is given by:

$$f(h) = (2h/H_{rms}^2)\exp-(2h/H_{rms}^2)$$

In discrete (histogram) form, with class interval size Δh and total number of waves N, then the number of waves (n) in each class interval is given by

$$n = f(h)N\Delta h$$

The integral equations defined in Section 3.3 can be applied to each wave height class in turn and a weighted average value of the field transport rate found as follows:

For a particular wave height h and associated period T_h, the time of action of these n waves is

$$T_n = nT_h$$

and the total time of action for all waves is

$$T_N = \Sigma T_n$$

Hence a weighted average value for the field transport rate is

$$Q_f = \Sigma \, (Q_h T_n)/T_N$$

To calculate this value appropriate wave periods need to be assigned to each wave height. The joint distribution of wave height and period can be established from the field measurements. Currently, this work has not been completed.From a previous study (7), the mean zero crossing period (for wind waves) was found to fit the equation

$$T_z = 0.925 Hs + C \quad \text{where } C = 3.136$$

In this instance the measured T_z was used to establish the constant C and the mean wave height in each class interval (h) substituted for Hs to find the wave period T_h.

3.5 Summary of results

The principles described in Sections 3.3 and 3.4 have used to calculate the field longshore transport rates from the trapped volumes. The data and results are given in Table 1.

To relate these values of field longshore transport rates to the causitive longshore wave power a second set of calculations were necessary. The wave climate was originally measured at the beach toe. To estimate the wave height, angle, and group wave speed at the breakpoint linear refraction and shoaling theory has been applied. The longshore wave power (P_1) has been calculated from the equation

$$P_1 = \rho g \sigma^2 \, C_{gb} \, \frac{\sin 2\theta_b}{2}$$

where σ^2 is the spectral equivalent of $H_{rmsb}^2/8$ (see (5))

C_{gb} is the group wave speed at the wave breakpoint

θ_b is the wave angle at the wave breakpoint

and C_{gb} and θ_b are calculated from the significant wave height breakpoint (for an explanation of this see Section 6.3)

The results of this analysis are given In Table 2.

4. Derivation of a longshore transport formula from the field measurements

Currently, various versions of the CERC energy equation are in common use. The CERC equation requires the use of a coastal constant which is a function, inter alia, of the beach material size. For shingle beaches, this constant has not been evaluated with any accuracy. In addition, the introduction of a threshold of motion term would be desirable.

Brampton and Motyka[8] have proposed the following form of equation

$$Q = \frac{K}{\gamma} \; P_1 \; (L/D)^{\epsilon} \; (1-8.1D/H)^{\beta}$$

where $\gamma = (\rho_s - \rho)ga'$

$$\begin{array}{ll} a' & =1/(1-e) \\ e & = \text{voids ratio} \\ D & = 90\% \text{ grain size} \\ H & = \text{significant wave height at breaking} \end{array}$$

and K, ϵ and β are constants to be evaluated from field measurements.

Given a sufficiently large data set containing a large range of the controlling variables, it should be possible to test the applicability of the above equation and, if satisfactory, to calibrate the three constants. This has not proved possible with the available Shoreham data.

However, the data has been plotted in the form of Q against P_1 as shown in Fig. 2. The equation of the best fit line to the 5 results measured when the SEM was operational results in the following relationship

$$Q=3.553 \times 10^{-6}(P_1-12.2) \qquad (m^3/s)$$

This equation can be written as

$$Q = \frac{K}{\gamma} (P_1 - P_0)$$

where $K = 0.0384$

$\gamma = 10807$

and P_0 is the threshold wave power. γ has been determined from the following data:

dry bulk density $\rho_{db} = 1800 \text{ kg/m}^3$

material density $\rho_s = 2650 \text{ kg/m}^3$

as $\rho_{db} = \rho_s/(1+e)$ then $e=0.47$

and $a' =1/(1+e) = 0.68$

therefore $\gamma = (\rho_s - \rho)ga' = 10807 \text{ kg/m}^2\text{s}^2$

5. The calibration of a numerical model of beach plan shape evolution

5.1 Background

A copy of a numerical model of beach plan shape evolution (BPSM) was provided by Sir William Halcrow and Partners.The BPSM describes the movement of the beach line in response to longshore transport caused by wave action. It uses two basic equations, one to describe the relationship between the incident waves and longshore transport and the other to describe the relationship between longshore transport and beach line movement.

$$Q = 1/2K\ C_o\ (Hrms\ Kr)^2\ \sin(2\theta_b) \qquad (m^3/day) \qquad (1)$$

where K = coastal constant

C_o = deep water wave speed

$Hrms$ = deep water root mean square wave height

Kr = refraction coefficient

θ_b = wave angle at breaking

The coastal constant is calculated from an equation derived by Swart for sand beaches given by-:

$K=10000\ \log(0.00146/D_{50})$

The continuity equation is used to describe the beach line movement, given by-:

$h\ dx/dt=dQ/dy$

where h=active beach height
x=distance along the beach
y=beach line movement perpendicular to the beach

The model also incorporates the effect of internal and external boundary conditions.

5.2 Calibration data

The Southern Water Authority (SWA) have been carrying out an annual aerial survey of the Sussex coastline since 1973. Table 3 summarises the data abstracted from the SWA records. It shows the mean beach line movements and area and volume changes recorded over the period 1973 to 1984 for sections 341 to 358. Inspection of Table 3 leads to the conclusion that a stable point exists between sections 345 and 346. In addition the harbour arm acts as a complete cut-off of the shingle transport. Given these two end conditions, the mean annual volume change between section 345 and the harbour arm (14539 m^3/annum) is indicative of the mean longshore transport rate. Hence to calibrate the beach plan shape model, the model prediction of volume change must be an accretion of about 14500 m^3/annum.

In addition, the model prediction of beach line movement should be in accordance with the aerial survey data. In this case, the beach line movement has been taken as that recorded at the 2.0m AOD contour. This is closest to the mean high water level (2.3m AOD), a commonly used reference line. Figure 3 illustrates the recorded beach line movements at different levels for SWA Section 350. These are very similar throughout the profile.

The input wave data to the model is a mean nearshore climate derived from wind records for 1980 to 1984 (this data was provided by Hydraulics Research Limited). Therefore the numerical model predictions could not be expected to reproduce the historical year to year variations. Instead, the model predictions of beach line movements have been compared with the long term mean accretion rates. This has been achieved by using the 1981 2.0m beach line as the initial line and estimating the expected beach line in 1984 from the long term mean accretion rates. The 1981 beach line was input to the numerical model which was then run for a 3 year period (using the mean wave climate for 1980-84) and its predicted beach line compared with the expected beach line.

5.3 Results

Full details of the calibration and sensitivity analysis of the BPSM are given in (9). Only a brief summary is given here. The BPSM predictions together with the recorded 1981 beach line and the expected 1984 beach line are shown in Fig. 4. The two principal calibration parameters were the coastal constant (K) and the active beach height (h).

In the calibrated model a D_{50} size of 1.381mm was used which produced a coastal constant of 242. This D_{50} size is obviously not indicative of a representative shingle size and can only be regarded as pure calibration constant. This was expected as the Swart equation was derived from sand beaches.

The calibrated model active beach height was 3.0m. Given that the mean beach line accretion rate is 1.06 m/annum over the 2km beach length and the mean volume accretion rate is 14500 m^3/annum then the mean active beach height must be 6.8m. Clearly, there is a large discrepancy between the expected and calibrated values. The explanation for this may be found in Figure 4. At both boundaries of the model, the predicted beach line is significantly different from the 1984 expected beach line. If the model predicted beach line had been closer to the expected beach line at either or both boundaries then the central section would have moved further seaward. Hence to maintain the position of the central section under these circumstances would have required a larger active beach height.

The model also highlighted the influence of wave reflection from the harbour arm on the beach line movements. It was found necessary to incorporate this effect in the model to obtain realistic results. Although none of the model runs completely reproduced the beach line near to the harbour arm, they did demonstrate the range of possible beach lines for the cases of full, partial and no reflection.

6. A comparison of the performance of five longshore transport equations

6.1 Methodolgy of comparisons

The central component of the BPSM is a longshore transport equation. In essence it is a simple matter to replace that transport equation (in the BPSM) with a variety of other transport equations, whose relative performance can then be evaluated. However, it proved possible to use a much simpler numerical model to compare the relative performance of different transport equations. The simple numerical model developed is called ANLST. It is based on the following principles:

- Beach assumed long and straight with a constant orientation.

- Nearshore wave climate transposed to wave climate at the breakpoint using linear refraction and shoaling theory.

- Mean annual longshore transport rate calculated as the sum of all the contributing wave heights, angles and frequencies.

The main differences between this model and the BPSM are that the BPSM takes account of internal and external boundary conditions and calculates the re-orientation of the beach line through time. In other respects the two models are similar.

6.2 Details of the longshore transport equations

Five equations have been compared using ANLST. The control results were produced using the equation (1) from the BPSM, the other four equations are as follows:

- Basic CERC equation

$$Q = \frac{K}{\gamma} P_1 \qquad\qquad (m^3/s) \qquad\qquad (2)$$

where

K = coastal constant

$$P_1 = E_b C_{gb} \frac{\sin 2\theta}{2}$$

$$E_b = \frac{\rho g Hrms_b^2}{8}$$

$$\gamma = (\rho_s - \rho)\, ga' \; : \; a' \; = 1/(1-e) \; : \; e = \text{voids ratio}$$

C_{gb} is the group wave speed at the wave breakpoint

θ_b is the wave angle at the wave breakpoint

and C_{gb} and θ_b are calculated from the significant wave height breakpoint (for an explanation of this see Section 6.3)

- Modified Delft equation

The Delft Hydraulics Laboratory has carried out a series of

investigations concerning the profile development and longshore transport of coarse material under regular and irregular wave attack. The findings of these investigations are summarised in four reports, Hijum(10), Hijim(11), Hijum & Pilarczyk(12) and Pilarczyk & Boer(13). These investigations included a literature survey of longshore transport of coarse material and a sequence of physical model studies of the same. Based on these model studies a longshore transport equation under irregular wave attack was proposed as follows:

$$Q = 0.000712 \ (gD_{90}^2 T)W(W-8.3) \sin \theta_v / \tanh(k_s h)_v$$

where

$$W = \frac{Hs_d (\cos\theta_v)^{0.5}}{D_{90}}$$

h = water depth

$k_s = 2 \pi / L_2$

Subscript d refers to deep water

Subscript v refers to beach toe

It can be seen that this equation uses a mixture of deep water properties and those measured at the toe of the model beach. This makes application of the equation to other sites unnecessarily difficult. In addition the tanh term has no real significance (as explained by Hijum & Pilarczyk) because $\sin \theta / \tanh(k_s h) = $ constant.

Consequently, the equation has been reformulated in terms of conditions at the breakpoint and the tanh term has been dropped. This has been achieved by taking all the original experimental results (measured at the beach toe) and applying linear refraction and shoaling theory to determine the break point conditions. The results are shown in Table 4. and Fig. 5. The best fit line to the modified data at the break point results in the following equation.

$$Q = 0.0013 \ (gD_{90}^2 T)W(W-8.3) \sin \theta_b \qquad (m^3/s) \qquad (3)$$

where $\quad W = \dfrac{Hs_b (\cos\theta_b)^{0.5}}{D_{90}}$

- CERC type equation with a threshold of motion term

A reduced form of the Brampton and Motyka equation given in Section 4 has been used by taking $\epsilon = 0$ and $\beta = 1$ giving

$$Q = \frac{K}{\gamma} P_1(1-8.1D/H) \qquad (m^3/s) \qquad (4)$$

391

- CERC type equation with a threshold of motion term derived from field measurement

$$Q = \frac{K}{\gamma} \cdot (P_1 - P_0) \qquad (m^3/s) \qquad (5)$$

where K $\quad= 0.0384$

$\quad\quad \gamma \quad= 10807$

and $P_0 \quad= 12.2$

6.3 The estimation of true transport rates using energy based equations

The same arguments apply here as already presented in Section 3.4. Assuming a CERC type equation holds true for an individual wave of given height and period, then the true transport rate(Q_t) for a complete distribution of wave heights and periods is given by

$$Q_t = \sum (Q_h T_n)/T_N$$

The true transport rate can then be compared to that predicted using either Hrms or Hs. This comparison has been carried out for a range of offshore wave angles using a computer program RAYL. The results are summarised in Table 5. The best approximation to the true transport rate was found to be that produced by using Hrms for wave energy, but Hs for the break point parameters.

6.4 Comparative results

The program ANLST has been used to compare the performance of the five transport equations . The complete set of results from this program are summarised in Table 6.

The BPSM equation (1) gives a mean annual transport of 15050 m^3. This is very close to that predicted by the BPSM program (cf 14500 m^3/annum), thus demonstrating the validity of ANLST. Hence, the other four transport equations have been compared with this base figure.

The coastal constants for the basic CERC equation (2) and the CERC threshold equation (4) have been set to produce the same annual transport as equation 1. These constants can therefore be considered to have been calibrated to the Shoreham site.

The modified Delft equation (3) and the CERC field equation (5) have not been calibrated, but included to gauge their predictive capacity. The Delft equation performs very badly, giving a transport prediction 4.4 times too large. The CERC field equation, calibrated by direct measurements, performs rather better giving a transport prediction of 0.7 times the true value.

7. Conclusions

7.1 Field measurements

- A new method for relating field measurements of shingle trapping to longshore transport rates has been developed, based on physical principles. Currently, this method relies on an a priori assumption regarding the distribution of transport flux across the swash zone and the use of a Rayleigh wave height distribution. It is intented to carry out further research into this aspect of the study.

- A new transport equation has been derived from the field measurements which gives encouraging results when used to predict the mean annual longshore transport rates. This supports, to some extent, the assumed distribution of the transport flux across the swash zone and the calibrated constants in the derived transport equation. However, better results would undoubtedly ensue with more field data.

7.2 Application of the BPSM

- The predictive capacity of a numerical model of beach evolution is critically dependent on its calibration from field data. An accurate calibration may require many years of data. The two principal parameters of active beach height and coastal constant (or eqivalent) should not be concomitantly calibrated unless the model predictions are in accord with recorded values along the whole beach. This will prove to be difficult if there is significant wave reflection and/or refraction from any adjacent coastal structures.

7.3 Longshore transport formulae

- The methodology adopted to compare the performance of various transport equations is a useful tool. It provides comparative results quickly and easily and can potentially be used to calibrate any new transport equation.

- In applying a CERC type transport equation, any wave height parameter can be selected provided that the coastal constant is is calibrated to it. However, it is preferable that the parameters used should be related as closely as possible to the true processes. Hence it is recommended that Hrms be used for calculating wave energy and Hs used to determine the break point.

- Three transport equations (1), (2) and (4) have been calibrated to the Shoreham site. This represents an advance in knowledge concerning suitable values for the coastal constant, as applied to shingle beaches. However, although equation (4) includes a threshold term related to grain size, its validity and hence universal applicability has not been proven.

- The modified Delft transport equation (3) did not give satisfactory results. This fact supports the argument that small scale laboratory model studies of shingle transport cannot easily be transposed to the prototype situation.

- The transport equation (5) derived from field measurements, gave encouraging results. This supports the argument that further progress in quantifying the longshore transport of coarse materials can be made from field studies. The field measurement principles established in the Shoreham Beach Project could be used to obtain a much larger data set. This should then enable better transport equations to be evaluated and calibrated for universal application.

8. Acknowledgements

The author would like to express his gratitude to the following organisations and persons:

Southern Water Authority (Sussex and Kent Divisions), Brighton Borough Council, Rother District Council and Havant Borough Council for sponsoring the field measurement programme.

Sir William Halcrow and Partners for providing the BPSM.

Hydraulics Research Limited for providing the inshore wave climate data at Shoreham.

Dr. C Fleming (of Sir William Halcrows & Partners), Dr. A Brampton (of HRL) and Professor D M McDowell for their helpful comments on the research programme.

9. References

1. McDowell D. M.:"The sea-face of estuaries". In:"Developments in Hydraulic Engineering V.". Amsterdam, Holland,Elsevier: In preparation.

2. McDowell D. M,and O'Connor B. A.: In"Hydraulic Modelling in Maritime Engineering". London, UK, Thomas Telford, 1981.

3. Institution of Civil Engineers:"Research requirements in coastal engineering". London, UK, Thomas Telford, 1985.

4. Chadwick A. J."Sediment dynamics on open shingle beaches: Report on the Shoreham Beach field measurement programme". Brighton, UK, Brighton Polytechnic, Dept. Civil Eng., 1987.

5. Chadwick A. J.:"The measurement and analysis of inshore wave climate using a micro computer based data logging and analysis system". Submitted to Proc. Inst. Civ. Eng. Part 2.

6. British Standard 6349.:"Maritime Structures". London, UK, British Standards Institute, 1984.

7. Chadwick A. J. and Pope D. J.:"The beach plan shape model. Application to Shoreham Beach". Brighton, UK, Brighton Polytechnic, Dept. Civil Eng., 1986.

8. Brampton A. and Motyka J. M.:"Modelling the plan shape of shingle beaches". Estuarine Studies,Vol 12, Springler Verlag, 1984.

9. Chadwick A. J.:"The calibration of a numerical model of beach plan shape evolution to Shoreham Beach". Brighton, UK, Brighton Polytechnic, Dept. Civil Eng., 1988.

10. van Hijum E.:"Equilibrium profiles of coarse material under wave attack". Delft, Holland, Delft Hydraulics laboratory, publication No.133, 1974.

11. van Hijum E.:"Equilibrium profiles and longshore transport of coarse material under wave attack". Delft, Holland, Delft Hydraulics laboratory, publication No. 174, 1977.

12. van Hijum E. and Pilarczyk K. W.:"Equilibrium profile and longshore transport of coarse material under regular and irregular wave attack". Delft, Holland, Delft Hydraulics laboratory, publication No. 274, 1982.

13. Pilarczyk K. W. and den Boer K.:"Stability and profile development of coarse materials and their application in coastal engineering". Delft, Holland, Delft Hydraulics laboratory, publication No. 293, 1983.

Table 1 Data and results for conversion of trapped volumes to longshore transport rates

Date	Hrmsb	Tz	beach slope angle	Rate of tidal rise	Trapped volume	Qr	Qrms	Qrms/Qr
	(m)	(s)	(deg)	(m/s)	(l)	m³/sx10⁻⁶	m³/s x10⁻⁶	
22/09/86	0.23	2.32	7.7	0.00037	26.9	47.2	49	1.04
22/09/86	0.24	2.32	7.7	0.00037	33.8	60.2	63	1.05
03/10/86	0.28	2.79	6.1	0.00037	106	299	322	1.08
04/10/86	0.32	2.90	6.1	0.00042	67.6	238	259	1.09
09/10/86	0.25	2.42	6.3	0.00032	32.9	68.9	73.1	1.06
Visually estimated wave data								
03/11/86	0.48	3.00	6.1	0.00042	84.5	372	411	1.10
07/11/86	0.60	2.90	6.1	0.00032	72.6	276	304	1.10

The units row shows: Qr in m³/sx10⁻⁶ ($m^3/s \times 10^{-6}$) and Qrms in m³/s x10⁻⁶ ($m^3/s \times 10^{-6}$).

Notes: Qr=longshore transport rate based on Rayleigh distribution
Qrms=longshore transport rate based on Hrms
Rates of tidal rise taken from tide gauge records
beach slope angle taken from recorded cross-sections
Lo calculated from Tz

Table 2 Data and results for conversion of wave parameters at the beach toe to the breakpoint

Date	Hrms	Hs	Tz	d	wave angle	Hrmsb	Hsb	wave angle at breakpoint	8γ² /Hrms	longshore wave power
	(m)	(m)	(s)	(m)	(deg)	(m)	(m)	(deg)		(W)
22/09/86	0.22	0.31	2.32	3.80	24	0.23	0.31	12.5	1.0	24.1
22/09/86	0.23	0.33	2.32	3.80	24	0.24	0.33	13.0	1.0	27.9
03/10/86	0.26	0.37	2.79	3.83	40	0.28	0.37	18.5	1.36	57.2
04/10/86	0.31	0.45	2.90	3.97	40	0.32	0.45	20.0	1.07	86.0
09/10/86	0.23	0.32	2.42	3.60	25	0.25	0.33	13.0	1.31	30.5
Visually estimated wave data										
03/11/86	0.43	0.56	3.0	2.2	21	0.48	0.60	14.0	----	158.5
07/11/86	0.55	0.77	2.9	1.8	10	0.60	0.80	8.0	----	155.4

Table 3 Summary of SWA aerial survey data for sections 341-358, 1973-84

Section No.	Spacing (m)	2.0m line movement (m/year)	Correlation coef.	Total area change (m²/year)	Correlation coef	Volume change (m³/year)
341		0.14	0.16 (i)			
342		-0.72	-0.62 (s)			
343		0.24	0.29 (i)			
344		-0.47	-0.80 (s)			
345		-0.33	-0.42 (i)	-0.3	-0.40 (i)	
	164					-369.0
346		0.10	0.11 (i)	-1.5	-0.20 (i)	
	135					202.5
347		0.77	0.54 (i)	4.5	0.45 (i)	
	190					1349.0
348		1.46	0.80 (s)	9.7	0.68 (s)	
	162					1401.3
349		1.24	0.92 (s)	7.6	0.85 (s)	
	178					1326.1
350		1.03	0.94 (s)	7.3	0.93 (s)	
	156					1318.2
351		1.27	0.94 (s)	9.6	0.81 (s)	
	150					1387.5
352		1.22	0.95 (s)	8.9	0.93 (s)	
	160					1296.0
353		1.17	0.83 (s)	7.3	0.74 (s)	
	110					885.5
354		1.35	0.85 (s)	8.8	0.71 (s)	
	142					1015.3
355		1.28	0.89 (s)	5.5	0.54 (i)	
	150					1102.5
356		1.54	0.97 (s)	9.2	0.84 (s)	
	150					1650.0
357		1.71	0.81 (s)	12.8	0.83 (s)	
	140					1435.0
358		0.99	0.61 (s)	7.7	0.65 (i)	
	70					539.0
Harbour arm				7.7 (assumed)		
					Total	14538.9

Note:
    (i) insignificant correlation at the 95% level
    (s) significant correlation at the 95% level

Table 4 Conversion of Delft experimental data for longshore transport

Delft experimental data					Conversion data		
Test No.	Transport function	Hsv (mm)	Ts (s)	Hsb (mm)	θb (deg)	Longshore wave power function		
						(1)	(2)	(3)
1	2.9	74	1.177	79	16.7	24.5	21.9	19.8
2	3.1	86	1.002	85	18.6	40.8	38.8	27.9
3	4.5	78	1.442	88	16.6	34.2	27.1	28.4
4	18.3	126	1.456	134	20.0	156.0	122.7	110.5
5	5.0	88	1.430	98	17.4	52.8	42.0	40.2
6	11.9	122	1.446	130	19.8	142.3	112.4	101.2
7	4.8	83	1.797	97	16.7	50.9	34.2	40.1
8	13.6	122	1.812	135	19.5	168.3	112.5	110.0
9	10.6	117	2.051	132	19.3	166.1	100.2	102.5
10	29.2	161	2.057	174	22.0	388.8	233.9	226.3

Notes: $D90=5.76$mm
$\theta v=30$ deg
water depth at model beach toe$=0.44$m
Transport function$=Q/gD90^2 Ts \times 10^2$ m³/s
Longshore wave power function (1)$=Wv(Wv-8.3)\sin\theta v/\tanh(Ksh)$
(2)$=Wv(Wv-8.3)\sin\theta v$
(3)$=Wb(Wb-8.3)\sin\theta b$
$W=Hs(\cos\theta)^{0.5}/D90$
v refers to model beach toe
b refers to break point

Table 5 Summary of results for longshore transport using a Rayleigh distribution

Deep water angle	Q	Qrms/Q	Qs/Q	Qrms,s/Q
45	5.05	0.83	1.93	0.97
30	4.51	0.82	1.92	0.96
15	2.65	0.82	1.93	0.97

Notes: Q=true transport based on Rayleigh distribution
Qrms=transport based on Hrms
Qs=transport based on Hs
Qrms,s=transport based on Hrms for energy, Hs for break
point

Table 6 Summary of comparisons of mean annual longshore transport for five transport equations

Equation	Parameters	Transport (m³/annum)	Transport/15050
1. BPSM	K=0.0028	15050	1.0
2. CERC	K=0.0527	15040	1.0
3. Modified Delft	D90=0.04	66580	4.4
4. CERC threshold	D90=0.04 K=0.0696	15050	1.0
5. CERC field	K=0.0384 Po=12.2	10640	0.7

Plate 1: View of Shoreham Beach looking East

Plate 2: Surface elevation monitor and visual wave pole array

Plate 3: Surface Mounted shingle trap in operation

Plate 4: Triangular wave pole array

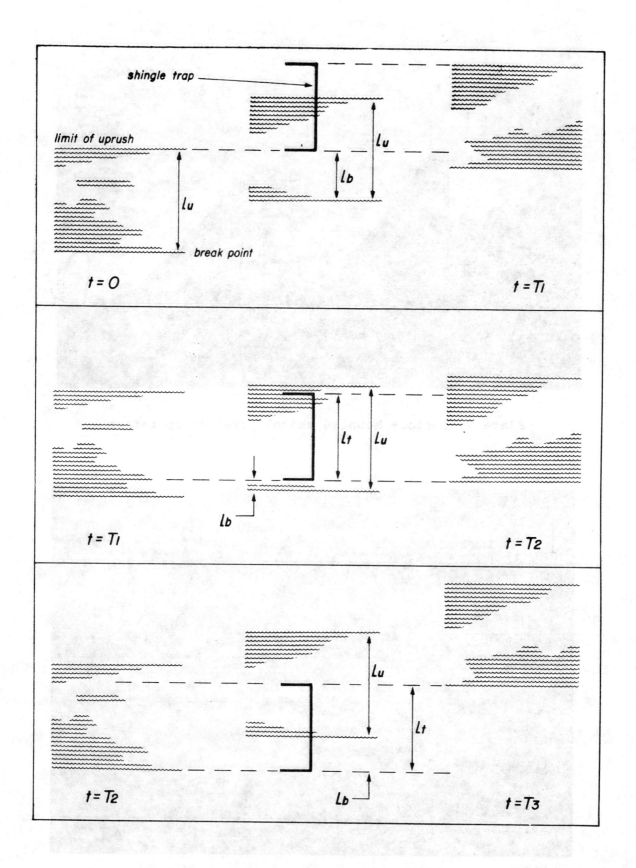

Figure 1: Progression of the swash zone through a trap

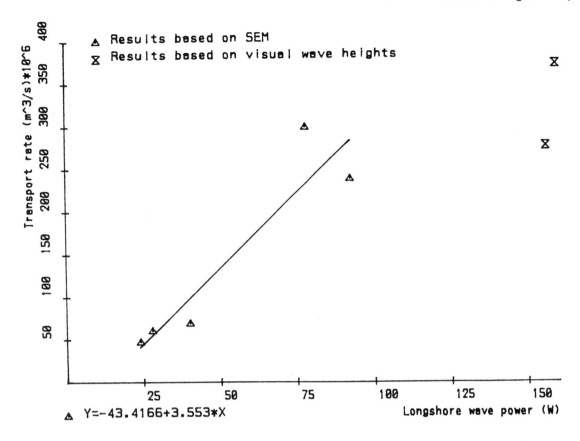

Figure 2: Plot of field longshore transport rates against wave power

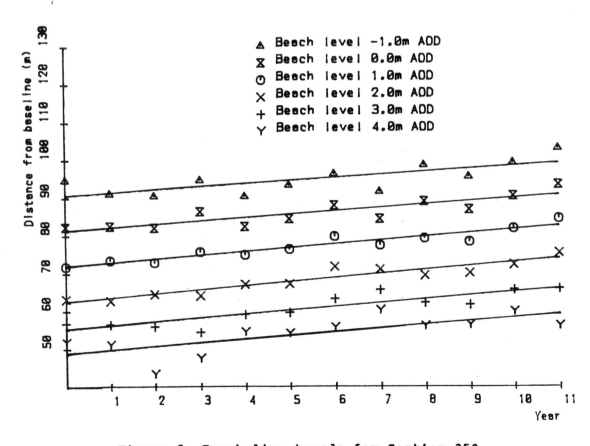

Figure 3: Beach line trends for Section 350

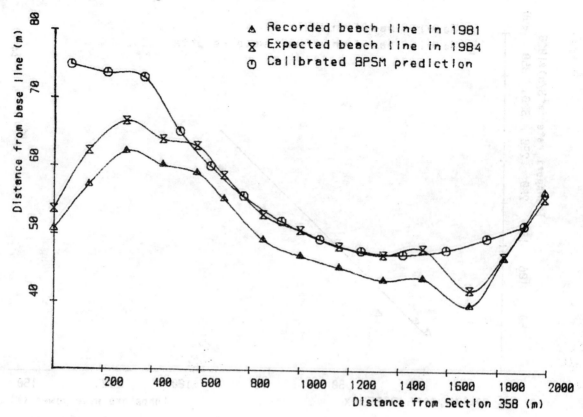

Figure 4: Plot of recorded and predicted beach line movements

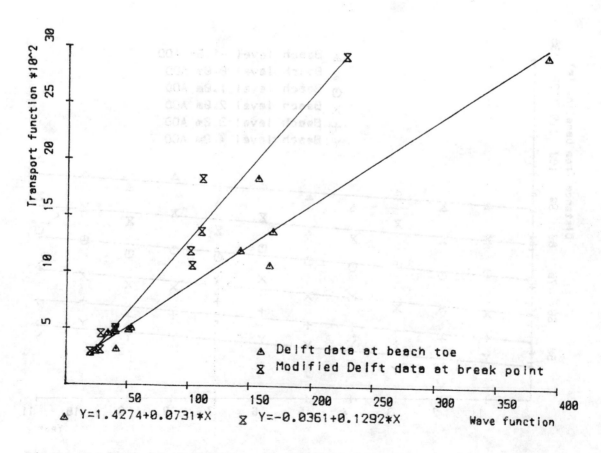

△ Y=1.4274+0.0731*X   Ⅹ Y=-0.0361+0.1292*X   Wave function

Figure 5: Plot of original and modified Delft longshore transport data